8th Joint European Magnetic Symposia (JEMS 2016)

AA005677

Journal of Physics: Conference Series Volume 903

Glasgow, United Kingdom
21 – 26 August 2016

ISBN: 978-1-5108-5062-0
ISSN: 1742-6588

Printed from e-media with permission by:

Curran Associates, Inc.
57 Morehouse Lane
Red Hook, NY 12571

Some format issues inherent in the e-media version may also appear in this print version.

Copyright© (2016) by the Institute of Physics
All rights reserved. The material featured in this book is subject to
IOP copyright protection, unless otherwise indicated.

Printed by Curran Associates, Inc. (2017)

For permission requests, please contact the Institute of Physics
at the address below.

Institute of Physics
Dirac House, Temple Back
Bristol BS1 6BE UK

Phone: 44 1 17 929 7481
Fax: 44 1 17 920 0979

techtracking@iop.org

Additional copies of this publication are available from:

Curran Associates, Inc.
57 Morehouse Lane
Red Hook, NY 12571 USA
Phone: 845-758-0400
Fax: 845-758-2633
Email: curran@proceedings.com
Web: www.proceedings.com

Table of contents

Volume 903

8th Joint European Magnetic Symposia (JEMS2016)
21–26 August 2016, Glasgow, UK

Accepted papers received: 27 September 2017
Published online: 28 October 2017

Preface

011001
OPEN ACCESS
8th Joint European Magnetic Symposia (JEMS 2016)

011002
OPEN ACCESS
Peer review statement

Biomagnetism and medical applications

012001
OPEN ACCESS
Ferromagnetic resonance in the ethmoid bones of salmon and silver carp

Svitlana Gorobets, Oksana Gorobets, Volodymyr Golub and Marina Gromnadska.....1

012002
OPEN ACCESS
The prediction of biogenic magnetic nanoparticles biomineralization in human tissues and organs

O Medviediev, O Yu Gorobets, S V Gorobets and V S Yadrykhins'ky.....4

Electronic correlations, superconductivity, superconducting spintronics

012003
OPEN ACCESS
Hedgehog-like Spin Fluctuation and the Boson Peak in Zn-Mg-Tb Quasicrystals

Y. Sekiyama, R. Maeda and I. Kanazawa.....7

Frustrated and disordered magnetism including spin ice

012004
OPEN ACCESS
Methods of dilatometric investigations under extreme conditions and the case of spin-ice compounds

M Doerr, S Granovsky, M Rotter, T Stöter, Z-S Wang, S Zherlitsyn and J Wosnitza.....10

012005
OPEN ACCESS
Spin Anisotropy in $Cu(tn)Cl_2$: A Quasi-Two-Dimensional $S = 1/2$ Spatially Anisotropic Triangular-Lattice Antiferromagnet

R Tarasenko, A Orendáčová, E Čižmár, M Orendáč, I Potočňák, S Zvyagin and A Feher.....14

012006
OPEN ACCESS
Energy surface and transition rates in a hexagonal element of spin ice

S Y Liashko, V M Uzdin and H Jónsson.....17

Magnetic memories and magnetic recording, sensors

012007
OPEN ACCESS
CoFe-microwires with stress-dependent magnetostriction as embedded sensing elements

M.M. Salem, M.G. Nematov, A. Uddin, L.V. Panina, M.N. Churyukanova and A. T. Marchenko.....20

012008
OPEN ACCESS
Noise study of magnetic field sensors based on magnetic tunnel junctions

M. Mouchel, A. Bocheux, C. Ducruet, Ph. Sabon, I.-L. Prejbeanu, Y. Conraux, J. Alvarez-Hérault, K. Mackay and C. Baraduc.....24

012009
OPEN ACCESS
Switching Field and Resolution of MFM Tips Prepared by Coating Fe/$Co_{50}Pt_{50}$ Magnetic Films

Yuma Tomita, Masaaki Futamoto, Mitsuru Ohtake, Fumiyoshi Kirino and Nobuyuki Inaba.....28

012010
OPEN ACCESS
Magnetostrictive Materials for aerospace applications

A. Al-Taher, R.W. Reiss, A.D. Lafferty, S.A. Hayes, N. Lupu, I. Murgulescu and N.A. Morley.....31

Magnetic shape memory, magnetoelastic and multifunctional materials

012011
OPEN ACCESS
Tunable microwave electric polarization in magnetostrictive microwires

L.V. Panina, D. P. Makhnovskiy, A. Dzhumazoda, S.V. Podgornaya, V. G. Kostishyn and H.X. Peng.....35

012012
OPEN ACCESS
Study of magnetic field distribution in anisotropic single twin-boundary magnetic shape memory (MSM) element in actuators

N Gabdullin and S H Khan.....39

012013
OPEN ACCESS
Change of magnetic domain structure by mechanically induced twin boundary motion in Ni-Mn-Ga single crystal

Vít Kopecký and Oleg Heczko.....42

Magnetic thin films, surface, interfaces and patterned thin films

012014
OPEN ACCESS
Magnetic and dielectric properties of o-$LuFeO_3$/$SrTiO_3$

D V Mamedov, T P Gavrilova, I V Yatsyk, I F Gilmutdinov, Z Y Seidov, M A Aljanov, M J Najafzade, I N Ibrahimov, V I Chichkov, N V Andreev, E Yu Koroleva and R M Eremina.....46

012015
OPEN ACCESS
Anisotropic magnetic property of nanocomposite $Nd_2Fe_{14}B$/Mo/α-Fe multilayer films

K Kobayashi, D Ogawa, K Koike, H Kato, M Oogane, T Miyazaki, Y Ando and M Itakura.....51

012016
OPEN ACCESS
First principles and metadynamics study of the spin-reorientation transition in Fe/Au(001) films

B. Nagyfalusi, L. Udvardi and L. Szunyogh.....54

012017
OPEN ACCESS
Charge transfer and magnetization of a MoS_2 monolayer at the Co(0001)/MoS_2 interface

T. Garandel, R. Arras, X. Marie, P. Renucci and L. Calmels.....57

012018
OPEN ACCESS
Magnetic properties of multilayers based on anisotropic Heisenberg films with dipolar long-range interaction

P V Prudnikov, V V Prudnikov, A P Soldusova and A N Purtov.....60

012019
OPEN ACCESS
Dependence of magnetic anisotropy on MgO sputtering pressure in $Co_{20}Fe_{60}B_{20}$/MgO stacks

A Kaidatzis, C Serletis and D Niarchos.....63

012020
OPEN ACCESS
Modification of magnetization orientation in Pt/Co/Pt ultrathin films by femtosecond laser pulses

J. Kisielewski, Z. Kurant, I. Sveklo, M. Tekielak, E. Dynowska, M. Jakubowski, A. Wawro and A. Maziewski.....66

012021
OPEN ACCESS
Thickness and angular dependent magnetic anisotropy of $La_{0.67}Sr_{0.33}MnO_3$ thin films by Vectorial Magneto Optical Kerr Magnetometry

S K Chaluvadi, P Perna, F Ajejas, J Camarero, A Pautrat, S Flament and L Méchin.....69

012022
OPEN ACCESS
Preparation of high-quality planar FeRh thin films for *in situ* TEM investigations

Trevor P. Almeida, Damien McGrouther, Yevheniy Pivak, Hector Hugo Perez Garza, Rowan Temple, Jamie Massey, Christopher H. Marrows and Stephen McVitie.....72

012023
OPEN ACCESS
Magnetic and structural properties of RTA-processed $L1_0$ FePt-Mn thin films

N. Y. Safonova, M. Riepp, O. Klein and M. Albrecht.....76

012024
OPEN ACCESS
Evolution of the structural and electronic properties of thin Bi films on Ge(111)

C Zucchetti, F Bottegoni, A Calloni, G Bussetti, L Duò, M Finazzi and F Ciccacci.....79

Magnetism and spin transport in graphene/h-BN, carbon based and organic materials

012025
OPEN ACCESS
Magneto-optical reflection spectroscopy on graphene/Co in the soft x-ray range

H.-Ch. Mertins, C. Jansing, M. Gilbert, M. Krivenkov, J. Sanchez-Barriga, A. Varykhalov, O. Rader, H. Wahab, H. Timmers, A. Gaupp, M. Tesch, A. Sokolov, D. Legut and P M. Oppeneer.....83

012026
OPEN ACCESS
Magnetization measurement of a solid CO_2 grain based on its field-induced translation under microgravity conditions

Wakana Yamaguchi, Keiji Hisayoshi and Chiaki Uyeda.....87

012027
OPEN ACCESS
Ab-Initio Calculation of the Magnetic Properties of Metal-Doped Boron-Nitrogen Nanoribbon

J. Rufinus.....91

Magnetism in alloys and intermetallics

012028
OPEN ACCESS
Spin density wave behaviour in the $(Cr_{98.4}Al_{1.6})_{100-y}Mo_y$ and $(Cr_{100-x}Al_x)_{95}Mo_5$ alloy series

B Muchono, ARE Prinsloo, CJ Sheppard, AM Venter and KD Liss.....95

012029
OPEN ACCESS
Bulk and local properties of intermetallic GdAgSn compound

K Łątka and A W Pacyna.....98

012030
OPEN ACCESS
An *ab initio* study of the effects of vacancies on the static and dynamic magnetic properties of Co_2MnSi

B. Pradines, R. Arras and L. Calmels.....102

012031
OPEN ACCESS
Uniaxial pressure effect on magnetization process of ferromagnetic domains in a tensile deformed Pt_3Fe antiferromagnet

Satoru Kobayashi and Yasunori Konno.....106

012032
OPEN ACCESS
Synthesis and magnetic properties of the thin film exchange spring system of MnBi/FeCo

S. Sabet, E. Hildebrandt and L. Alff.....109

012033
OPEN ACCESS
Structure and magnetic properties of $Sm_{1-x}Zr_x Fe_{10}Si_2$ (x=0.2-0.6) alloys

M. Gjoka, C. Sarafidis, V. Psycharis, E. Devlin, D. Niarchos and G. Hadjipanayis.....113

012034
OPEN ACCESS
Defects and magnetic structure of CuMnSb

František Máca, Josef Kudrnovský, Václav Drchal and Ilja Turek.....116

012035
OPEN ACCESS
Study and characterization of soft magnetic materials for beam intensity monitors at CERN

S Aguilera, H Hofmann and P Odier.....119

Magneto-transport, spin electronics, topological insulators

012036
OPEN ACCESS
Circularly polarized light interaction in topological insulators investigated by time-resolved ARPES

D Bugini, H Hedayat, F Boschini, H Yi, C Chen, X Zhou, C Manzoni, C Dallera, G Cerullo and E Carpene.....123

012037
OPEN ACCESS
Magnetotransport properties of nanogranular Bi and Co(c=0.2)Bi(1-c) thin films

P Athanasopoulos, Th Speliotis and C Christides.....127

012038
OPEN ACCESS
Three-dimensional Dirac semimetal and magnetic quantum oscillations in Cd_3As_2

Filip Orbanić, Mario Novak, Nikola Biliškov, Sanda Pleslić and Ivan Kokanović.....130

Materials for energy (permanent magnets, magnetocalorics, soft magnetic materials, Heusler alloys)

012039
OPEN ACCESS
A semi-empirical approximation of static hysteresis for high flux densities in highly grain-oriented silicon iron

C Carrander and G Engdahl.....133

012040
OPEN ACCESS
Development of High-Speed Switched Reluctance Motor for Electric Power Tools

Kenji Nakamura, Yuya Kumasaka and Osamu Ichinokura.....136

012041
OPEN ACCESS
Synthesis of magnetocaloric $LaFe_{11.6}Si_{1.4}$ alloy by spark plasma sintering

P. Shamba, G. Kerridge, N.A. Morley, I. M. Reaney and W. M. Rainforth.....139

012042
OPEN ACCESS
Thermal stability of $Sm_2Fe_{17}N_3$ magnet powders

N Imaoka, K Takagi, K Ozaki, T Iriyama and H Kato.....143

012043
OPEN ACCESS
First-order antiferromagnetic to ferromagnetic transition in Mn(Co,Fe)P probed by x-ray absorption experiments

F. Guillou, A. Hen, F. Wilhelm and A. Rogalev.....147

012044
OPEN ACCESS
Influence of La substitutions on the magnetic after-effect in hexaferrites and garnets

P Hernández-Gómez, X Liu, J M Muñoz, C Torres and O Alejos.....151

012045
OPEN ACCESS
Ni-Mn-Sn Heusler: milling and annealing effect on structural and magnetic properties

Florin Popa, Traian Florin Marinca, Horea Florin Chicinaş, Olivier Isnard and Ionel Chicinaş.....154

012046
OPEN ACCESS
Kinetics of heat flux avalanches at the first order transition in La(Fe-Mn-Si)$_{13}$-H$_{1.65}$ compounds

M Piazzi, C Bennati and V Basso.....157

012047
OPEN ACCESS
Magnetic Circuit Model combined with Play Model Obtained from Landau-Lifshitz-Gilbert Equation

H Tanaka, K Nakamura and O Ichinokura.....162

012048
OPEN ACCESS
Accuracy Improvement of Magnetic Hysteresis Calculated by LLG Equation

H Tanaka, K Nakamura and O Ichinokura.....165

012049
OPEN ACCESS
Effect of long-term thermal aging on magnetic hysteresis for low-alloy pressure vessel steel

Satoru Kobayashi, Ferenc Gillemot, Ákos Horváth, Marta Horváth and Almásy László.....168

012050
OPEN ACCESS
Tailoring the ultra-soft magnetic properties of sputtered FineMET thin films for high-frequency power applications

Ansar Masood, Paul McCloskey, Cian Ó Mathúna and Santosh Kulkarni.....171

012051
OPEN ACCESS
Influence of operating frequency on design of power electronic transformers

V C Valchev, A Van den Bossche and T P Todorova.....174

Micromagnetics, magnetization processes

012052
OPEN ACCESS
Micromagnetic simulation of nanomagnets with geometry-tuned domain wall nucleation

W. Kaiser, M. Kiechle, G. Žiemys, D. Schmitt-Landsiedel and S. Breitkreutz-v. Gamm.....177

012053
OPEN ACCESS
Angular dependence of the domain wall depinning field in sensors with segmented corners.

D. Heinze, B. Borie, H. Grimm, J. Wahrhusen and M. Kläui.....180

012054
OPEN ACCESS
Magnetization of a Single Carbonaceous Grain Obtained by Field-Induced Acceleration

Keiji Hisayoshi and Chiaki Uyeda.....186

012055
OPEN ACCESS
Probing domain walls in cylindrical magnetic nanowires with electron holography

M Staňo, S Jamet, J C Toussaint, S Bochmann, J Bachmann, A Masseboeuf, C Gatel and O Fruchart.....191

012056
OPEN ACCESS
New calculations of Gilbert damping in ferromagnetic transition metals

A. Umerski and D. M. Edwards.....194

Nanoparticles and interfaces , nanomaterials and molecular magnetism

012057
OPEN ACCESS
Liquid-liquid phase separation and cluster formation at deposition of metals under inhomogeneous magnetic field

O Yu Gorobets, Yu I Gorobets, V P Rospotniuk, V I Grebinaha and A A Kyba.....197

Optically driven spin excitations, magneto-optics and magnetoplasmonics

012058
OPEN ACCESS
Terahertz excited plasmon-magnon interaction and magnetoplasmon-enhanced energy transfer within the framework of generalized spin Hamiltonian

Vyacheslav S. Gritzaenko, Dmitry I. Bazhanov and Oleg V. Farberovich.....201

012059
OPEN ACCESS
Magnetic response of surface plasmon polaritons in silver / $Co_2FeSi_{0.6}Al_{0.4}$ Heusler alloy / silver films

Yoshito Ashizawa, Yuya Fukumoto and Katsuji Nakagawa.....207

012060
OPEN ACCESS
Development of in situ magneto-ellipsometry for studying correlation between the optical and magneto-optical properties of ferromagnetic thin films

O A Maximova, N N Kosyrev, S N Varnakov, S A Lyashchenko, I ATarasov, I AYakovlev, O M Maximova, D V Shevtsov and S G Ovchinnikov.....210

Perovskites, Multiferroics, artificial/composite multiferroics

012061
OPEN ACCESS
Influence of Mn diffusion on IrMn thickness threshold for the onset of exchange bias in IrMn/Co bilayers

G Vinai, L Frangou, C Castan-Guerrero, V Bonanni, B Gobaut, S Auffret, I L Prejbeanu, B Dieny, V Baltz and P Torelli.....213

Spin orbitronics, spintronics in antiferromagnets and skyrmions

012062
OPEN ACCESS
On Berezinskii-Kosterlitz-Thouless transition in monoaxial chiral helimagnets

Igor Proskurin, Alexander S Ovchinnikov and Jun-ichiro Kishine.....217

Spin waves, magnonics and dynamics

012063
OPEN ACCESS
Correlation between Gilbert damping and electric resistivity in $Fe_{94}Co_6$/GaAs single crystal thin films: A role of electron scattering

Akihiro Iwasaki, Takahiko Suzuki, Nobuyuki Inaba, Yutaka Takahashi, Fumiyoshi Kirino, Mitsuru Ohtake and Masaaki Futamoto.....225

IOP Conf. Series: Journal of Physics: Conf. Series **903** (2017) 011001 doi:10.1088/1742-6596/903/1/011001

JEMS 2016

The 8th Joint European Magnetic Symposia (JEMS 2013) were held in Glasgow, United Kingdom, on August 21-26, 2016. JEMS 2016 was an EMA sponsored conference jointly organized with the Institute of Physics. The Joint European Magnetic Symposia are the most important and comprehensive conference on magnetism in Europe. Previous JEMS conferences took place in Grenoble (2001), Dresden (2004), San Sebastian (2006), Dublin (2008), Krakow (2010), Parma (2012) and Rhodes (2013).

The success of the conference was made possible by the quality and diversity of the contributions as well as by the lively discussions that followed both the oral and the poster sessions. JEMS2016 was attended by 620 delegates, including 198 students (31%), originating from more than 40 countries. The 17 Symposia, which attracted ~ 800 contributions, were focused on a broad range of topics embracing fundamental aspects and application aspects of magnetism, as well as novel magnetic materials.

The conference programme consisted of 5 plenary, 10 semi-plenary, 16 symposia oral and 3 poster sessions (for all 16 symposia); we had also 38 invited talks, 244 oral presentations and over 500 poster presentations. Special sessions were dedicated to early career researchers, flash posters, and women in Physics. An afternoon tutorial session was held, sponsored by the Graduate School of Excellence "Materials Science in Mianz".

The symposia at JEMS 2016 and the names of their organizers were as follows:

- **Biomagnetism and medical applications**
  Symposium organisers: Russell Cowburn and Alessandro Lascialfari
- **Electronic correlations, superconductivity, superconducting spintronics**
  Symposium organisers: Steve Hayden and Marco Aprilli
- **Frustrated and disordered magnetism including spin ice**
  Symposium organisers: Steve Bramwell and Francois Montaigne
- **Deliberate Design of Exotic Phases in Nano-Magnets via Unusual, Dedicated Geometries**
  Symposium organiser: Cristiano Nisoli
- **Magnetic memories and magnetic recording, sensors**
  Symposium organisers: Jim Miles and Hélène Béa
- **Magnetic thin films, surface, interfaces and patterned thin films**
  Symposium organisers: Dan Allwood, Francesca Casoli and Jeffrey McCord
- **Magnetic shape memory, magnetoelastic and multifunctional materials**
  Symposium organisers: Manfred Kohl and Volodymir Chernenko
- **Magnetism and spin transport in graphene/h-BN, carbon based and organic materials**
  Symposium organisers: Ivan Vera-Marun and Saroj Dash
- **Magnetism in alloys and intermetallics**
  Symposium organisers: Katayun Barmak and Ludwig Schultz
- **Magneto-transport, spin electronics, topological insulators**
  Symposium organisers: Astufumi Hirohata and Riccardo Bertacco

Content from this work may be used under the terms of the Creative Commons Attribution 3.0 licence. Any further distribution of this work must maintain attribution to the author(s) and the title of the work, journal citation and DOI.
Published under licence by IOP Publishing Ltd

IOP Conf. Series: Journal of Physics: Conf. Series **903** (2017) 011001 doi:10.1088/1742-6596/903/1/011001

- **Materials for energy (permanent magnets, magnetocalorics, soft magnetic materials, Heusler alloys)**
  Symposium organisers: Leslie Cohen and George Hadjipanayis
- **Micromagnetics, magnetization processes**
  Symposium organisers: Hans Fangohr and Thomas Schrefl
- **Nanoparticles and interfaces , nanomaterials and molecular magnetism**
  Symposium organisers: Mark Murrie and Gregory Chaboussant
- **Optically driven spin excitations, magneto-optics and magnetoplasmonics**
  Symposium organisers: Roy Chantrell, Oksana Chubykalo and Paolo Vavassori
- **Perovskites, Multiferroics, artificial/composite multiferroics**
  Symposium organisers: Neil Mathur, C Ross and Caf Vaz
- **Spin orbitronics, spintronics in antiferromagnets and skyrmions**
  Symposium organisers: Chris Marrows and Yuriy Mokrousov
- **Spin waves, magnonics and dynamics**
  Symposium organisers: Volodymyr Kruglyak and Giovanni Carlotti

The success of JEMS 2016 owes everything to the efforts of the Symposium Organizers, the Programme Committee, chaired by Franca Albertini, the International Advisory Committee, chaired by Dino Fiorani and the Publications Committee, chaired by Stephen McVitie.

The Local organizing committee did an excellent job in organizing a flawless meeting. Also the excellent facilities and catering offered by the conference made an essential contribution. The meeting arrangements were effectively managed by the Institute of Physics.

We would like to acknowledge the institutional support received by:
Cryogenic Ltd, Cryomech, Inc., ICE Oxford, Graduate School of Excellence Materials Science in Mainz, Institute of Physics, IOP Magnetism Group ,IOP Publishing, Lake Shore Cryotronics, LOT-Quantum Design Ltd, Mantis Deposition Ltd, Metrolab Technology SA, NanoMagnetics Instruments Ltd, Oxford Instruments Nanoscience, Rohde & Schwarz UK Ltd, local government agency (Subvention), The UK Magnetics Society.

Finally, some contributors made the additional effort of turning their presentations into a high quality written report, which we now offer to the research community in the form of these proceedings published open access and available online. This should further strengthen the dissemination of the research presented at this conference. The turnout of JEMS2016 demonstrated that magnetism continues to be an exciting field of research with bright prospects for the future. The next JEMS 2018 edition will be held in Mainz, Germany, August or September. Additionally JEMS 2019 is planned to be held in Uppsala, Sweden.

The JEMS 2016 Conference Chairs:
Prof. Robert Stamps, University of Glasgow, UK.
Prof. Stephen Lee, University of St Andrews, UK.

IOP Conf. Series: Journal of Physics: Conf. Series **903** (2017) 011002 doi:10.1088/1742-6596/903/1/011002

Peer review statement

All papers published in this volume of *Journal of Physics: Conference Series* have been peer reviewed through processes administered by the proceedings Editors. Reviews were conducted by expert referees to the professional and scientific standards expected of a proceedings journal published by IOP Publishing.

Content from this work may be used under the terms of the Creative Commons Attribution 3.0 licence. Any further distribution of this work must maintain attribution to the author(s) and the title of the work, journal citation and DOI.
Published under licence by IOP Publishing Ltd

Ferromagnetic resonance in the ethmoid bones of salmon and silver carp

Svitlana Gorobets[1], Oksana Gorobets[1], Volodymyr Golub[1], Marina Gromnadska[2]

[1] Institute of Magnetism of NASU and MESU, Kyiv 03142, Ukraine
[2] Department of General and Experimental Physics, National Technical University of Ukraine "Kyiv Polytechnic Institute", Kyiv 03056, Ukraine

E-mail: pitbm@ukr.net

Abstract. The detection of biogenic magnetic nanoparticles (BMN) with different magnetic properties in biological material was done using magnetic resonance (MR) spectroscopy. MR spectra of biological material of ethmoid bone of salmon (containing ferritin and BMN), bacteria *E. coli* K13 (containing ferritin and without BMN), yeast *S. cerevisiae* (without ferritin or BMN) and ethmoid bone of silver carp (containing ferritin and not investigated for the presence of BMN) were investigated. The analysis of MR spectra shows that *S. cerevisiae* cells produce much lower signal MR than samples of ethmoid bones of salmon and silver carp which is confirming conclusions about the presence of BMN and ferritin in the ethmoid bones of fishes. The narrow MR linewidth indicates that the magnetic particles in the ethmoid bones of salmon and silver carp are in monodisperse state. The presence of a broad line and the absence of a narrow peak in MR spectrum of *E. coli* K13 cells are typical for ferritin.

1. Introduction

In 1962, Lowenstam first discovered biochemically precipitated magnetite as a capping material in the radula (tongue plate) teeth of chitons (marine mollusks of the class Polyplacophora) [1]. In 1975, Blakemore discovered magnetotactic bacteria (MTB) containing chains of BMN. Now magnetotactic bacteria are the most intensively studied biomagnetic systems [1].

Since that BMNs have been detected in a bunch of organisms which are related to all three domains: prokaryotes, archaea and eukaryotes [2]. A number of organisms such as insects, shellfish, fish, birds, mammals can produce biogenic magnetite and/or iron oxide compounds [3]. Biogenic magnetite particles were also found in many human organs [2]. Moreover, some studies have shown the increase of BMN concentrations in the affected area during neurodegenerative diseases and cancer [4]. Thus it is very important to determine the origin of magnetic nanoparticles and develop methods of detection of BMN in living organisms.

2. Experimental procedure

Different biological materials were selected as the objects of the study, namely ethmoid bone of salmon, bacteria *Escherichia coli* K13, yeast *Saccharomyces cerevisiae*, and biological material of ethmoid bone of silver carp.

The choice of biological materials was done based on the analysis of publication data about the presence of BMN and iron-containing proteins, including ferritin (table 1). It is known that *E.coli* has

Content from this work may be used under the terms of the Creative Commons Attribution 3.0 licence. Any further distribution of this work must maintain attribution to the author(s) and the title of the work, journal citation and DOI.
Published under licence by IOP Publishing Ltd

at least two iron-containing proteins, namely ferritin and bakterioferritin [5] but doesn't contain BMN, while ferritin and BMN presents in ethmoid bone of salmon. *S. cerevisiae* cells do not contain ferritin or BMN. So these cells were used as control samples.

Table 1. Samples of biological material for MR.

The biological material where BMN and ferritin was found (migratory fish)	The biological material containing ferritin and free BMN	The biological material that contains neither ferritin nor BMN	The biological material containing ferritin, but was not investigated for the presence of BMN (non-migratory fish)
Ethmoid bone of salmon [1]	*E.coli* [5]	*S. cerevisiae* [6]	Ethmoid bone of silver carp

Magnetic resonance spectra were recorded for 0.01 g dry biological materials using BrukerBioSpin ELEXSYS E500 spin resonance spectrometer.

3. Results and discussion

MR studies have shown that resonance spectra and intensities of lines for investigated samples of biological materials are quite different. The results of the measurements for the samples of *E. coli* K13, biological material of ethmoid bones of salmon and silver carp, and samples of *S. cerevisiae* are presented in figure 1. The MR spectrum intensity for *S. cerevisiae* cells is much lower than that for other samples. This is in good correlation with literature data: yeast does not contain magnetic nanoparticles. Spectra of biological material of ethmoid bone of salmon and ethmoid bone of silver carp contain a narrow peak around 3.5 kG while there is a quite intense broad peak in *E. coli* K13 spectrum. It should be noted that the intensity of the spectrum of *E. coli* K13 is an order of magnitude lower than that for ethmoid bone of salmon and ethmoid bone of silver carp. The possible explanation of this is the presence of both ferritin and BMN in ethmoid bones of salmon and silver carp.

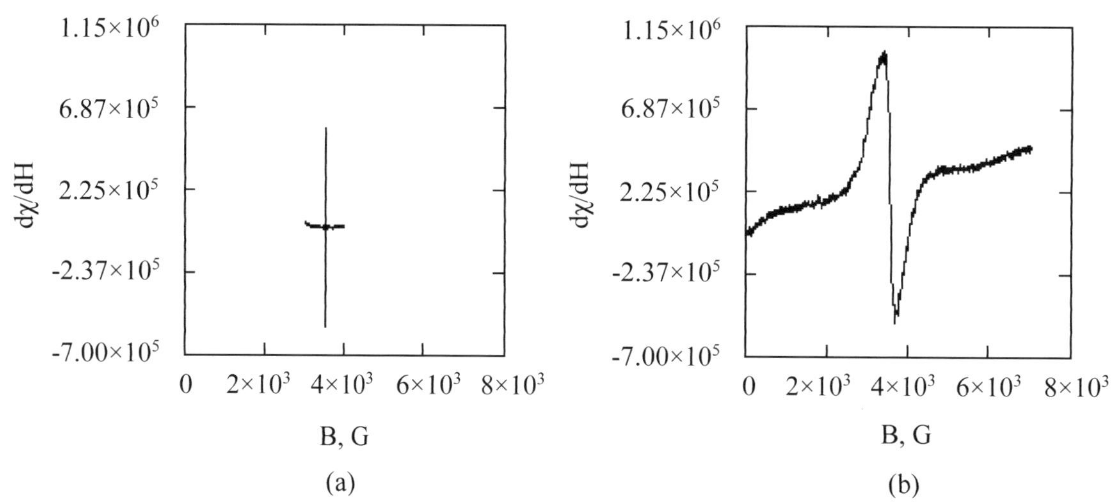

Figure 1. MR spectra for the ethmoid bone of salmon (a), bacteria *E. coli* K13 (b).

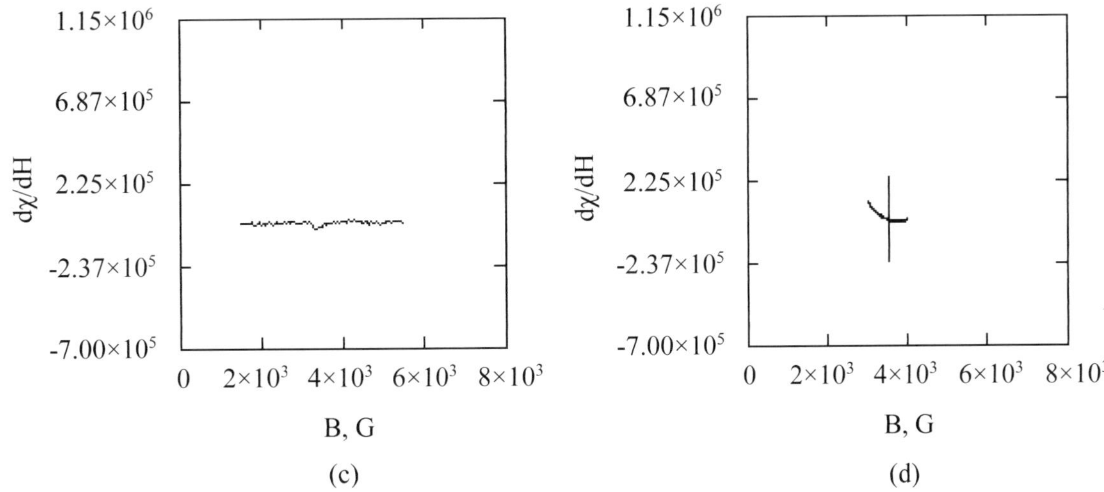

Figure 1. Continued. MR spectra for the yeast *S. cerevisiae* (c), ethmoid bone of silver carp (d).

4. Conclusions

In the paper it has been shown that MR spectroscopy is a power tool to identify ferritin and biogenic magnetic nanoparticles in biological material, which has been confirmed by comparison of MR spectra of different biological material. MR spectrum of biological material of ethmoid bones of salmon and silver carp contains a narrow peak. The appearance of this peak is probably related to the presence of BMN. The spectrum of *E.coli* with ferritin contains much broader peak. The presence of broad peak is characteristic for the samples containing ferritin [7] while the presence of a narrow peak on MR spectra is characteristic feature of monodisperse magnetite nanoparticles.

This project has received funding from the European Union's Horizon 2020 research and innovation programme under the Marie Skłodowska-Curie grant agreement No 644348 (MagIC).

References

[1] Šafařík I, Šafaříková M 2002 *Monatshefte für Chemie/Chemical Monthly* **133** 737
[2] Kirschvink J L, Kobayashi-Kirschvink A, Diaz-Ricci J C and Kirschvink S J 1992 *Bioelectromagnetics* **13** 101
[3] Posfai M and Dunin-Borkowski R E 2009 *Elements* **5** 235
[4] Komeili A 2012 *FEMS Microbiol Rev.* **36** 232
[5] Hudson A J, Andrews S C, Hawkins C, Williams J M, Izuhara M, Meldrum F C, Mann S, Harrison P M, Guest J R 1993 *Eur. J. Biochem.* **15** 985
[6] Hassett R F, Romeo A M, Kosman D J 1998 *J. Biol. Chem.* **273** 7628
[7] Wajnberg E, El-Jaick L J, Linhares M P, Esquivel D M. 2001 *J. Magn. Reson.* **153** 69

The prediction of biogenic magnetic nanoparticles biomineralization in human tissues and organs

O Medviediev[1], O Yu Gorobets[1], S V Gorobets[1] and V S Yadrykhins'ky[1]

[1] National Technical University of Ukraine "Kyiv Polytechnic Institute", 37 Prospect Peremohy, Kyiv 03056, Ukraine

E-mail: will.be.psychedelic@gmail.com

Abstract. In this study, human homologs of magnetosome island proteins basing on pairwise and multiple alignment of amino acid sequences were found. The expression levels of genes, which encode magnetosome island proteins of *M. gryphiswaldense* MSR-1, that were cultured under oxygen deficiency conditions and also under microaerobic conditions were compared to the expression levels of genes that encode the relevant homologs in human organism. The possibility of BMN biomineralization in human tissues and organs, in which BMN were not experimentally found before, was predicted.

1. Introduction

Biogenic magnetic nanoparticles (BMN) are ferrite nanocrystals (magnetite, maghemite, greigite) formed during the genetically controlled biosynthesis. This mechanism is called biomineralization. BMN biomineralization is the most studied in magnetotactic bacteria (MTB). Magnetic nanoparticles are situated inside special membrane structures - magnetosome islands (MI). Magnetite biomineralization is induced under microaerobic conditions. There is a distinct correlation between the extracellular levels of oxygen and magnetite quantity in *Magnetospirillum gryphiswaldense*. Magnetite biomineralization occurs at the oxygen levels below the threshold of 2 kPa (2%) and the largest number of BMN was found inside MI at 25 Pa (0.025%). Higher level of oxygen completely inhibit BMN formation [1].

BMN presence is experimentally confirmed in different human tissues [2, 3]: liver, heart, spleen [4] adrenal glands, slatted bone and brain [5]. Moreover, several studies have shown increased BMN concentrations during neurodegenerative diseases [6, 7, 8] and cancer [9, 2] in the affected area.

The main objective of this study is the prediction and detection of human organs and tissues in which BMN biomineralization occurs by using bioinformatics methods.

Since the mechanism of biomineralization BMN is the same for all three kingdoms of organisms, the prediction of biomineralization could be based on finding MI MTB homologous proteins in human and relevant gene analysis [10].

2. Experimental procedure

Mechanism of BMN biomineralization is based on a set of MI MTB proteins. They are mamA, mamB, mamM, mamE, mamO, mamN. In this paper, only the first three proteins, acting as protein-protein interaction and transport of cations Co^{2+}, Zn^{2+}, Cd^{2+}, were analyzed respectively.

Content from this work may be used under the terms of the Creative Commons Attribution 3.0 licence. Any further distribution of this work must maintain attribution to the author(s) and the title of the work, journal citation and DOI.
Published under licence by IOP Publishing Ltd

Homologs were found by means of pairwise and multiple alignment of amino acid sequences methods using BLAST program of National Center for Biotechnology Information (NCBI). Proteome of *M. gryphiswaldense* MSR-1 was used as a microorganism for the original sequences of mam proteins obtaining, as the most researched. MamA, mamB and mamM proteins were chosen because their expression levels were experimentally obtained for *M. gryphiswaldense* MSR-1 cultured under oxygen deficiency conditions as well as under microaerobic conditions [1].

Evaluation on the similarity was performed basing on the I and E-value indices with the thresholds > 18% and < 0.05, respectively. I (%) - number of identical amino acid residues in comparing proteins. E-value is an indicator that reflects the statistical significance of the alignment, reducing the value that indicates a lower level of coincidence chance [11].

The comparison of the gene expression levels was conducted by using an average of FPKM (Fragment per Kilobase of Exon per Million Fragments Mapped). The data were obtained using The Human Protein Atlas.

3. Results and discussion

In this research human homologs of MI MTB proteins were found. They are: pex-5 protein as a mamA homolog, SLC30A9, SLC39A4, SLC39A3 as mamB homologs and SLC30A9, SLC39A4 as mamM homologs. In addition, the expression levels of genes that encode MI MTB proteins and the expression levels of genes that encode the relevant homologs in human organism were compared. The results of the genes expression levels comparison are represented in the table 1.

Table 1. The results of the comparison of the expression levels of genes, encoding MI MTB proteins and the expression levels of genes, encoding the relevant homologs in human organs and tissues. "+" - the expression level of genes, encoding homologs in human organism is higher than the relevant expression level of genes in MTB, "-" is respectively lower.

	mamA	mamB			mamM	
	Pex-5	SLC30A9	SLC39A4	SLC39A3	SLC30A9	SLC39A4
Endometrium	+	+	-	-	+	+
Cerebral cortex	+	+	-	+	+	-
Lung	+	+	-	-	+	+
Ovary	+	+	-	-	+	+
Prostate	+	+	-	-	+	+
Stomach	+	+	+	-	+	+
Thyroid gland	+	+	-	+	+	+
Testis	+	+	-	+	+	+
Kidney	+	+	+	-	+	+
Smooth muscle	+	+	-	+	+	-
Duodenum	+	+	+	-	+	+
Gall bladder	+	+	-	-	+	+
Bone marrow	+	-	-	+	+	+
Tonsil	+	+	-	-	+	+
Urinary bladder	+	+	-	+	+	-
Adrenal glands	+	+	-	+	+	+
Esophagus	+	+	-	-	+	+
Small intestine	+	+	+	-	+	+
Colon	+	+	+	-	+	+
Fallopian tube	+	+	-	+	+	+
Rectum	+	+	-	-	+	+
Salivary gland	+	+	-	-	+	+

BMN biomineralization can be predicted in human tissues and organs, in which the expression levels of three mam homologs is higher than the expression levels of mam proteins in M. gryphiswaldense MSR-1, cultured under microaerobic conditions, as it is represented in the table 1.

4. Conclusions

In this paper, the possibility of BMN biomineralization in human tissues and organs, in which BMN were not experimentally found earlier was predicted. This tissues and organs are endometrium, cerebral cortex, lung, ovary, prostate, stomach, thyroid gland, testis, kidney, smooth muscle, duodenum, gall bladder, bone marrow, tonsil, urinary bladder, esophagus, small intestine, colon, fallopian tube, rectum, salivary gland.

This project has received funding from the European Union's Horizon 2020 research and innovation programme under the Marie Skłodowska-Curie grant agreement No 644348 (MagIC). The prospect of the further research is MI MTB homologs gene expression levels analysis in human tumor tissues.

References

[1] Schubbe S, Wurdemann C, Peplies J, Heyen U, Wawer C, Glockner F O and Schuler D 2006 *Appl. Environ. Microbiol.* **72** 5757

[2] Brem F, Hirt A M, Winklhofer M, Frei K, Yonekawa Y, Wieser H-G and Dobson J 2006 *J. R. Soc. Interface* **3** 833

[3] Moos T and Morgan E H 2004 *Ann. NY Acad. Sci.* **1012** 14

[4] Schultheiss-Grassi P P, Heller F and Dobson J 1997 *Biometals* **10** 351

[5] Kirschvink J L, Kobayashi-Kirschvink A, Diaz-Ricci J C and Kirschvink S J 1992 *Bioelectromagnetics* **13** 101

[6] Bartzokis G, Sultzer D, Mintz J, Holt L E, Marx P, Phelan C K and Marder S R 1994 *Biological Psychiatry* **35** 480

[7] Bartzokis G, Cummings J L, Markham C H, Marmarelis P Z, Treciokas L J, Tishler T A, Marder S R and Mintz J 1999 *oMagnetic Resonance Imaging* **17** 213

[8] Beyhum W, Hautot D, Dobson J and Pankhurst Q A 2005 *J. Physics: Conf. Ser.* **17** 50

[9] Kobayashi A, Yamamoto N, Kirschvink J 1997 *J. Jpn. Soc. Powder Metall.* **44** 294

[10] Gorobets O Yu., Gorobets S V and Gorobets Yu I 2014 *Biogenic Magnetic Nanoparticles: Biomineralization in Prokaryotes and Eukaryotes, In Dekker Encyclopedia of Nanoscience and Nanotechnology* (New York: CRC Press) pp 300-308

[11] Li W, Pio F, Pawlowski K and Godzik A 2000 *Bioinformatics* **16** 1105

IOP Conf. Series: Journal of Physics: Conf. Series **903** (2017) 012003 doi:10.1088/1742-6596/903/1/012003

Hedgehog-like Spin Fluctuation and the Boson Peak in Zn-Mg-Tb Quasicrystals

Y.Sekiyama, R.Maeda, I.Kanazawa

Department of Physics, Tokyo Gakugei University, Koganei, Tokyo 184-8501, Japan

Kanazawa@u-gakugei.ac.jp

Abstract. We have discussed localized collective fluctuations of short-range ordered spin in a dodecahedral spin cluster in Zn-Mg-Tb icosahedral quasicrystal and the relation to the boson peak in the topological glasses.

1. Introduction

Quasicrystals are characterized by sharp Bragg reflections with an anomalous point-symmetry such as fivefold or tenfold symmetry. Taking into account the Bragg reflection and point-symmetry, quasiperiodicity has been proposed and is widely accepted[1]. Icosahedral quasicrystals can most easily be obtained in real alloys such as Al-Pd Mn [2] and Al-Li-Cu [3]. Then, a new type of icosahedral quasicrystals was found in the Zn-Mg-RE (RE: rare earth) system[4].

Since this system has localized 4f magnetic moments on the RE sites, it provides an opportunity to study magnetism on the quasiperiodic lattice. Ordering and excitations of quasiperiodically arranged magnetic moments are fundamental open problems, despite the intensive efforts continuously made since the discover of the quasicrystals[5].

The Zn-Mg-RE (RE: rare earth) icosahedral quasicrystals [6] are the most extensively studied magnetic quasicrystals.

Sato et al.[7] have studied the low-temperature spin dynamics in the face-centered icosahedral Zn-Mg-Tb quasicrystal around its spin-glass-like freezing temperature (T_f~5.8K) by inelastic neutron scattering. Furthermore, a single broad inelastic peak, centered at $\hbar\omega \sim 2.5meV$, was observed in the inelastic scattering spectrum at temperature ranging from the base temperature up to ~20K. This broad peak can be interpreted as localized collective fluctuations of short-range-ordered spin in a dodecahedral spin cluster. Also, they gave very important suggestions.

That is, they indicate a possible close relation between the broad inelastic spin-excitation peak and the so-called boson peak in topological glasses[8,9].

In this study, we will argue localized collective fluctuation of short-range ordered spin in a dodecahedral spin cluster in Zn-Mg-Tb quasicrystal and the relation to the boson peak in the topological glasses, using field-theoretical formula[10-15].

2. Localized collective spin-fluctuations around dodecahedral clusters

Content from this work may be used under the terms of the Creative Commons Attribution 3.0 licence. Any further distribution of this work must maintain attribution to the author(s) and the title of the work, journal citation and DOI.
Published under licence by IOP Publishing Ltd

IOP Conf. Series: Journal of Physics: Conf. Series **903** (2017) 012003 doi:10.1088/1742-6596/903/1/012003

We adopt the parameter $\rho(u,r) \equiv \rho^a(a=1,2,3,4)$ which is similar to that in the Sachdev and Nelson model [11]. Where r and u are the spatial axis and the perpendicular ones in the quasicrystal. The ρ^a is specified by the rotation, which is related to the gauge fields SO(4) of A_μ^a [10,11]. S^i is the spin of Tb as shown in Fig.1. To represent the dodecahedral cluster, we set the symmetry breaking $\langle 0|\rho^a|0\rangle = \langle 0,0,0,v\rangle$ of the parameter field ρ^a in the Lagrange density as follows,

$$\begin{aligned} L = &\frac{1}{2}(\partial_i S^j - g_1 \varepsilon_{ijk} A_i^a S^k)^2 \\ &+\frac{1}{2}(\partial_\mu \rho^a - g_4 \varepsilon_{abc} A_\mu^b \rho^c)^2 \\ &-\frac{1}{4}(\partial_\nu A_\mu^a - \partial_\mu A_\nu^a + g_3 \varepsilon_{abc} A_\mu^b A_\nu^c)^2 \\ &-\lambda^2(\rho^a\rho^a - v^2)^2 \end{aligned} \tag{1}$$

After the symmetry breaking $\langle 0|\rho^a|0\rangle = \langle 0,0,0,v\rangle$, we can obtain the effective Lagrange density.

$$\begin{aligned} L_{eff} = &\frac{1}{2}(\partial_i S^j - g_1 \varepsilon_{ijk} A_i^a S^k)^2 \\ &-\frac{1}{4}(\partial_\nu A_\mu^a - \partial_\mu A_\nu^a + g_3 \varepsilon_{abc} A_\mu^b A_\nu^c)^2 \\ &+\frac{1}{2}(\partial_\mu \rho^a - g_4 \varepsilon_{abc} A_\mu^b \rho^c)^2 \\ &+\frac{m_1^2}{2}[(A_\mu^1)^2 + (A_\mu^2)^2 + (A_\mu^3)^2] \\ &+m_1(A_\mu^1 \partial_\mu \rho^2 - A_\mu^2 \partial_\mu \rho^1) \\ &+m_1(A_\mu^2 \partial_\mu \rho^3 - A_\mu^3 \partial_\mu \rho^2) \\ &+m_1(A_\mu^3 \partial_\mu \rho^1 - A_\mu^1 \partial_\mu \rho^3) \\ &+gm_1\{\rho^4[(A_\mu^1)^2 + (A_\mu^2)^2 + (A_\mu^3)^2] \\ &-A_\mu^4[\rho^1 A_\mu^1 + \rho^2 A_\mu^2 + \rho^3 A_\mu^3]\} \\ &-\frac{m_1^2}{2}(\rho^4)^2 - \frac{m_2 g}{2m_1}\rho^4(\rho^a)^2 \\ &-\frac{m_2^2 g^2}{8m_1^2}(\rho^a\rho^a)^2 \end{aligned} \tag{2}$$

where $m_1 = v \cdot g_4$, and $m_2 = 2(2)^{\frac{1}{2}}\lambda \cdot v$. The effective Lagrange describes three massive gauge fields A_μ^1, A_μ^2, A_μ^3 and one massless gauge field A_μ^4 . Since massive gauge fields are created through the Anderson-Higgs mechanism by introducing the dodecahedral cluster, the fields A_μ^1, A_μ^2, and A_μ^3

are localized around the dodecahedral cluster within the length of $\sim 1/|m_2| \equiv R_c$. R_c approximately corresponds to the radius of the dodecahedral cluster.

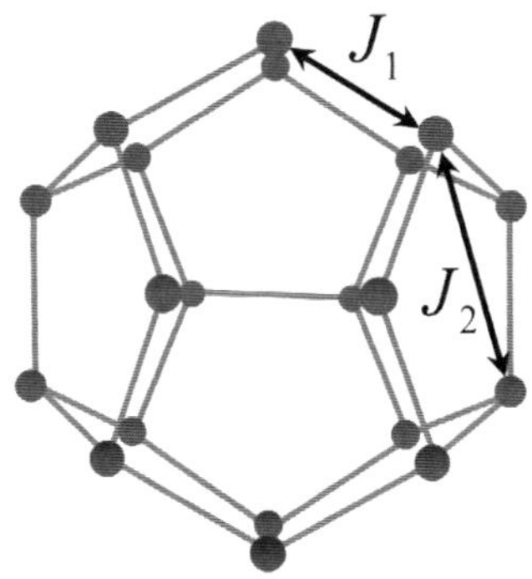

Fig.1 Dodecahedral spin cluster model. A typical ground-state spin configuration is shown by the thick solid lines. The nearest-neighbor (J_1) and next-nearest neighbor (J_2) interactions are also indicated by arrows.

From first term in eq.(2), it is shown that the massive gauge fields $\boldsymbol{A}^1_\mu$, $\boldsymbol{A}^2_\mu$, and $\boldsymbol{A}^3_\mu$ induce localized collective fluctuation of spins in a dodecahedral spin cluster, taking into account the short-range spin fluctuation mechanism by massive gauge fields [16,17].

It should be noticed that the massive gauge fields $\boldsymbol{A}^1_\mu$, $\boldsymbol{A}^2_\mu$, and $\boldsymbol{A}^3_\mu$ around the dodecahedral clusters in eq.(2) are much similar to massive gauge fields around the hedgehog-like clusters (locally preferable clusters) in topological glasses, which might correspond to the so-called boson peak [13-15].

3.Conclusion

Using field-theoretical formula, with parameter which is similar to that in Sachdev and Nelson model, we have argued analytically tractable localized collective fluctuation of short-range ordered spin in a dodecahedral spin cluster in Zn-Mg-Tb quasicrystal and the relation to the boson peak in the topological glasses.

References

[1] Yamamoto A, 1994 *Acta Crystallogr.* **Sect.A52**, 509
[2] Tsai A P et al. 1990 *Mater. Trans.* **JIM 31**, 98
[3] Dubost B et al. 1986 *Nature* **324**, 48
[4] Luo Z et al. 1993 *Scr. Metall. Mater.* **28**, 1513
[5] Fukamichi K , in Physical Properties of Quasicrystals, edited by Stadnik Z M (Springer-Verlag, Berlin, 1999), PP.295-326
[6] Niikura A et al. 1994 *Phil.Mag.Lett.* **69**, 351
[7] Sato T J et al, 2006 *Phys. Rev.* **B73,** 054417
[8] Frick B and Richter D, 1995 *Science* **267**, 1939
[9] Kanaya T and Kaji K, 2001 *Adv. Polym. Sci.* **154**, 87
[10] Sethna J P, 1983 *Phys. Rev. Lett.* **51**, 2198
[11] Sachdev S and Nelson D R, 1984 *Phys. Lev. Lett.* **53**, 1947
[12] Kanazawa I, 1997 *Prog. Theor. Phy.Suppl.* **126**, 393
[13] Kanazawa I, 2001 *J.Non-Cryst. Solids*, **293-295**, 615
[14] Kanazawa I, 2002 *J.Non-Cryst. Solids*, **312-314**, 608
[15] Suzuki H and Kanazawa I, 2010 *Intermetallics* **18**, 1809
[16] Kanazawa I, 2003 *J.Phys.* **A36**,9371
[17] Kanazawa I and Sasaki T, 2015 *Phys.Scr.* **T165**, 014038

Methods of dilatometric investigations under extreme conditions and the case of spin-ice compounds

M Doerr[1], S Granovsky[1,*], M Rotter[1,2], T Stöter[1], Z-S Wang[1,3], S Zherlitsyn[3], and J Wosnitza [1,3]

[1] Technische Universität Dresden, Institut für Festkörperphysik, D-01062 Dresden, Germany.
[2] *McPhase*-project (www.mcphase.de) Dresden, Germany.
[3] Helmholtz-Zentrum Dresden-Rossendorf, Hochfeld-Magnetlabor, D-01314 Dresden, Germany.
[*] also at Faculty of Physics, M.V.Lomonossov Moscow State University, Moscow, Russia.

E-mail: `doerr@physik.tu-dresden.de`

Abstract. We give an overview on how dilatometric methods have been developed in the last decade. The concept of capacitive dilatometry was successfully adapted to dilution refrigerators with a resolution of 10^{-9}. Miniaturized dilatometers with an overall diameter of 18 mm or less are optimally suited for measuring longitudinal and transversal components of the striction tensor. Going to another extreme, to the highest (pulsed) fields, optical methods, such as the FBG technology, were developed for investigations up to 100 T.

As examples for utilizing dilatometry at low temperatures we show results for the spin-ice materials $Dy_2Ti_2O_7$ and $Ho_2Ti_2O_7$. To characterise the magneto-elastic coupling in these materials, we investigated the thermal expansion and magnetostriction between 80 mK and 15 K and in magnetic fields aligned along the [111] direction and found field-induced phases and strong correlations below 500 mK. Our data demonstrate, that the formation of the field-induced phase is strongly influenced by lattice distortions: any change in interatomic distances will result in a variation of the exchange couplings.

1. Introduction

Dilatometric investigations, *i.e.* measurements of the changes in sample dimensions or crystal symmetries, can shed light on how the lattice couples to electronic degrees of freedom. Another aspect is the connection to the elastic properties such as elastic moduli etc. Besides the common thermal expansion, the effects in magnetic fields (molecular fields or external ones) shifted into the scope of research and application. Basics, physical models and experimental results of magnetostriction were published in review articles, *e.g.* [1].

Magnetic materials change their lattice parameters when exposed to a magnetic field, but the effects can even beyond the experimental resolution. This requires to improve the sensitivity of existing methods. Nowadays, length variations in the sub-Å range can be resolved. Additionally, the recent interest in quantum criticality and magnetic frustration has made it necessary to extend the temperature range to the millikelvin region. On the other hand, measurements in fields up to 100 T can be utilized now to overcome exchange forces and to study magnetoelastic interactions.

A focus of current research in solid-state physics is on highly frustrated magnetism, mainly driven by competing interactions in a geometrically frustrated lattice. Long-range magnetic

Content from this work may be used under the terms of the Creative Commons Attribution 3.0 licence. Any further distribution of this work must maintain attribution to the author(s) and the title of the work, journal citation and DOI.
Published under licence by IOP Publishing Ltd

IOP Conf. Series: Journal of Physics: Conf. Series **903** (2017) 012004 doi:10.1088/1742-6596/903/1/012004

order is suppressed and new ground states and exotic excitations, as for example magnetic monopoles [2, 3], are studied. Prominent examples are pyrochlores $R_2M_2O_7$ (R = rare earth; M = Ti,Zr,Hf,Si,Ge,Sn). In these materials, highly degenerated spin-ice or quantum spin-ice ground states exist and new kinds of excitations emerge in magnetic fields and in the millikelvin range [4]. In this context, magnetostriction is an important analytical method because it detects lattice expansions or distortions which are strongly related to the magnetic exchange couplings. Simulation of the magnetostrictive behaviour, *e.g.* with the program package *McPhase*, may help to elucidate if the lattice is affected either by crystal-field or exchange striction.

2. Measurement techniques

Magnetostriction of three-dimensional solids (bulk) can be measured, in principle, either by A) electromechanical or B) optical methods. Method A) comprises the commonly used capacitive dilatometers and the resistive strain-gauges whereas in method B) interferometry, lately also using fibre-bragg gratings (FBG, see below) and optical transmission are included.

To extend the measurements to millikelvin temperatures, the existing capacitive dilatometers, where the sample is connected to a flexible plate of a capacitor, are mounted into dilution refrigerators. The first report of millikelvin-range experiments was given in [5], followed by a number of other sophisticated designs ([6] and others). The developments boomed especially in the last decade. Some designs are published in [7, 8, 9]. To realize a sensitivity of 10^{-9}, the temperature stability and homogeneity over the dilatometer cell is an important issue. An appropriate material with high thermal conductivity has to be used. Additionally, in the case of magnetostriction eddy currents should be reduced. Geometrically, conducting loops perpendicular to the field have to be as small as possible. Again, such effects can be reduced by the right material choice, *i.e.* copper beryllium alloys or brass. Mechanical stability is another important ingredient.

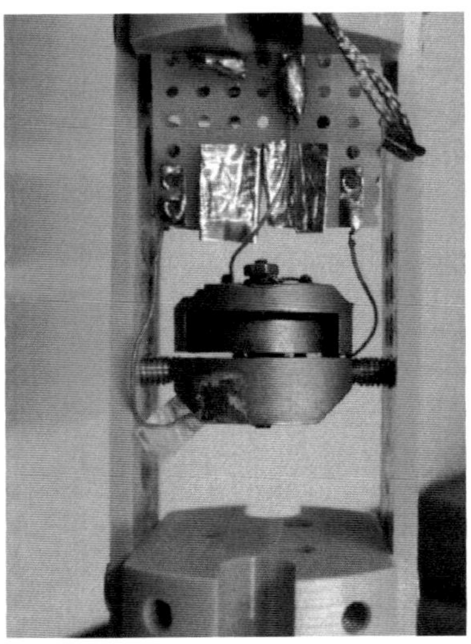

Figure 1. Millikelvin dilatometer ready for installation into a dilution refrigerator.

In our case, a tilted-plate miniature dilatometer [10] (material: brass) was mounted inside a top-loading dilution refrigerator equipped with a superconducting 15 T magnet, see fig. 1. The sample is mounted inside the capacitor and the overall outer diameter is 18 mm. All components of the striction tensor can be measured. In contrast to other devices, the dilatometer is located in the mixing chamber, *i.e.* immersed directly into the ^{3}He-^{4}He mixture. Although a dielectric fluid is sometimes discussed to cause problems due to evaporation and generation of gas bubbles inside the capacitor, we never encountered such problems. The capacitance indeed is stable and the sensitivity is in the range of 10^{-9} over the full temperature range 0.015 K to 2 K.

In general, all existing capacitive dilatometers are applicable in steady-field magnets, but the available field is limited to 45 T. To extend the field range to 100 T the pulsed-magnet technique with pulse durations between 10 ms and 1 s is state of the art. The fast field-sweep rates make eddy-current effects a critical issue. Thereby, not only the thermalisation is a crucial challenge, but also the induced magnetic moments. As a first try [11] a push-rod design was used with the capacitor outside the magnet. However, the not temperature stabilized push-rod reduced the resolution. Next, an attempt was made with nonmetallic miniaturised capacitors using electrically insulating glass plates coated with slitted metallic films [12, 13, 14]. However, temperature instabilities

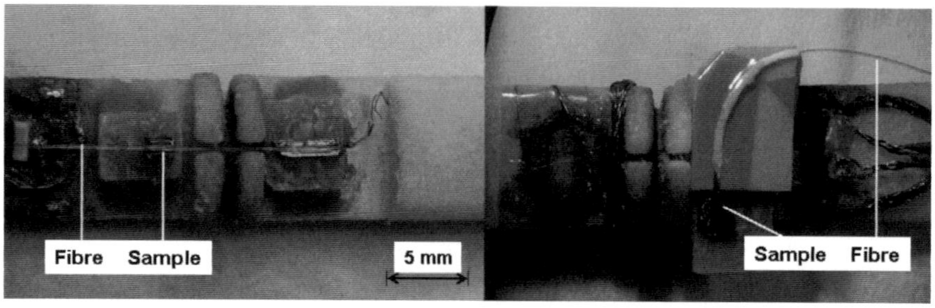

Figure 2. Details of the FBG setup. Left: Longitudinal geometry, fibre coming from left. Right: Transversal geometry, fibre bent, coming from right. Sample is small part in black.

and, more severely, mechanical noise limited the sensitivity to 10^{-5}. A real breaktrough was achieved using the optical fibre-bragg gratings. It is based on the modulation of the refractive index in an optical fibre. A length change of the fibre is detected by a change of the wavelength of reflected light. The fibre can be glued directly to the sample surface ([15] and fig. 2). This technique improved the sensitivity to 10^{-7} and beyond [16, 17].

The next step to measure dilatometry in even more extreme experimental conditions is to apply high pressure. The FBG technique is perfectly suited to be coupled to pressure cells [18]. Compact size piston-cylinder cells give the ability to perform dilatometry studies at pressures up to 3 GPa and at temperatures down to 25 mK.

3. Results

The magnetostriction of the spin-ice compounds $Dy_2Ti_2O_7$ and $Ho_2Ti_2O_7$ was measured longitudinally with the miniature dilatometer shown in fig. 1 in order to study the coupling of the lattice to the magnetic properties. The external field was applied along the [111] direction. To separate magnetodielectric effects [19] the sample was completely shielded by a metallic foil and/or a metallic shield installed into the dilatometer.

The magnetostriction of $Dy_2Ti_2O_7$ at lowest temperatures (fig. 3) shows a steady increase of the sample length up to about 1.2 T followed by a sudden drop and saturation above 2 T. The anomalies can be understood by a change from the 2-in/2-out ground state to the Kagomé-ice state. Above 1 T, this configuration switches to the 3-in/1-out (or vice-versa) configuration visible by the negative expansion. In zero field, the spin-ice state breaks down at temperatures above 1.5 K into a thermally excited non-ice-rule configuration [20, 21]. Our magnetostrictive investigations agree with the phase diagram reported in [22].

The situation for $Ho_2Ti_2O_7$ (fig. 4) is similar. The relative lattice changes are again in the order of 10^{-4}. A clear maximum in the magnetostriction is visible about 2.0 T that separates the Kagomé-ice state from the 3-in/1-out configuration. The data at 0.1 K and 0.3 K are highly hysteretic in field, consistent with [23]. This behavior is related to the slow spin dynamics [24]. Again, at higher temperatures the maximum of the magnetostriction anomaly shifts to higher fields characteristic for predominantly ferromagnetic correlations.

In a first simulation using a mean-field theory (program *McPhase*, www.mcphase.de) the behaviour can be explained by a dominant crystal-field striction process where two-ion striction effects seem to be negligible.

In conclusion, we may state that capacitive dilatometry is a versatile technique that allows magnetoelastic investigations even in the millikelvin range. As examples, we showed some results of such measurements for the spin-ice materials $Dy_2Ti_2O_7$ and $Ho_2Ti_2O_7$. These compounds have comparable features in the magnetostriction. At the transition from the Kagomé-ice state to the 3-in/1-out configuration clear anomalies in the magnetostriction signal are observed. The strong lattice effects indicate the influence of lattice distortions and interatomic distances.

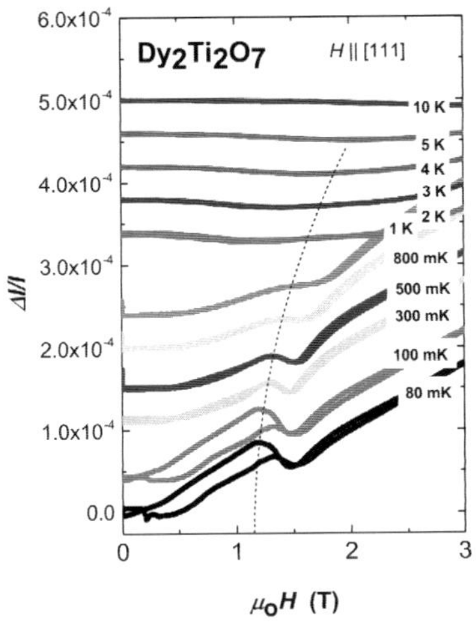

Figure 3. Magnetostriction of $Dy_2Ti_2O_7$. (The data are shifted for a better visibility).

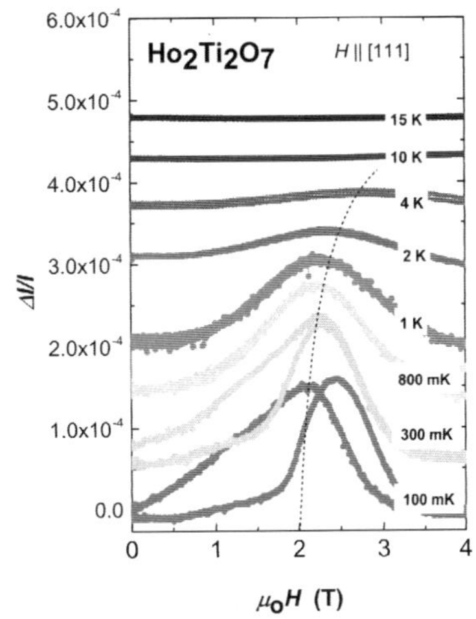

Figure 4. Magnetostriction of $Ho_2Ti_2O_7$. (The data are shifted for a better visibility).

Acknowledgments
We acknowledge support by the Deutsche Forschungsgemeinschaft within CRC 1143. The work was also supported by HLD at HZDR, member of the European Magnetic Field Laboratory (EMFL). We wish to thank O. Petrenko (University of Warwick, UK) as well as C. Wiebe and H. Zhou (University of Winnipeg, Canada) providing us with the single-crystalline samples.

References

[1] Ekreem N, Olabi A, Prescott T, Rafferty A and Hashmi M 2007 *Mat. Proc. Techn.* **191** 96
[2] Castelnovo C, Moessner R and Sondhi S 2008 *Nature* **451** 42–45
[3] Bramwell S, Giblin S, Calder S, Aldus R, Prabhakaran D and Fennell T 2009 *Nature* **461** 956
[4] Gingras M and Clarty P M 2014 *Rep. Prog. Phys.* **77** 056501
[5] Lang M 1991 *Dissertation, TH Darmstadt*
[6] Hill R, Janssen T, Meeson P, Cuthbert M, Springford M and Wasserman A 1996 *Czech. J. Phys.* **4, S4** 2077
[7] Schmiedeshoff G, Lounsbury A, Luna D and et al 2006 *Rev. Sci. Instrum.* **77** 123907
[8] Küchler R, Bauer T, Brando M and Steglich F 2012 *Rev. Sci. Instrum.* **83** 095102
[9] Abe S, Sasaki F, Oonishi T and et al 2012 *Cryogenics* **52** 452
[10] Rotter M, Müller H, Gratz E, Doerr M and Loewenhaupt M 1998 *Rev. Sci. Instrum.* **69** 2742–2746
[11] Heremans J, Michenaud J, Iye Y and et al 1983 *J. Phys. E: Sci. Instrum.* **16** 382
[12] Kido G 1989 *Physica B* **155** 199
[13] Sakon T, Takaha A, Matsuoka Y and et al 2004 *Jap. J. of Appl. Phys.* **43** 7467
[14] Doerr M, Lorenz W, Neupert T and et al 2008 *Rev. Sci. Instrum.* **79** 063902
[15] Daou R, Weickert F, Nicklas M, Steglich F, Haase A and Doerr M 2010 *Rev. Sci. Instrum.* **69** 033909
[16] Daou R, Haase A, Doerr M and et al 2011 *J. Phys.: Conf. Ser.* **273** 012111
[17] Jaime M, Daou R, Crooker S and et al 2012 *PNAS* **109** 12404
[18] Grockowiak A, Graf D, Coniglio W and et al 2016 *MAR16 APS meeting* **unpublished**
[19] Katsufuji T 2004 *Phys. Rev. B* **69** 064422
[20] Saito M, Higashinaka R and Maeno Y 2005 *Phys. Rev. B* **72** 144422
[21] Hiroi Z, Matsuhira K, Takagi S, Tayama T and Sakakibara T 2002 *J. Phys. Soc. Japan* **72** 411
[22] Kolland G, Valldor M, Hiertz M, Frielingsdorf J and Lorenz T 2013 *Phys. Rev. B* **88** 054406
[23] Krey C, Legl S, Dunsinger S and et al 2012 *Phys. Rev. Lett.* **108** 257204
[24] Erfanifam S, Zherlitsyn S, Yasin S and et al 2014 *Phys. Rev. B* **90** 064409

IOP Conf. Series: Journal of Physics: Conf. Series **903** (2017) 012005 doi:10.1088/1742-6596/903/1/012005

Spin Anisotropy in $Cu(tn)Cl_2$: A Quasi-Two-Dimensional $S = 1/2$ Spatially Anisotropic Triangular-Lattice Antiferromagnet

R Tarasenko[1], A Orendáčová[1], E Čižmár[1], M Orendáč[1], I Potočňák[2], S Zvyagin[3] and A Feher[1]

[1] Institute of Physics, Faculty of Science, P.J. Šafárik University, Park Angelinum 9, 041 54 Košice, Slovak Republic
[2] Institute of Chemistry, Faculty of Science, P.J. Šafárik University, Moyzesova 11, 0414 54 Košice, Slovak Republic
[3] Dresden High Magnetic Field Laboratory (HLD), Helmholtz-Zentrum Dresden-Rossendorf and TU Dresden, D-01314 Dresden, Germany

E-mail: robert.tarasenko@upjs.sk

Abstract. The angular dependence of electron paramagnetic resonance spectra of $Cu(tn)Cl_2$ single crystals has been studied in the X-band frequency range in the temperature range from 2.5 to 300 K. The presence of an easy-plane exchange anisotropy with $J_z/J_{x,y} < 1$ and g-factor anisotropy, $g_z/g_{x,y} > 1$ was found from analysis of the variation of the resonance fields with temperature.

1. Introduction

$Cu(tn)Cl_2$ (tn = 1,3-diaminopropane) has been previously identified as a potential realization of a spatially-anisotropic triangular Heisenberg antiferromagnet with spin 1/2, intralayer exchange coupling, $J/k_B \approx -3$ K, and interlayer exchange coupling, $J' \approx 0.001\ J$. Previous specific-heat studies in zero magnetic field did not show a phase transition to long-range magnetic order down to 60 mK. These studies indicated a field-induced anomaly forming below 1 K in magnetic fields up to 7 T and this anomaly was assigned to a Berezinskii-Kosterlitz-Thouless phase transition [1].

In this paper, we focus on the analysis of angular dependence of electron paramagnetic resonance spectra (EPR) of $Cu(tn)Cl_2$ single crystals with the aim to indicate the type of a spin anisotropy in the studied system, which plays an important role in stabilization of magnetic order.

2. Experimental details

The crystal structure of $Cu(tn)Cl_2$ determined at 150 K is orthorhombic, space group $Pna2_1$ with the lattice parameters a = 17.956 Å, b = 6.859 Å, c = 5.710 Å and Z = 4 [2]. The structure consists of covalent ladders running along the c axis, while adjacent ladders are linked through hydrogen bonds (Figure 1a).

Content from this work may be used under the terms of the Creative Commons Attribution 3.0 licence. Any further distribution of this work must maintain attribution to the author(s) and the title of the work, journal citation and DOI.
Published under licence by IOP Publishing Ltd

The 4 + 2 local coordination can be considered to be axially symmetric to a good approximation. The axial symmetry permits choosing a local (*xyz*) coordinate system with the *z* axis parallel with the crystallographic *c* axis and *x*, *y* axes parallel with the directions of the chemical bonds within the equatorial plane. The angle between the *x*, *y* and *a*, *b* axes is about 45° (Figure 1b).

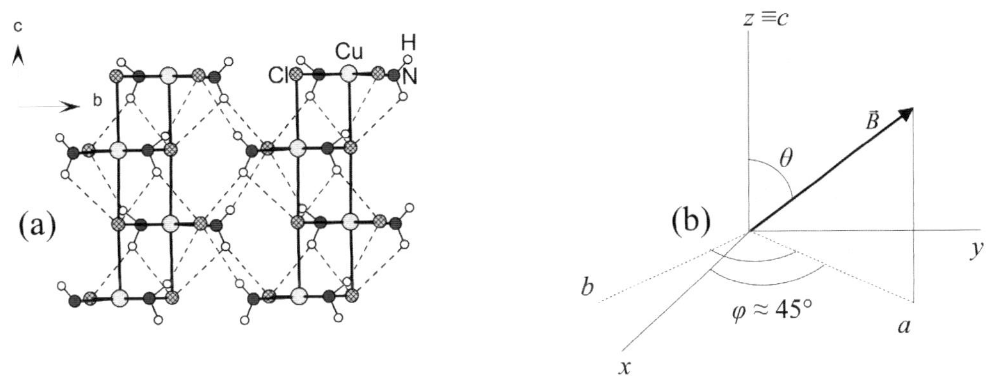

Figure 1. (a) Crystal structure and a unit cell of $Cu(tn)Cl_2$ projected to the *bc* plane. Carbon atoms are omitted for the sake of clarity. The dashed lines represent hydrogen bonds. (d) Orientation of local anisotropy axes *x*, *y*, *z* with respect to crystallographic *a*, *b*, *c* axes.

Using a modified method published in Ref. [2], $Cu(tn)Cl_2$ single crystals were prepared in the form of blue flat plates with typical dimensions $a' \times b' \times c' = 1.5 \times 1 \times 0.5$ mm^3. Using X-ray study the longest crystal edge a' was identified with the crystallographic *c* axis, b' edge is parallel to the *b* axis and the shortest crystal edge c' is parallel to the *a* axis.

EPR angular measurements of a single crystal within *ac* plane have been performed at Dresden High Magnetic Field Laboratory using an X-band spectrometer (Bruker ELEXSYS E500) at the fixed frequency of 9.4 GHz and magnetic fields up to 0.5 T.

3. Results and discussion

Angular variation of the *g*-factor of $Cu(tn)Cl_2$ in the *ac* plane in the temperature range from 2.5 to 300 K is presented in Figure 2a. Experimental data were fitted using the standard relation

$$g^2 = g_z^{\ 2}\cos^2\theta + g_x^{\ 2}\sin^2\theta\cos^2\varphi + g_y^{\ 2}\sin^2\theta\sin^2\varphi, \qquad (1)$$

where g_x, g_y, and g_z are the *g*- factors corresponding to the local anisotropy axes *x*, *y*, and *z*. θ is the angle between the magnetic field and the local *z* axis and φ is the angle between the projection of the magnetic field in the *xy* plane and the local *x* axis (Figure 1b).

A possible deviation of the (*xyz*) frame from the actual local anisotropy axes was introduced by free fitting parameters, p_θ and p_φ, taking $\theta = \Gamma + p_\theta$ and $\varphi = p_\varphi$ in Eq. (1). Γ expresses angle of rotation within *ac* plane. We found the nearly temperature-independent parameter $p_\theta = (90 \pm 0.5)^\circ$. The parameter p_φ changed monotonously from 41° at 2.5 K to 52° at room temperature, which is rather close to the value 45°. The obtained corrections correspond to the choice of the coordinate system as depicted in Figure 1b.

The fits also provided g_x, g_y, and g_z parameters, which systematically vary with temperature (Figure 2b). The increase of g_z and decrease of g_x, g_y parameters at low temperatures, is typical for the presence of an effective exchange anisotropy. At higher temperatures, an increase of the *g* parameters appears, probably as a result of a thermal lattice expansion. The fits performed in the temperature interval from 10 to 100 K using Eq. (5) in Ref. [3] and Curie paramagnetic temperature, Θ_C = - 4 K from Ref. [2], yield the high-temperature values g_x^0 = 2.035, g_y^0 = 2.064, g_z^0 = 2.241 and effective spin-anisotropy parameters $J_{xx}/k_B \approx 0$ K, $J_{yy}/k_B \approx -0.02$ K, and $J_{zz}/k_B \approx -0.17$ K. The latter approximate easy-plane anisotropy in Cu(tn)Cl_2. Nearly zero value of J_{xx}/k_B suggests a presence of easy axis within the easy plane, parallel to the *x*-axis.

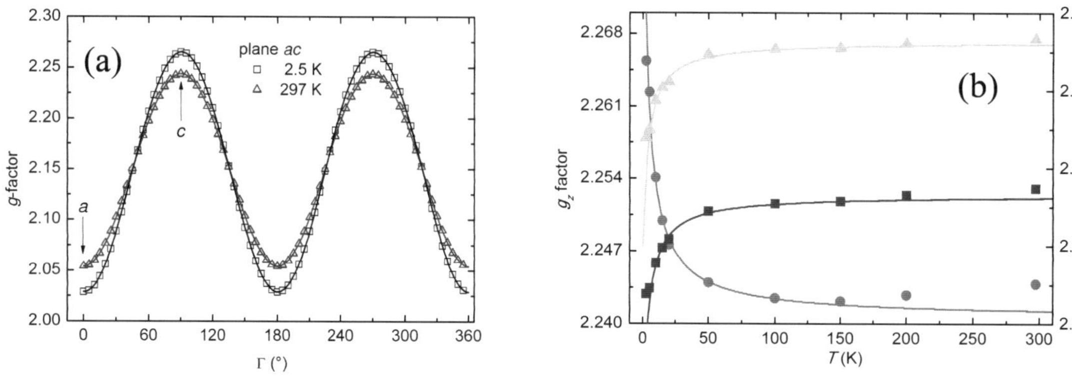

Figure 2. (a) Angular variation of the *g*-factor of Cu(tn)Cl_2 in the *ac* plane. For clarity, only data at selected temperatures are shown. The solid lines represent the fits to Eq. (1). (b) Temperature dependence of g_x (square), g_y (triangle), and g_z (circle) parameters as obtained from the fits of *g*-factor variation within the *ac* plane. The solid lines represent fitting curves (see text).

4. Conclusions

The analysis of EPR spectra of Cu(tn)Cl_2 in the *ac* plane at various temperatures revealed the presence of a spin exchange anisotropy with the easy-plane character. In future, the presence of easy-plane anisotropy will be verified by single crystal measurement of magnetic susceptibility at temperatures below 0.1 J/k_B [4].

Acknowledgments

This work was supported by the project APVV LPP-0202-09, VEGA grant No. 1/0143/13 and ERDF EU project. No. ITMS26220120005. Material support from U.S. Steel Košice s.r.o. is greatly acknowledged.

References

[1] Orendáčová A *et al* 2009 *Phys. Rev.* B **80** 144418
[2] Zeleňák V *et al* 2006 *Inorg. Chem.* **45** 1774
[3] Krug von Nidda H A *et al* 2002 *Phys. Rev.* B **65** 134445
[4] Cuccoli A *et al* 2003 *Phys. Rev.* B **67** 104414

IOP Conf. Series: Journal of Physics: Conf. Series **903** (2017) 012006 doi:10.1088/1742-6596/903/1/012006

Energy surface and transition rates in a hexagonal element of spin ice

S Y Liashko[1,2], **V M Uzdin**[2,3], **H Jónsson**[1,4]
[1]Science Institute and Faculty of Physical Sciences, VR-III, University of Iceland, Reykjavík, Iceland
[2]Department of Natural Sciences, University ITMO, St. Petersburg, 197101 Russia
[3]Department of Physics, St. Petersburg State University, St. Petersburg, 198504 Russia
[4]Department of Applied Physics, Aalto University, FIN-00076 Espoo, Finland

Abstract. The energy surface of a hexagonal element of Kagome spin ice consisting of six prolate magnetic islands is investigated as a function of applied magnetic field. Minimum energy paths for magnetic reversals are determined to estimate energy barriers and the transition rates estimated using harmonic transition state theory for magnetic systems. The overall transition rate between equivalent ground states is calculated using the stationary state approximation including all possible transition paths. The calculated transition rates are in close agreement with reported experimental measurements taken in the absence of an applied field. Predictions are made for the change in the energy landscape and transition rates as a magnetic field is applied.

1. Introduction

Artificial spin ice systems make it possible to study new magnetic phenomena at various spatial and time scales ranging from atomic to micromagnetic dimensions. Temperature induced magnetic switching of magnetization has been observed experimentally with typical lifetime ranging from seconds to days as temperature changes from 300 K to 400 K [1]. Here, we present results of calculations of the rate of magnetisation reversals using rate theory and analysis of the multidimensional energy surface. We study a hexamer of islands arranged as a unit of a Kagome lattice. Since the size of a typical island in spin ice is a few nanometers, the interaction between islands is dominated by the dipole-dipole interaction. The total energy of the hexamer is written as

$$E = \sum_i^6 V\left(\vec{M}_i \vec{K}_{out}\right)^2 - \sum_i^6 V\left(\vec{M}_i \vec{K}_{in}\right)^2 - \frac{1}{2}\sum_{i\neq j}^6 \left(\frac{V^2(\vec{M}_i \vec{M}_j)}{|\vec{r}_{ij}|^3} - \frac{3V^2(\vec{M}_i \vec{r}_{ij})(\vec{M}_j \vec{r}_{ij})}{|\vec{r}_{ij}|^5}\right) - \sum_i^6 V(\vec{M}_i \vec{H}) \quad (1)$$

where $\vec{K}_{out}$ and $\vec{K}_{in}$ are respectively the in-plane and out-of-plane shape anisotropy. The dimensions of an island are chosen in such a way that the volume is $V = 470 \times 170 \times 3$ nm^3, the saturation magnetization $M_s = 166 \times 10^3$ A/m and the anisotropy $K_{in} = 4550$ erg/cm^3 $K_{out} = 17991$ erg/cm^3 to mimic an experimentally measured system [1]. The values of K_{in} and K_{out} are obtained from micro-magnetic simulations given the volume and saturation magnetization. H is the external magnetic field applied in plane, perpendicular to the long side of the top and bottom islands. The applied magnetic field strongly changes the energy surface both the location of energy minima and the height of energy barriers between states. The geodesic nudged elastic band (GNEB) method [2] is used to find minimum energy paths (MEPs) for various values of H.

Content from this work may be used under the terms of the Creative Commons Attribution 3.0 licence. Any further distribution of this work must maintain attribution to the author(s) and the title of the work, journal citation and DOI.
Published under licence by IOP Publishing Ltd

IOP Conf. Series: Journal of Physics: Conf. Series **903** (2017) 012006 doi:10.1088/1742-6596/903/1/012006

2. Lifetime calculation

By identifying all the possible paths, the reversal of the total magnetization of the hexamer and the transition rate for each of the elementary steps, the lifetime of the system in a particular ground state can be estimated. The rate is estimated using harmonic transition state theory (HTST) for magnetic systems [3]. The rate constant of an elementary transition, k, is given by the Arrhenius law:

$$\mathrm{k} = A\,exp\left(-\frac{E_a}{k_B T}\right) \tag{2}$$

The steady state approximation is applied to determine the lifetime of the ground state as a function of external field strength. The set of master equations, i.e. differential equations for the variation of occupation probabilities of the various states (see figure 1) as a function of time, is

$$\begin{cases} \frac{dn_1}{dt} = -6W_{1,2}n_0 + 6W_{2,1}n_2 \\ \frac{dn_2}{dt} = -(2W_{2,3} + W_{21})n_2 + W_{2,1}n_1 + 2W_{2,3}n_3 \\ \frac{dn_i}{dt} = -(2W_{i,i+1} + 2W_{i,i-1})n_i + 2W_{i,i-1}n_{i-1} + 2W_{i,i+1}n_{i+1} \\ \dots \\ \frac{dn_6}{dt} = -W_{6,7}n_6 - 2W_{6,5}n_6 + 2W_{5,6}n_5 \end{cases} \tag{3}$$

where n_i is the occupation probability of a particular state. $W_{i,j}$ is the transition rate between states i and j. In order to find the overall transition rate between ground states of the hexamer, eqn. (3) is solved in terms of n_0 to give the equation $\frac{dn_1}{dt} = -n_1(G)$, where the G is a combination of the $W_{i,j}$. Each elementary transition rate, $W_{i,j}$, is given by eqn. (2) with E_a obtained from calculated MEPs and A obtained from HTST [3]. From the temperature dependence of the overall rate, an effective activation energy and pre-exponential factor can be extracted.

3. Results

The applied field strongly affects the energy surface as can be seen in figure 1. The energy of the various local minima and the energy barriers between them change. Some of the intermediate energy minima disappear when an even stronger field is applied.

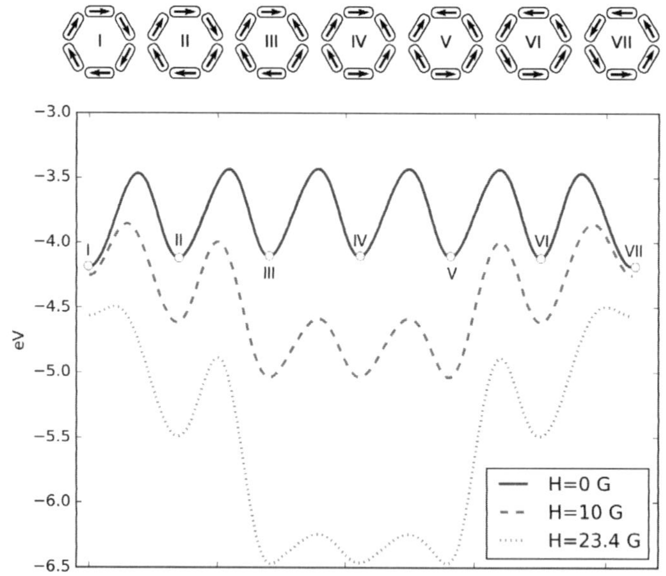

Fig. 1. Minimum energy paths between the clockwise and anti-clockwise ground states for various values of the applied magnetic field, H.

By using the methodology described above, we first studied the case where an applied field is not present. The lifetime of a ground state of the kagome ring was estimated to be 12 seconds at 420K but several days at 300K. These values are in close agreement with the experimental measurements [4]. The effective pre-exponential factor turns out to be ca. 10^8 s^{-1} and the effective activation energy 0.74 eV. These values are, however, quite different from what has been used in previous modeling of the system [1]. When the external field is included, the lifetime of the various intermediate states as well as that of the ground states changes. Some of the states become less stable, as shown in figure 1, and even completely disappear as local minima on the energy surface when the field is strong.

It can be important to take into account the internal magnetic structure of the islands during spin transitions. This is schematically illustrated in figure 2a.

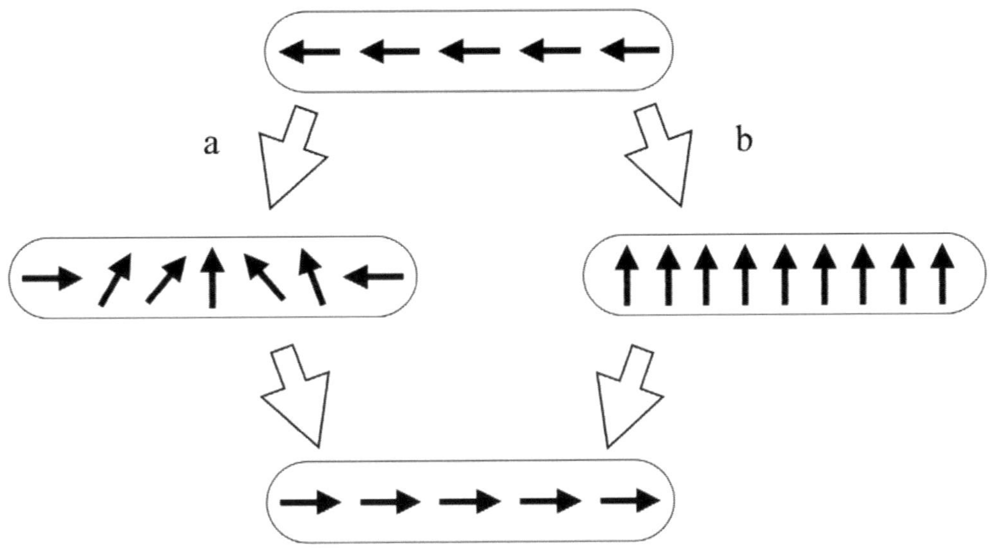

Fig. 2. A schematic comparison of different magnetization reversal mechanisms: (a) Each island is described as a set of interacting magnetic moments that can rotate in different ways; (b) Each island is described as a macrospin undergoing coherent rotation. The former description is needed for long enough islands.

When an island is long compared with the length scale characterizing the exchange coupling of spins within the island, the macrospin approximation illustrated in figure 2b overestimates the energy barrier for the transition. The magnetization reversal of an island then occurs via a temporary domain wall mechanism, as has recently been illustrated for Fe islands on a substrate [3]. This effect will in future work be included in long time scale simulations of spin ice systems using the AKMC method [5].

References:

[1] Farhan A, Derlet P M, Kleibert A, Balan A, Chopdekar R V, Wyss M, Anghinolfi L, Nolting F, Heyderman L J 2013 Nature Physics **9** 375.

[2] Bessarab P F, Uzdin V M, Jónsson H 2015 Comp. Phys. Commun. **196**, 335.

[3] Bessarab P F, Uzdin V M, Jónsson H 2013 Phys. Rev. Letters **110** 020604; 2012 Phys. Rev. B **85**, 184409.

[4] Farhan A, Derlet P M, Kleibert A, Balan A, Chopdekar R V, Wyss M, Anghinolfi L, Nolting F, Heyderman L J 2013 Nature Physics **9** 375.

[5] Pedersen A, Jónsson H 2009 Acta Materialia 57 4036; Jónsson H 2011, PNAS 108 944.

Acknowledgment

This work was financially supported by Government of Russian Federation, Grant 074-U01, RFBR grant №14-02-00102 and the Icelandic Research Fund.

CoFe-microwires with stress-dependent magnetostriction as embedded sensing elements

M.M. Salem[1,3], M.G. Nematov[1], A. Uddin[1], L.V. Panina[1,2], M.N. Churyukanova[1] and A. T. Marchenko[1]

[1] National University of Science and Technology, MISiS, Moscow 119991, Russia
[2] Institute for Design Problems in Microelectronics RAS, Moscow 124681, Russia
[3] Physics Department, Faculty of Science, Tanta University, Tanta 31527, Egypt

E-mail: elshshtawy@science.tanta.edu.eg, lpanina@plymouth.ac.uk

Abstract. Testing internal stress/strain condition of polymer composite materials is of high importance in structural health monitoring. We are presenting here a new method of monitoring internal stresses. The method can be referred to as embedded sensing technique, where the sensing element is a glass-coated ferromagnetic microwire with a specific magnetic anisotropy and stress-dependent magnetostriction. When the microwire is remagnetized the sharp voltage is induced which is characterized by high frequency harmonics. The amplitude of these harmonics sensitively depends on various stresses. The microwire of composition $Co_{71}Fe_5B_{11}Si_{10}Cr_3$ with the metallic core diameter of 22.8 μm show abrupt transformation of the magnetization process under applied tensile stress owing to the stress-dependent magnetostriction.

1. Introduction.

Magnetostrictive materials are widely used for stress and strain sensing. When a mechanical force is applied to a magnetostrictive material, its magnetization state changes. This can be measured and inversely translated into a corresponding indication of stress or strain change. In many cases it is assumed that a good magnetostrictive material should have a large magnetostrictive constant λ_s as Terfenol-D ($Tb_xDy_{1-x}Fe_{2-y}$, $x \approx 0.3,\ 0 < y < 0.2$) with $\lambda_s \approx 1.6 \cdot 10^{-3}$[1]. The situation can be different for amorphous or nanocrystalline materials where the intrinsic crystalline anisotropy is small. In this case, the main source of the anisotropy is due to anisotropic stress-magnetostriction coupling effect. It is well known that the magnetic structure in amorphous microwires with small magnetostriction $\lambda \approx 10^{-7} - 10^{-6}$ can be easily changed by a moderate stress [2-4]. Therefore, the magnetization process will be highly stress-dependent which can be used for sensitive stress sensing, for example, based on magnetoimpedance effect in near-zero magnetostrictive wires [5-6].

The stress sensitivity of the magnetization state in low magnetostrictive metallic glasses may be further enhanced by the dependence of the saturation magnetostriction on the applied tensile stress σ_{ex} [7-9]. Typically, the behavior is described by a linear relationship:

$$\lambda_s(\sigma_{ex}) = \lambda_s(0) - \alpha\sigma_{ex}$$

where the parameter α is found to be in the range of $(1-6) \cdot 10^{-10}$MPa^{-1}. Similar effects on stress dependence of the saturation magnetostriction were recently observed in nanocrystalline alloys [10].

A number of mechanisms were proposed to explain such magnetostriction behavior. In many cases, the phenomenon of stress-dependent magnetostriction can be attributed to local atomic

Content from this work may be used under the terms of the Creative Commons Attribution 3.0 licence. Any further distribution of this work must maintain attribution to the author(s) and the title of the work, journal citation and DOI.
Published under licence by IOP Publishing Ltd

rearrangements resulting in the redistribution of local anisotropy axes. The coexistence of two amorphous phases, elastically coupled, with different magnetostriction to each other may also lead to the stress dependence of magnetostriction. The purpose of this paper is not to elucidate a particular mechanism of the stress dependence of magnetization, but to demonstrate how this phenomenon can be used in embedded stress sensors. The sensing principle is based on the stress- sensitivity of the harmonic spectrum originated by the wire remagnetization process.

2. Experimental details

Glass-coated $Co_{71}Fe_5B_{11}Si_{10}Cr_3$ amorphous wires with small magnetostriction constant have been studied. The wires were produced by the modified Taylor-Ulitovski technique [11,12] based on direct casting from the melt. The wires with different geometry were used: having the total diameter of 29.5 and 41.5 μm and the metallic core diameter of 23.9 and 36.3 μm, respectively for samples labeled as No (1) and No (2). For structural characterization, differential scanning calorimetry (DSC) measurements were performed using (DSC 204 F1 Netzsch) calorimeter in Ar atmosphere at a heating rate of 10 K/min. The mass of the samples was 10-15 mg. The extent of crystallization, Curie temperature and crystallization temperature were estimated from DSC curves using standard IT application.

The magnetization curves were measured by an inductive method with two differential coils under the effect of external load by hanging weights up to 35 g. The magnetization coil was excited by a current of frequency 500 Hz, which produced a magnetizing field with the amplitude of 1000 A/m. The induced voltage vs time was digitally integrated to obtain a hysteresis loop. The wire was inserted into a narrow magnetization/detection coil having the inner diameter of 3 mm and the length of 5 cm. This stress is redistributed between the metal core and glass shell which have different Young's moduli ($E_m = 130\ GPa, E_g = 70 GPa$) and cross-sections ($A_m, A_g$), for metal core and glass shell, respectively. The stress σ_{ex} in the metal core can be estimated by using the method of consistent deformations caused by a gravitational force F:

$$\sigma_{ex} = \frac{E_m F}{E_m A_m + E_g A_g}$$

Magnetostriction was measured by small-angle magnetization rotation technique [13,14].

The harmonics of the voltage generated at remagnetization are measured with the help of a lock in amplifier (Signal Recovery 5210) and function generators [15].

3. Result and discussion

The DSC curves are shown in Figure 1. It is seen that the crystallization proceeds in two stages for both types of samples; however, for sample (1) with smaller diameter the main peak is much narrower. This indicates a partial crystallization of sample (2). From the ratio of the peak areas the proportion of the crystalline phase has been evaluated as 12.5%. The Curie temperature is defined from the position of the inflexion point which is clearly seen for sample (1) at 364.4 °C as shown in insert of Figure 1. For sample (2) inflexion is not noticeable which also supports the conclusion about its partial crystallization. The partial crystallization of sample (2) can be attributed to its larger diameter.

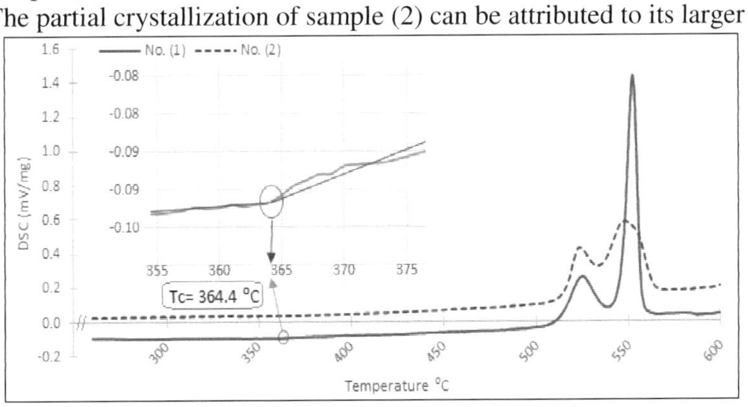

Figure 1: DSC curves for samples (1) (solid red curve) and (2) (dashed blue curve). Insert shows the inflexion point on the DSC curve for sample (1).

The results for the saturation magnetostriction are shown in Figure 2. For sample (2) the value of λ_s is positive and much larger owing to partial crystallization. For sample (1), the magnetostriction is negative and the absolute value increases with increasing σ_{ex}. For linear portion, the gradient of this increase is $2.7 \cdot 10^{-10}\ MPa^{-1}$ which is consistent with previously reported results [16,17]. The value of the magnetostriction near zero stress is questionable since reliable measurements to avoid the effect of the off-axis anisotropy require the application of a certain stress. However, linear extrapolation gives a value close to zero (smaller than 10^{-8}). The sign is unclear but considering the hysteresis loops without stress suggests that λ_s is positive for $\sigma_{ex} = 0$.

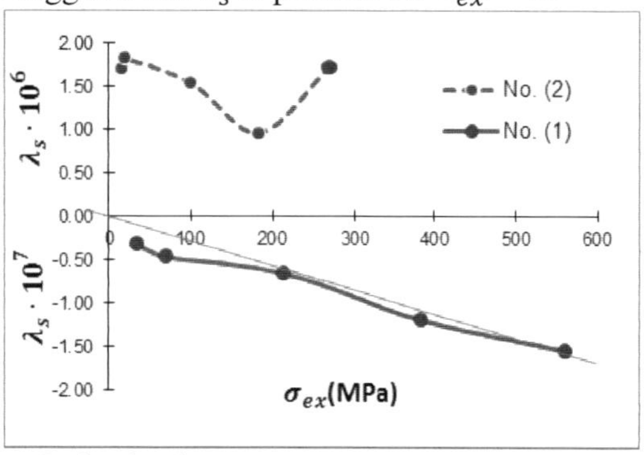

Figure 2: Saturation magnetostriction as a function of applied tensile stress for as-prepared $Co_{71}Fe_5B_{11}Si_{10}Cr_3$ wires two different diameters. Linear extrapolation (shown by blue line) for sample 1 gives λ_s close to zero at $\sigma_{ex} = 0$.

Indeed, when no stress is applied the hysteresis loops for both samples are of bistable type and almost identical with small coercitivity of about 25 A/m. The value of remanence close to saturation magnetization confirms the existence of the axial anisotropy nearly in the entire wire. Applying a tensile stress results in sharp transformation of the hysteresis for sample (1): for stress higher than 200 MPa the loop becomes inclined. For sample (2), the loop remains a rectangular form but the coercivity increases. The change in remanence and coercivity under the effect of tensile stress is shown in Figure 3.

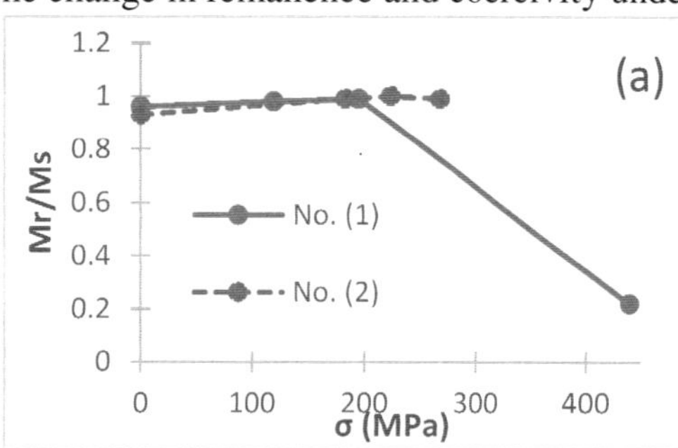

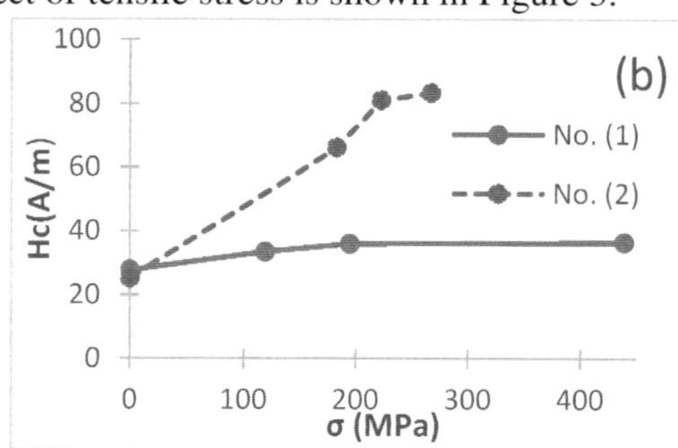

Figure 3: Effect of tensile stress on remanence (a) and coercivity (b) of $Co_{71}Fe_5B_{11}Si_{10}Cr_3$ wires with different diameters.

In all types of wire samples, the voltage induced by the wire re-magnetization is characterized by the generation of high harmonics. However, only for wire sample (1) their amplitudes sensitively change under the effect of stress as shown in Figs. 4 (a, b) and the sensitivity with respect to stress increases for higher harmonics. It is interesting to notice that for wire sample (2) with a bistable type of hysteresis for all applied stresses and the coercivity increasing with σ_{ex}, there is no dependence of the harmonic spectrum on σ_{ex}.

For practical applications, it is interesting to investigate the stress dependence of the harmonics ratio since this parameter is not dependent on a particular measurement scheme. This was done for sample (1) which demonstrates a considerable stress sensitivity of the harmonic spectra. Figure 5 shows the stress dependences of higher frequency harmonics normalized to the amplitude of the third harmonic. The normalized harmonics change by 10-15% per 100 MPa in the stress sensitive region ($\sigma_{ex} >$ 200 MPa) depending on the harmonics number. The change in the induced harmonic spectra may give information about the strain/stress state inside materials which otherwise is very difficult to measure. The methods to avoid insensitive stress-range should be further investigated.

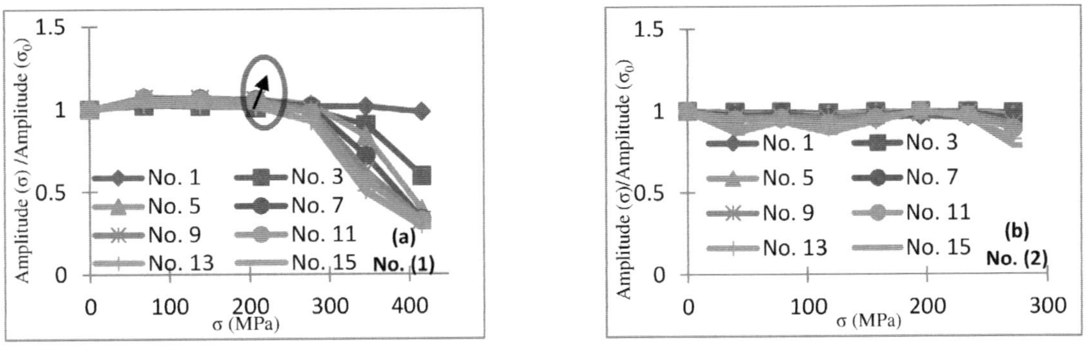

Figure 4: Amplitudes of high harmonics as a function of applied stress: (a) for wire sample (1) and (b) for wire sample (2). The amplitudes are normalized by their values at zero applied stress.

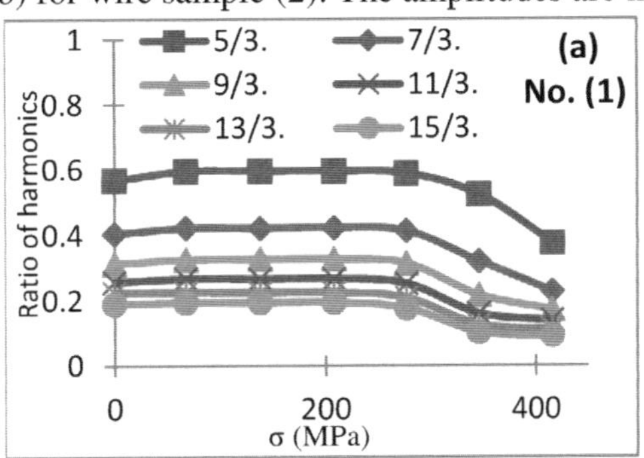

Figure 5: Ratio of higher harmonics amplitudes to the amplitude of the third harmonic as a function of applied stress for wire sample (1).

In summary, we have proposed to use the effect of stress-dependent magnetostriction for the development of embedded wireless stress sensors which is based on sharp transformation of the hysteresis loops and stress-sensitive harmonic spectrum.

L. V. Panina acknowledges the support for this work under Russian Federation State Contract for organising a scientific work.

References

[1] D.C. Jiles, J.B. Thoelke 1994 *J. Magn. Magn. Mat.* V **134, No 1**, pp. 143-160
[2] J. Olivera, M. González, J. V. Fuente, R. Varga, et al, 2014 *Sensors*, V. **14**, pp. 19963-19978
[3] V. Zhukova, A. Zhukov, J.M. Blanko, et al, 2003 *J. appl. Phys.* V. **93**, pp. 7208-7210
[4] F. X. Qin, H. X. Peng, V. V. Popov, L. V. Panina, M. Ipatov, V. Zhukova, A. Zhukov, and J. Gonzalez 2011 *J. Appl. Phys.* V. **109**, **No 7**, Article ID 07A310
[5] K. Mandal, S. Puerta, M. Vazquez, et al,2000 *IEEE Trans. Magn.* V. **36**, **No 5**, pp. 3257-3259
[6] V. V. Popov, V. N. Berzhansky, et al, 2013. *J. Appl. Phys.* V. **1139**, **No 17**, Article ID 17A326.
[7] H. Szymczak 1987 *J. Magn. Magn. Mat.* V. **67**, **No 2**, p.227-231
[8] J. M. Barandiarán, A. Hernando, V. Madurga, O. V. Nielsen, et al, 1987, *Phys. Rev.* B **35**, 5066
[9] A.Zhukov, Churyukanova, et al, 2016 *J. Electronic Materials,* V. **45, No 1**, pp. 226-234
[10] G. Herzer, S. Flohrer, C. Polak, 2010. *IEEE Trans. Magn.* V.**46, No 2**, p.341-344
[11] A.V. Ulitovski, 1950, Author certification, USSR patent, No. 128427, 3.9
[12] S.A. Baranov, et al, 1989 Phys. Met. Metall. **V 67**, pp.73
[13] J. Yamasaki, Y. Ohkubo, and F. B. Humphrey, 1990 *J. Appl. Phys.* V **67, No 9**, pp. 5472-5474
[14] R. Grössinger, et al, 2012 *Journal of Electrical Engineering,* V. **63, No 7s**, pp. 9-14
[15] T. Shirane , M. Ito. 2012. *IEEE Trans. Magn.* V**. 48, No 4,** pp. 1437-1440
[16] V. Zhukova et al, 2003 *Journal of Magnetism and Magnetic* Materials, **V**.254–255 pp.94–96
[17] A. Zhukov, et al, 2000 *J. Appl. Phys.*, V.87, pp. 1402-1408

Noise study of magnetic field sensors based on magnetic tunnel junctions

M. Mouchel[1,2,3,4], A. Bocheux[5], C. Ducruet[1], Ph. Sabon[2,3,4], I.-L. Prejbeanu[2,3,4], Y. Conraux[1], J. Alvarez-Hérault[1], K. Mackay[1] and C. Baraduc[2,3,4]

[1] Crocus Technology, 4 Place Robert Schuman, 38025 Grenoble cedex, France
[2] Univ. Grenoble-Alpes, INAC-SPINTEC, F-38000 Grenoble, France
[3] CEA, INAC-SPINTEC, F-38000 Grenoble, France
[4] CNRS, SPINTEC, F-38000 Grenoble, France
[5] LPC2E, 3A, avenue de la Recherche Scientifique, 45071 Orléans cedex 2, France

E-mail: myckael.mouchel@cea.fr

Abstract. Low frequency noise has been studied for two types of magnetic field sensors based on magnetic tunnel junctions (MTJ). The first structure, composed of a few large MTJs, is designed for low noise applications; the second one, composed of hundreds of small MTJs, is designed for general purposes. At low frequency, both structures exhibit $1/f$ noise, but with very different amplitudes. The sensors for general purposes show a much higher noise level compared to the low-noise sensors. However, the sensitivity of the low noise sensors is much smaller compared to the other ones. Thus, the limit of detection, defined as the ratio of noise and sensitivity, turns out to be roughly the same for both technologies. Using the advantages of each sensor could help to design a sensor with an improved limit of detection.

1. Introduction

Magnetic field sensors based on Magnetic Tunnel Junctions (MTJs) could become a serious alternative to other technologies like Hall effect devices [1], thanks to their low power consumption, good sensitivity and robustness against radiations. These junctions are composed of two ferromagnetic layers separated by a tunnel barrier and their resistance depends on the relative orientation of the magnetizations of both layers [2]. To use them at low-frequency, such magnetic field sensors need to have low noise levels. Moreover, at low frequency, $1/f$ noise may restrict the limit of detection of the sensors [3]. This noise can be expressed as $\frac{\alpha_{Hooge}V^2}{Af}$, where V is the voltage across the MTJ, A its area, f the frequency and α_{Hooge} the Hooge coefficient. This noise is usually high compared to *Johnson* or *shot noise* and has to be reduced in order to improve the performances of MTJ-based sensors at low frequencies. Its proportionality factor α_{Hooge} called Hooge coefficient [4] characterizes the noise level. It can be understood as the *noise quality factor* of the junction and is a good parameter to compare different technologies or solutions [5]. Another important characteristics of a magnetic sensor is its limit of detection, *i.e.* the lowest measurable field under operating conditions. It is defined as the ratio between noise and sensitivity and is expressed in $T/\sqrt{\text{Hz}}$. It is frequency-dependent due to $1/f$ noise [6]. Its value is therefore estimated at a given frequency.

Content from this work may be used under the terms of the Creative Commons Attribution 3.0 licence. Any further distribution of this work must maintain attribution to the author(s) and the title of the work, journal citation and DOI.
Published under licence by IOP Publishing Ltd

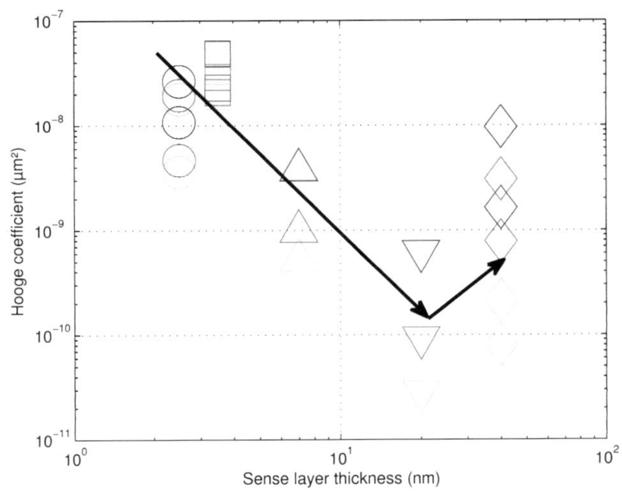

Figure 1. Hooge coefficients in parallel state (green), anti-parallel state (red), near operating point (blue) as a function of sense layer thickness. The following stacks (thicknesses in nm): □ (CoFeB/NiFe)3.5; △ (CoFeB/NiFe)7; ○ CoFeB2.5; ▽ CoFeB20; ◇ CoFeB40. The black line draws the global observed trend. The Hooge coefficient decreases as a function of sense layer thickness up to a point where the thickness becomes too high and where we suspect magnetic properties are deteriorating.

2. Samples and experiments

Two types of sensors were studied. The first type (Sensors A) consists of rectangular junctions with a surface from 4 μm^2 to 20 μm^2, specifically designed for low-noise purposes [7]. In these sensors, there are no more than 3 junctions in series. The second type of sensors (Sensors B) consists of arrays of small circular junctions with a surface of $\approx$ 0.05 μm^2. These arrays are composed of respectively, 104, 480 and 600 MTJs, in a serial / parallel configuration with respectively N_s and N_p junctions. In both cases, the sense layer is either a single CoFeB layer or a CoFeB/NiFe bilayer. Noise has been measured using a SR780 Stanford Research Spectrum Analyzer that can perform noise measurements up to 100 kHz. The noise signal is filtered and amplified using a Stanford Research SR560 low-noise pre-amplifier. The devices are biased with a battery and a potentiometer [8]. Custom coils and electromagnets are used to apply a magnetic field on the sample. Noise measurements are corrected from the measurements performed at zero bias voltage (background noise). Note that the *Johnson noise* is included in the background noise.

3. Noise in sensor A

Noise was studied for a wide range of magnetic fields along the easy and hard axes leading to a full noise characterization of these samples. Several samples were studied with different sense layer compositions and thicknesses, and different junction sizes. The results are summarized in figure 1. The lowest noise levels are recorded when the magnetizations of the reference and sense layers are parallel. Noise in anti-parallel state is always higher than in parallel state, which is consistent with previous observations showing that the noise in anti-parallel state is roughly twice the noise in parallel state [5, 9]. Nevertheless, there is no constant factor between both noise levels in our case. All other things being equal, the Hooge coefficient is expected to decrease inversely to the sense layer thickness [10]. One can see in figure 1 that the Hooge coefficient is decreasing with the sense layer thickness (see black line) until a certain point where it seems to increase again. The reversal point, which is the best sensor in terms of noise, corresponds to a sensor with a sense layer thickness of 20 nm (CoFeB20). After this limit, increasing the thickness seems to deteriorate the noise level. However, the aspect ratio varies with the junction size which has an important impact on magnetic properties and thus on noise [11, 12]. Further studies are necessary to distinguish the impacts of the junction size, aspect ratio and magnetic

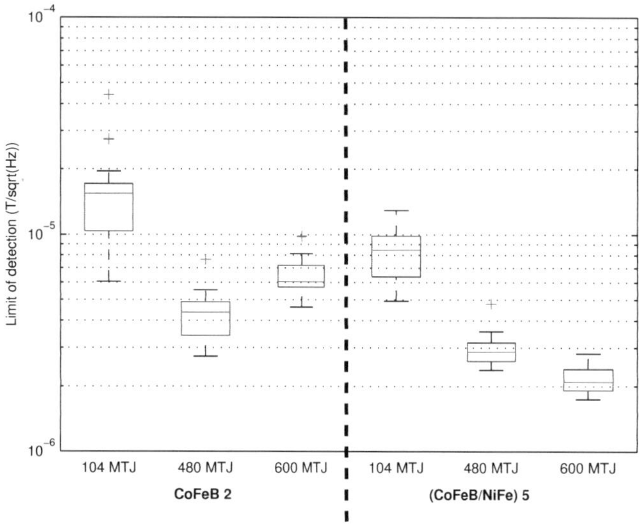

Figure 2. Limit of detection (at 1 Hz) as a function of number of MTJs in sensors B for two sense layers. Each box comprises 18 samples. Increasing the sense layer thickness and/or increasing the number of MTJs improves the limit of detection. Note that for 480-MTJ-CoFeB2 sensors, the median of limits of detection ($4.4 \times 10^{-6}\ T/\sqrt{Hz}$) is better than expected ($7 - 8 \times 10^{-6}\ T/\sqrt{Hz}$)

volume. However, minimizing noise is not sufficient for obtaining a high performance sensor. The sensitivity must also be high. To optimize the properties of a sensor requires to find the best compromise between noise and sensitivity, which means to lower the limit of detection.

4. Noise in MTJs arrays – Sensors B

We performed a statistical study of noise in sensors composed of arrays of hundreds of MTJs. Three types of sensors with different numbers of MTJs have been considered. Sensors with two different sense layers are compared: 2 nm thick CoFeB layer or 5 nm thick (CoFeB/NiFe) bi-layer. All other things being equal, the Hooge coefficient should decrease as the inverse of the sense layer thickness [10]. In our case, the expected reduction should be 2.5 which is larger than the observed reduction factor of 1.71 ± 0.24. This can be explained by the fact that the sense layer material was also changed: we compare a pure CoFeB sense layer and a bi-layer of CoFeB and NiFe. Taking into account the magnetizations of CoFeB and NiFe (respectively 1250 emu/cm^3 and 800 emu/cm^3), we can estimate a thickness of 3.92 nm for an equivalent layer composed only of CoFeB. This value would lead to a noise decrease by a factor 1.96, close to the factor observed experimentally. We can further analyze these results and consider the limit of detection of these sensors as a function of the sense layer composition and number of junctions. For each sensor we measured its sensitivity and its highest noise level inside the sensitive region, leading to an over-estimation of the sensor limit of detection. Figure 2 shows a clear dependence of the limit of detection as a function of the number of junctions and as a function of the sense layer thickness. We have calculated that the limit of detection should decrease as the square root of the number of junctions $\sqrt{N_s N_p}$. By comparison to the 104-MTJ sensors, the limit of detection of 480-MTJ and 600-MTJ sensors should decrease respectively by a factor 2.15 and 2.4. Experimentally observed reduction factor are 3.55 and 2.56 for sensors with CoFeB2, and 2.94 and 4.04 for sensors with (CoFeB/NiFe)5. These reduction factors are higher than expected. However, this discrepancy can be partially explained by the error bars on these data. Let us now compare arrays with an identical number of junctions and different sense layers. The reduction factors of the limit of detection for 104-MTJ, 480-MTJ and 600-MTJ sensors are respectively 1.83, 1.52 and 2.89, to be compared to $\sqrt{1.96} = 1.4$ as calculated above.

These two successive comparisons between experimental results and theoretical expectations show that 480-MTJ-CoFeB2 and 600-MTJ-(CoFeB/NiFe)5 sensors have better performances than expected, probably linked to fabrication details or particular micromagnetic states.

5. Conclusion

In this article we studied two different sensor technologies. The first one is based on micron-sized junctions specifically designed for low-noise applications. This goal is only partially achieved. Although, the Hooge coefficient measured both in saturated states and in sensitive region (10^{-10} to 10^{-9} μm^2 for the best sensors) is very low, the sensitivity of these sensors is rather low leading to a relatively high limit of detection (around 2 $\mu T/\sqrt{Hz}$ at 1 Hz for the best sensor A). The second technology studied was based on arrays of small MTJs with a high sensitivity comprised between 50 and 150 $V/V/T$. They exhibit higher noise with a Hooge coefficient of the order of 10^{-6} μm^2. Nevertheless, thanks to their sensitivity, their limit of detection is quite similar to sensor A (down to 1 $\mu T/\sqrt{Hz}$ at 1 Hz for 600-MTJ-(CoFeB/NiFe)5). Future work will combine the advantages of both types of sensors (A+B). For example, sensors A can be designed with junctions having a smaller aspect ratio or another shape in order to increase the sensitivity, while for sensors B, we can increase the junction size while keeping a large number of junctions, thus reducing the noise level. Further improvements of magnetic materials and anisotropy could also be implemented for both technologies.

Acknowledgments

This work has been supported by French *ANRT* through a *CIFRE* thesis grant. This work has been partially founded by *CNRS*-Défi Instrumentation aux Limites 2015.

References

[1] S. Tumanski. Modern magnetic field sensors a review. *Przeglad Elektrotechniczny*, 0033-2097(R.89), 2013.
[2] M. Julliere. Tunneling between ferromagnetic films. *Physics Letters*, 54A(3):225–226, 1975.
[3] C. Fermon and M. Pannetier-Lecoeur. Noise in GMR and TMR Sensors. In *Giant Magnetoresistance Sensors*, volume 6, pages 47–70. Springer, springer edition, 2013.
[4] F. N. Hooge. 1/f Noise. *Physica B*, 83:14–23, 1976.
[5] Z. Q. Lei, G. J. Li, W. F. Egelhoff, P. T. Lai, and P. W. T. Pong. Review of noise sources in magnetic tunnel junction sensors. *IEEE Transactions on Magnetics*, 47(3):602–612, 2011.
[6] R. Guerrero, M. Pannetier-Lecoeur, C. Fermon, S. Cardoso, R. Ferreira, and P. P. Freitas. Low frequency noise in arrays of magnetic tunnel junctions connected in series and parallel. *Journal of Applied Physics*, 105(11):113922, 2009.
[7] A. Bocheux, C. Cavoit, M. Mouchel, C. Ducruet, R. Fons, P. Sabon, I.-L. Prejbeanu, and C. Baraduc. High sensitivity magnetic field sensor for spatial applications. *To be published*, 2016.
[8] N. A. Stutzke, S. E. Russek, D. P. Pappas, and M. Tondra. Low-frequency noise measurements on commercial magnetoresistive magnetic field sensors. *Journal of Applied Physics*, 97(10):14–17, 2005.
[9] J. M. Almeida, P. Wisniowski, and P. P. Freitas. Low-frequency noise in MgO magnetic tunnel junctions: Hooge's parameter dependence on bias voltage. *IEEE Transactions on Magnetics*, 44(11):2569–2572, 2008.
[10] S. Kogan. *Electronic noise and fluctuations in solids*. Cambridge University Press, cambridge edition, 1996.
[11] J. M. Almeida, R. Ferreira, P. P. Freitas, J. Langer, B. Ocker, and W. Maass. 1f noise in linearized low resistance MgO magnetic tunnel junctions. *Journal of Applied Physics*, 99, 2006.
[12] R. Ferreira, E. Paz, P. P. Freitas, J. Wang, and S. Xue. Large area and low aspect ratio linear magnetic tunnel junctions with a soft-pinned sensing layer. *IEEE Transactions on Magnetics*, 48(11):3719–3722, 2012.

IOP Conf. Series: Journal of Physics: Conf. Series **903** (2017) 012009 doi:10.1088/1742-6596/903/1/012009

Switching Field and Resolution of MFM Tips Prepared by Coating Fe/$Co_{50}Pt_{50}$ Magnetic Films

Yuma Tomita[1], Masaaki Futamoto[1], Mitsuru Ohtake[1,2], Fumiyoshi Kirino[3] and Nobuyuki Inaba[4]

[1] Faculty of Science and Engineering, Chuo University, Tokyo 112-8551, Japan
[2] Faculty of Engineering, Kogakuin University, Tokyo 192-0015, Japan
[3] Graduate School of Fine Arts, Tokyo University of the Arts, Tokyo 110-8714, Japan
[4] Faculty of Engineering, Yamagata University, Yamagata, 992-8510

Email: y-tomita@futamoto.elect.chuo-u.ac.jp

Abstract. Magnetic force microscope (MFM) tips are prepared by coating Fe(10 nm)/$Co_{50}Pt_{50}$(x nm)/Ru(5 nm) (at. %) films at 300 °C on Si-base tips of 4 nm radius. The thickness of CoPt layer, x, is varied in a range of 10 – 200 nm. The effects of coating thickness on spatial resolution and switching field of MFM tip are investigated. With increasing the thickness, the resolution deteriorates from 9.4 to 12.1 nm. The switching field increases from 0.675 to 2.025 kOe with increasing the coating thickness from 10 to 200 nm. A tip prepared by coating 100 nm thick $Co_{50}Pt_{50}$ film which has high resolution and high switching field property is successfully applied to the observation of magnetization structure of high-K_u magnetic thin film with $L1_0$-ordered structure.

1. Introduction

Magnetic force microscopy (MFM) has been used to investigate the magnetization structures of magnetic samples. MFM tips are generally prepared by coating non-magnetic sharp tips with magnetic materials [1]. As the areal density of hard-disk-drive (HDD) medium is approaching 1 Tb/in^2, where the bit length is becoming narrower than 30 nm, MFM resolution around 10 nm is necessary. Furthermore, in the observation of materials with high K_u like permanent magnets and recording media, high switching field (H_{sw}) property is required so that the tip magnetization does not change during the observation process. Recently, we have shown that high H_{sw} can be realized by coating $Co_{50}Pt_{50}$ alloy films on Si base-tips at temperatures around 300 °C [2, 3]. A high H_{sw} of 1.5 kOe is observed for a tip coated with 20 nm-thick CoPt film which is due to high K_u property of $L1_1$ ordered phase of the film.

To further increase the H_{sw} while keeping high resolution property, Fe/CoPt films are deposited on Si base tips with varying the CoPt film thickness up to 200 nm. Top coating of Fe film is introduced to enhance the detection sensitivity of MFM tip which depends on the saturation magnetization. The effects of coating thickness of CoPt film on the resolution and the switching field are investigated.

2. Experimental procedure

MFM tips were prepared by coating base-Si tips of 4 nm radius with films by employing a radio-frequency magnetron sputtering system. Ru, $Co_{50}Pt_{50}$, and Fe targets of 3 in diameter were used. $Co_{50}Pt_{50}$ (x nm), and Fe (10 nm) films were sequentially deposited on Si tips. The thickness of $Co_{50}Pt_{50}$ layer, x, is varied in a range between 10 and 200 nm. The film growth temperature is fixed at 300 °C,

Content from this work may be used under the terms of the Creative Commons Attribution 3.0 licence. Any further distribution of this work must maintain attribution to the author(s) and the title of the work, journal citation and DOI.
Published under licence by IOP Publishing Ltd

where $L1_1$-CoPt phase with high-K_u can be prepared. MFM tips were magnetized along the tip axis by applying a magnetic field of 10 kOe so that the tip top possessed the south magnetic pole.

MFM observation was carried out at room temperature under pressures lower than 0.1 Pa by using a scanning probe microscope. A perpendicular medium recorded at linear densities from 500 to 2000 kFCI and an HDD perpendicular medium with an areal density of 163 Gb/in^2 were employed for estimation of spatial resolution and H_{sw}, respectively. In order to estimate H_{sw}, MFM observation was repeated for an MFM tip by applying a magnetic field opposite to the initial magnetization. Magnetic field application was carried out by using an electromagnet after removing the tip from the MFM system, and then the tip was put again in the system for MFM observation. This procedure was repeated by increasing the magnetic field in a stepwise of 0.05 kOe until the MFM image contrast reversed. H_{sw} was estimated from the magnetic field where the MFM image contrast reversed.

3. Results and discussion

MFM observations were carried out for the perpendicular medium to estimate the resolution. Three informations, MFM image, line intensity profile, and power spectrum were used for the estimation. Figures 1(a-1)–(a-3) show the MFM images observed by using an Fe(10 nm)/CoPt(50 nm)/Ru(5 nm) coated tip. As the linear density is increased, the sharpness of image is degraded. Figures 1(b-1)–(b-3) are the signal profiles measured along the dotted lines in figures 1(a-1)–(a-3). Figures 1(c-1)–(c-3) are the power spectra obtained by fast Fourier transformation of the MFM image data. The tip resolution is determined as a half the bit length of recorded linear density until which the bits can be discerned in the MFM image information. They are observable up to the linear density of 1200 kFCI (bit length: 21.2 nm) but not observable at 1300 kFCI (bit length: 19.5 nm). The resolution is thus estimated to be between 21.2/2 = 10.6 nm (1200 kFCI) and 19.5/2 = 9.75 nm (1300 kFCI), i.e. 10.2 ± 0.4 nm.

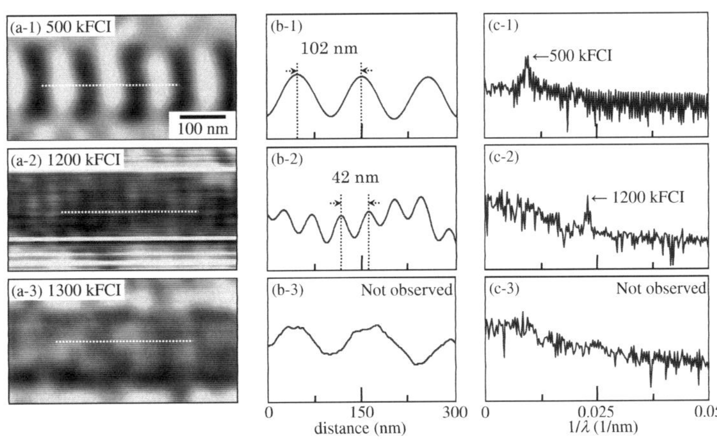

Figure 1. (a-1)–(a-3) MFM images of a perpendicular medium recorded at (a-1) 500, (a-2) 1200, and (a-3) 1300 kFCI observed by using an Fe(10 nm)/CoPt(50 nm)/Ru(5 nm) coated tip. (b-1)–(b-3) Signal profiles along the dotted lines in (a-1)–(a-3), respectively. (c-1)–(c-3) Power spectra for the magnetic bit images of (a-1)–(a-3), respectively.

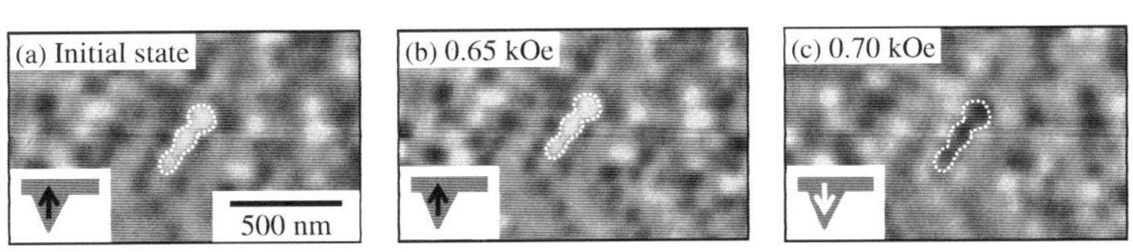

Figure 2. MFM images of a same area of HDD perpendicular medium observed by using an Fe(10 nm)/CoPt(50 nm)/Ru(5 nm) coated tip (a) before and (b, c) after applying magnetic fields of (b) 0.65 and (c) 0.70 kOe.

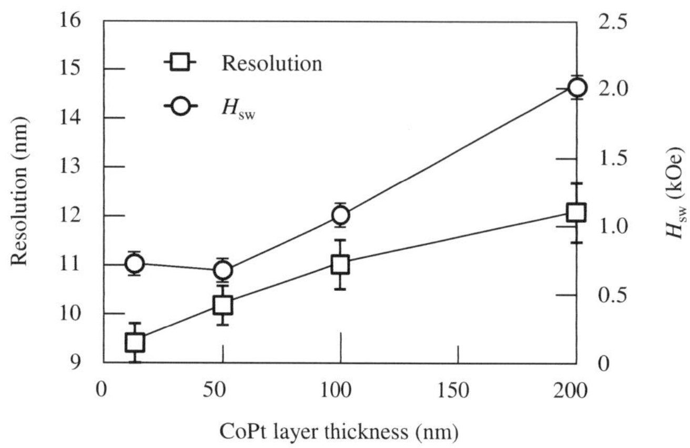

Figure 3. Effects of CoPt layer thickness on the resolution and the H_{sw} of MFM tip.

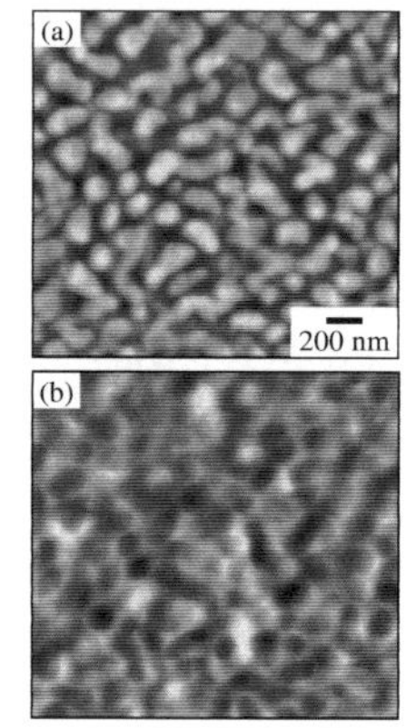

Figure 4. (a)AFM image and (b) magnetization structure of epitaxial $L1_0$-FePt film (5 nm) formed on MgO(001) substrate observed by using a tip coated with Fe/CoPt(100 nm)/Ru film.

MFM observations were carried out by using an HDD medium (163 Gb/in^2) sample to measure the H_{sw} values. Figure 2 shows the MFM images of a same area HDD medium observed by using an Fe(10 nm)/CoPt(50 nm)/Ru(5 nm) coated tip before and after applying magnetic fields. When a field of 0.65 kOe was applied to the tip, the contrast of MFM image was similar to the initial magnetization state as shown in figure 2(b). When a field of 0.70 kOe was applied, the contrast reversed as shown in figure 2(c). Thus, the H_{sw} of Fe(10 nm)/CoPt(50 nm)/Ru(5 nm) coated tip is determined to be between 0.65 and 0.70 kOe, i.e. 0.675 ± 0.025 kOe. Figure 3 shows the dependences of resolution and H_{sw} as a function of CoPt layer thickness. The resolution deteriorates from 9.4 to 12.1 nm with increasing the thickness of CoPt, which is related to the tip radius increase. With increasing the thickness, H_{sw} increases from 0.675 to 2.025 kOe. The experimental result indicates that it is possible to increase the magnetic switching field while keeping good spatial resolutions around 10 nm by employing an Fe/CoPt film structure for MFM tip preparation. Figure 4 is an example of MFM image for an epitaxial $L1_0$-FePt (5 nm) film formed on MgO(001) substrate observed by using an Fe(10 nm)/CoPt(100 nm)/Ru(5 nm) coated tip. A detailed remanent magnetization structure of high K_u $L1_0$-FePt film is clearly visualized.

4. Conclusion

MFM tips are prepared by coating Si tips with Fe and CoPt alloy films at 300 °C. The influences of coating CoPt layer thickness on the spatial resolution and the switching field are investigated. With increasing the thickness, the spatial resolution worsens from 9.4 to 12.1 nm, while the switching field increases from 0.725 to 2.025 kOe. An Fe/CoPt film coated tip is applied to the observation of high K_u magnetic material of epitaxial $L1_0$-ordered magnetic thin film.

References

[1] den Boef A. J. 1990 *Appl. Phys. Lett.* **56** 2045
[2] Ishihara S, Ohtake M and Futamoto M 2013 *J. Magn. Soc. Jpn.* **37** 255
[3] Nagatsu R, Ohtake M, Futamoto M, Kirino F and Inaba N 2016 *AIP Advances* **6** 056503

IOP Conf. Series: Journal of Physics: Conf. Series **903** (2017) 012010 doi:10.1088/1742-6596/903/1/012010

Magnetostrictive Materials for aerospace applications

A. Al-Taher[1], R.W. Reiss[1], A.D. Lafferty[3], S.A. Hayes[1], N. Lupu[2], I. Murgulescu[2] and N.A. Morley[1]

[1]Department of Materials Science and Engineering, University of Sheffield, Sheffield, S1 3JD, UK

[2]National Institue of R&D in Technical Physics, Iasi, Romania

[3] Institute for Materials and Processes, University of Edinburgh, Sanderson Building, Edinburgh EH9 3JF, UK

n.a.morley@sheffield.ac.uk

Abstract. Structural health monitoring of composite structures to detect barely visible damage is vitally important for the aerospace industry. This research has investigated amorphous magnetostrictive wires ($Fe_{77.5}Si_{7.5}B_{15}$ and $Co_{72.5}Si_{12.5}B_{15}$), as a possible solution to monitoring aerospace composites. The different amorphous wires were either embedded into the composite or epoxied on to the surface. How the wires effected the structure of the composite along with ultimate tensile strength was studied. Inductance measurements were used to study the strain within the composite, which provided a non-intrusive method of monitoring the composite.

1. Introduction

Composites are an ever-increasing material of choice across a wide variety of industries (aerospace, automotive and medical), as they possess superior mechanical properties combined with relatively low weight, compared with their metallic counterparts. However in the presence of a defect the composite is liable to severely deteriorate [1] this drives research into damage sensors. Thus to ensure the composite material integrity throughout the materials lifetime, various sensing techniques have been developed in the past decade, with the demand turning it into a multi-billion dollar industry [2].

There are two primary failure modes for composites: fibre rupture in tension or fibre buckle in compression [3], both of which result in an increase stress in the material. Thus the stress state of the composite material is the most important determinant of the structure safety [4]. However, it is difficult to obtain a direct reading of the stress within the material, so a measure of the strain, which is the deformation of a solid due to stress is used instead. There are a number of different sensors being developed for composite damage detection, these include self-sensing in the carbon fibres [5], glass fibre optical sensors [6] and piezoelectric sensors [7]. Each method has advantages and disadvantages, i.e. self-sensing in carbon fibres requires no added materials, but is limited to carbon composites, while piezoelectric sensors require the sensors to be attached to the composite surface, but multiple readings can be obtained simultaneously.

Magnetostrictive sensors using embedded magnetostrictive wires within the composite are also a promising option. Cristopolous et al [8] demonstrated that it was possible to embed a single magnetostrictive wire within composite material and detect a strain of 0.25 mStrain using a transducer. This paper presents studies of how two different amorphous magnetostrictive wires (FeSiB and CoSiB) embedded in the composite or epoxied onto the surface change the structural properties. Along with inductance measurements to determine the strain sensitivity of the wires on the composite surface.

Content from this work may be used under the terms of the Creative Commons Attribution 3.0 licence. Any further distribution of this work must maintain attribution to the author(s) and the title of the work, journal citation and DOI.
Published under licence by IOP Publishing Ltd

2. Experimental procedure

For magnetostrictive sensing, two different amorphous wires were studied, these were $Fe_{77.5}Si_{7.5}B_{15}$ (FeSiB) and $Co_{72.5}Si_{12.5}B_{15}$ (CoSiB). For both wire types uncoated and glass-coated wires were studied. They were produced via the melt spinning method with the glass-coated wires being produced using the improved glass-coating melt spinning method developed at the National Institute of Research and Development for Technical Physics in Iasi, Romania [9]. The uncoated FeSiB wires were 122μm in diameter and the glass coated wires were 20μm in diameter both with a magnetostriction constant of 30ppm. For the CoSiB wires, the uncoated wires were 110μm in diameter and the coated wires were 5μm in diameter, with a magnetostriction constant of -2.6ppm.

The composite samples were made from a 0°, 90° woven pre-impregnated carbon fibre epoxy system (VTC401®) from SHD Composites. It consisted of T300 carbon fibres in a 3k tow, indicating 3,000 fibres in each cluster combined with an epoxy based resin system. All the samples tested were 150mm x 50mm x 1mm in dimensions, which corresponded to 4 pre-preg layers. The magnetostrictive wires were evenly spaced, either 26mm, 13mm or 6.5mm between the second and third layers (Fig. 1a) or laid on top. The samples were then cured in the autoclave under vacuum, for 60 mins at 80° and then 60 mins at 120°. The samples produced contained either 2, 4 or 8 wires of each different wire type, along two samples with 2 uncoated wires epoxied to the surface and one sample with no wires as a reference.

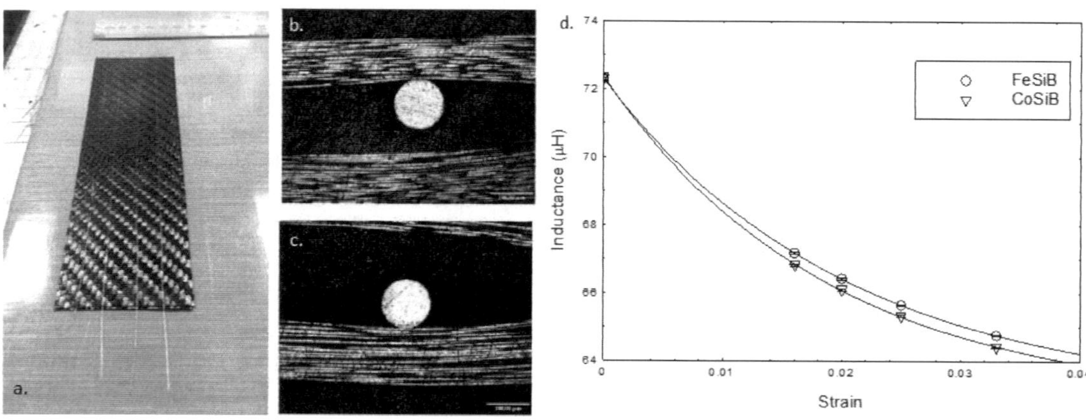

Figure 1a. Image of the magnetostrictive wires being placed within the composite. **1b&c.** Optical images of CoSiB wires within the composite at 50x magnification. **1d**. Inductance vs strain for two magnetostrictive wires epoxied to the composite surface. The lines are a guide for the eye.

To determine whether the embedded wires changed the structural integrity of the composite, static tensile testing using a Hounsfield machine with a 100 kN load cell was performed. Along with an interlaminar shear strength measurement using the short-beam method [10] for three sets of five samples. The samples were composite with no wires, composite with a wire along the sample length and composite with a wire perpendicular to the sample length. This test investigates the modes of failure due to external forces and whether they are acceptable (i.e. delamination) or not (tension or compression). The Young's modulus of the uncoated wires were also determined. Optical microscopy using a Nikon Eclipse LV150 microscope was used to image the embedded wires within the composite (Figs 1b&c). The strain sensitivity of the magnetostrictive wires in/on the composite was measured using an inductance method. A 112 turn pick-up coil connected to an Atlas LCR45 analyser was used to measure the inductance of the composite and the wires. To detect the response to strain, the composite was strained over a series of known bend radii (300, 400, 500, 600mm), which gave an uniform strain within the composite via the beam bending method. Using a 800 turn coil placed on the composite surface with the LCR45 analyser, the change in inductance was measured as a function of applied strain. This was repeated five times to check repeatability.

3. Results and Discussion

From the tensile tests, it was found that the ultimate tensile strength (UTS) increased with the number of wires within the composite, i.e. for 2 CoSiB wires UTS = 306 ± 1 MPa, while for 8 CoSiB wires UTS = 451 ± 1 MPa. All the composite samples with 4 or more wires embedded had a larger UTS than the composite only sample (UTS = 405 ± 1 MPa), suggesting that the wires helped to strengthen the composite. The CoSiB wires embedded in the composite gave higher UTS compared to the FeSiB wires, independent of whether they were glass-coated or uncoated. This is probably due to the CoFeSi wires having a smaller diameter and larger Young's modulus (E = 137 ± 2 GPa) than the FeSiB wires (E = 115 ± 5 GPa). For the interlaminar shear strength measurements, it was found that all three samples failed due to tension, with the composite failing at the centre of the underside. This was due to the manufacturing method of the samples. The breaking point for the samples with no wire was 920 ± 5 N compared to 860 ± 5 N for the wires parallel to the sample length and 830 ± 10 N for the wires perpendicular to the sample length. This is because the perpendicular wire is in the same direction as the loading force, so an increased volume of the wire lies in the same orientation as the induced load, which will increase the chances of failure.

Figs 1b and 1c shows a CoSiB magnetostrictive wire within the composite laminate. It is observed that there are two scenarios, the first (Fig. 1b) shows that the longitudinal fibres surround the wire, so increasing the fibre volume fraction around the wire. The second scenario (Fig. 1c) shows that the fibres have not followed the curvature of the wire leaving a relatively large area of epoxy matrix. No voids were observed in the images, meaning that the epoxy formed a sound bond between the wire and the fibres, thus allowing the stress to dissipate through the laminate.

For the inductance measurements, it was found for the same number of wires in the composite the FeSiB wires had a higher inductance compared to the CoSiB wires, i.e. for 2 FeSiB glass coated wires, L = 29±0.5 µH, while for 2 CoSiB glass coated wires, L = 27±0.5 µH. As expected the more wires within the composite the higher the inductance measured, i.e for 2 CoSiB uncoated wires L = 27.6±0.5 µH, and for 8 CoSiB uncoated wires L =34.8±0.5 µH. Thus as 8 wires embedded in the composite increased the UTS along with the inductance, this suggests that a spacing of 6.5 mm between the wires could be used for strain detection in composites, without being detrimental to the structural integrity. From Fig. 1d, it is observed that there is a decrease in inductance with an increase with strain, with a similar non-linear behaviour to that measured by Cristopolous [8]. The error bars (Fig. 1d) show the repeatability of the inductance measurement as a function of strain. The FeSiB wires gave a higher sensitivity compared to CoSiB wires, with the FeSiB wires able to detect a change strain of 500 µStrain, using a simple inductance technique. This demonstrates that the wires can measure a global strain to the composite. For structural health monitoring sensors, failure such as delamination produces a local strain within the composite. These local strains will change the magnetisation within the magnetostrictive wires, which will be detected using a handheld sensor such as a transducer [8]. Other structural health monitoring strain sensors, include surface mounted fibre optics, which can measure strains up to ±5000 µstrain with a resolution of ±1 µstrain [11], and surface mounted piezoelectric sensors, which have a measurement range of ±150 µstrain with a resolution of ±5 µstrain [12]. Thus further work has to be carried out to achieve magnetostrictive strain sensors with a resolution within the ±10 µstrain.

4. Conclusions

Magnetostrictive wires were embedded/epoxied onto composites to determine whether they could be used to measure the internal strain. It was found that the embedded wires improved the ultimate tensile strength of the composite with CoSiB wires increasing the UTS greater than the FeSiB wires. Thus the wires helped the structural integrity of the composite laminate. It was also found that there were no voids around the magnetostrictive wires, due to either the composite fibres or the epoxy matrix surrounding them. Measuring the inductance of the wires was found to be a reliable method to measure the strain within the composite, with the FeSiB wires having a higher inductance compared to CoSiB wires. A change of strain of 500 µStrain was measured using this technique for the FeSiB wires, making this a promising method to measure the internal strain in composites.

Acknowledgments
The results presented here have been partially supported by JTI-Cleansky2 project SHERLOC JTI-CS-2009-01-GRA-01-005

References

[1] Backman, B F, 2005, *Composite structures, design, safety and innovation*, (Amsterdam: Elsevier)
[2] Ristic R L R, 1994, *Sensor technology and devices*, (Boston: Artech House)
[3] Hashin Z, 1980 *J. Appl. Mech*, **47**, 329-334
[4] Duan Y, Zhang R, Zhao Y, Or S, Fan K and Tang Z, 2011, *J. Zhejiang Univ. Sci. A*, **12**, 12 895-901
[5] Abry J C, Choi Y K, Chateauminois A, Dalloz B, Giraud G and Salvia M, 2001, *Compos. Sci. Technol.* **61**, 855-864
[6] Fernando G F, 2005, *Reinf. Plast*, **49**, 12, 41-49
[7] Bowen C R, Dent A C E, Nelson L J, Stevens R, Cain M G and Stewart M, 2006, *Proc. Inst. Mech. Eng. Part C J. Mech. Eng. Sci*, 220, C11, 1655-1663
[8] Christoploulos A, Hristoforou E, Koulalis I and Tsamasphyros G, 2014, *Smart Mater. Struc.* 23, 8, 085035
[9] Ovari T-A, Lupu N and Chiriac H, 2015, *Magnetic Nano-and Microwires*, (Cambridge: Woodhead publishing)
[10] BS EN ISO14130 "Fibre-reinforced plastic composites determination of apparaent interlaminar shear strength by short beam mrthod", 3, 1108, 1-9, 1998
[11] Di Sante R, 2015, *Sensors*, **15**, 18666-18713
[12] Sirohi J and Chopra I, 2000, *J. Int. Mat Sys. Struc.* **11**, 246-257

Tunable microwave electric polarization in magnetostrictive microwires

L.V. Panina[1,2], D. P. Makhnovskiy[3], A. Dzhumazoda[1], S.V. Podgornaya[1], V. G. Kostishyn[1], H.X. Peng[4]

[1] National University of Science and Technology, MISiS, Moscow 119991, Russia
[2] Institute for Design Problems in Microelectronics RAS, Moscow 124681, Russia
[3] Robat Ltd, UK
[4] Zhejiang University, PR China
Email: drlpanina@gmail.com

Abstract. In this work, the induced electric dipole moment of ferromagnetic microwires dependent on the wire magnetic structure is investigated with the aim of sensing applications. The wire behaves as micro antenna exhibiting a resonance at half wavelength condition. At the vicinity of the resonance the wire polarization is sensitive to all the losses including magnetic losses. The transfer mechanism is based on the magnetoimpedance effect (MI) which requires a specific magnetic structure of a helical type. We demonstrate that scattering from a ferromagnetic microwire showing large MI effect at microwaves can be modulated with a low frequency magnetic field and the amplitude of this modulation near the antenna resonance can sensitively change in the presence of the external stimuli such as a dc magnetic field. The modelling is based on solving the scattering problem from a cylindrical wire with the impedance boundary conditions. The modelling results agree well with the experimental data.

1. Introduction.

Miniature sensors with remote wireless operation are required in many technological areas including structural health monitoring, smart composites, embedded biosensors, etc. Microwave technologies have a high potential for remote sensing applications. Recently, the research topic on microwaves interacting with ferromagnetic structures is intensively developing [1] and one of the main areas of applications is identified as sensor materials [2-3]. Here we investigate the performance of magnetic microwires as embedded sensors operating at microwave frequencies. The sensor operation is based on tunable electric polarization due to large change of the wire surface impedance (known as magnetoimpedance (MI) effect [4-6]).

In order to realize efficient tunable properties, magnetic microwires of CoFeSiB composition in amorphous state are considered. The absence of the crystalline structure is very useful since it is possible to induce a required magnetic anisotropy through the magnetoelastic interaction. For microwave sensing applications, the magnetic structure in the outer shell of a circumferential or helical type is preferable, which can be established in Co-rich alloys with negative magnetostriction. For such wires, the remagnetisation behaviour in an axial magnetic field less than the effective anisotropy field is almost linear, without a hysteresis and with a very high susceptibility. The external stress and temperature can substantially modify the magnetic anisotropy and the magnetization processes.

Content from this work may be used under the terms of the Creative Commons Attribution 3.0 licence. Any further distribution of this work must maintain attribution to the author(s) and the title of the work, journal citation and DOI.
Published under licence by IOP Publishing Ltd

IOP Conf. Series: Journal of Physics: Conf. Series **903** (2017) 012011 doi:10.1088/1742-6596/903/1/012011

The use of metal inclusions in microwave applications is often justified by their larger electric polarisability in comparison with dielectrics. This enables a strong response even from a single inclusion. Here we demonstrate theoretically that scattering from a ferromagnetic microwire showing large MI effect at microwaves can be modulated with low frequency magnetic field and the amplitude of this modulation near the antenna resonance can sensitively change in the presence of the external stimuli. The modelling is based on solving the scattering problem from a cylindrical wire with the impedance boundary conditions [7-8]. The modelling results agree well with the available experimental data [9-10].

2. Electric polarization of a ferromagnetic wire

The wire of finite length is interrogated by the microwave field with the electric field e_0 along the wire (z-axis). The field e_0 induces the current i in the wire with some distribution along the wire length l. Zero boundary conditions must be imposed at the wire ends: $i(\pm l/2) = 0$. This condition means the concentration of electric charge at the ends and the generation of a dipole moment $\mathcal{P}$ in the wire, which can be calculated using the continuity equation: $\partial i(z)/\partial z = j\omega\rho(z)$ and integrating by parts (ρ is a charge density per unit length, ω is a frequency, j is imaginary unit). The dipole moment is calculated as:

$$\mathcal{P} = \frac{j}{\omega}\int_{-l/2}^{l/2} i(z)dz \qquad (1)$$

The current distribution in a thin wire can be found in the antenna approximation [11]. However, this approach should be further developed when considering a ferromagnetic metallic wire and the effect of its magnetic structure on $i(z)$ and dipole moment [7-8]. The scattering problem is solved imposing the impedance boundary conditions at the wire surface

$$\bar{\boldsymbol{e}}_z = \hat{\varsigma}(\bar{\boldsymbol{h}}_\varphi \times \boldsymbol{n}) \qquad (2)$$

Here $\hat{\varsigma}$ is the surface impedance tensor, **n** is the unit radial vector directed inside the wire, $\bar{\boldsymbol{e}}_z$ and $\bar{\boldsymbol{h}}_\varphi$ are the longitudinal electric field and circumferential magnetic field at the wire surface, which include both the scattered and external fields.

The scattering problem is formulated in terms of the vector potential **A** which obeys the Helmholtz equation:

$$\Delta\boldsymbol{A} + k^2\boldsymbol{A} = \boldsymbol{i}\,, \quad k = (\omega/c)\sqrt{\varepsilon_d} \qquad (3)$$

c is the velocity of light and ε_d is the permittivity of the dielectric medium around the wire. The solution of Eq. (3) can be represented using the convolution of $\boldsymbol{i}(z)$ with the Green function $G(r) = \exp(jkr)/4\pi r$, r is radial coordinate. Then, the electric and magnetic fields are expressed in terms of integro-differential operators using the convolutions with the current. Imposing boundary condition (2) we obtain the integro-differential equation for the current density in the ferromagnetic wire (generalized antenna equation):

$$\frac{\partial^2(G * i)}{\partial z^2} + k^2(G * i) + \frac{j\omega\varepsilon_d\varsigma_{zz}}{2\pi ac}(G_\varphi * i) = \frac{j\omega\varepsilon_d}{4\pi}e_0 \quad (4)$$

Here a is the wire radius, G_φ is a sort of a Green function occurring due to the induced circular magnetic field:

$$G_\varphi = \frac{a^2(1 - jkr)\exp(jkr)}{2r^3} \qquad (5)$$

Equation (4) is completed with the zero boundary condition for the current at the wire ends. The internal losses are given by the impedance ς_{zz} and the convolution $(G_\varphi * i)$, whereas the imaginary part of $(G * i)$ determines the radiation losses. The solution of equation (4) can be obtained by the iteration method. We have compared the current distribution calculated in zero approximation and in the first approximation. If the skin effect is not very strong ($a \sim \delta_m$, δ_m is the skin-depth) the difference between the two approximations is less than few percent. The zero approximation obtained by putting imaginary parts of G and G_φ to zero corresponds to neglecting the radiation losses which are small for a moderate skin effect. Similar condition of a moderate skin-effect is needed to realize magnetic tuning of the electric polarization. Therefore, in further analysis a simplified analytical solution obtained in zero approximation is used.

The formalism for calculating the matrix $\hat{\varsigma}$ for a ferromagnetic wire with an arbitrary helical anisotropy was developed in [12]. In the case of a sufficiently strong skin-effect which is expected at GHz frequencies and considering that the surface layer of the wire has a helical magnetization with the constant angle θ (with respect to the wire axis), the component ς_{zz} is defined as:

$$\varsigma_{zz} = \frac{c(1-j)}{4\pi\sigma\delta}\left(\sqrt{\tilde{\mu}}\cos^2\theta + \sin^2\theta + \frac{\delta(1+j)}{4a}\right), \delta = \frac{c}{\sqrt{2\pi\sigma\omega}} \qquad (6)$$

In equation (6) σ is the wire conductivity, θ is the angle between the dc magnetization and the wire axis, $\tilde{\mu}$ is a permeability parameter composed of the components of internal permeability tensor and has a meaning of a circular permeability with respect to the dc magnetization.

3. Result and discussion

As it follows from Equation (6) the impedance may be sensitive to the change in the dc magnetization and ac permeability. For soft amorphous wires at GHz frequencies the parameter $\tilde{\mu}$ weekly depends on the moderate external magnetic field since this range of parameters corresponds to the tail of the ferromagnetic resonance. In the case of a circumferential anisotropy the impedance shows strong variations in the range of small magnetic field H_{ex} applied along the wire when the static magnetization changes its direction from circumferential to axial (the magnetization angle θ changes from $\pi/2$ to 0), as shown in Figure 1. The experimental results for MI at GHz frequencies confirm such behavior.

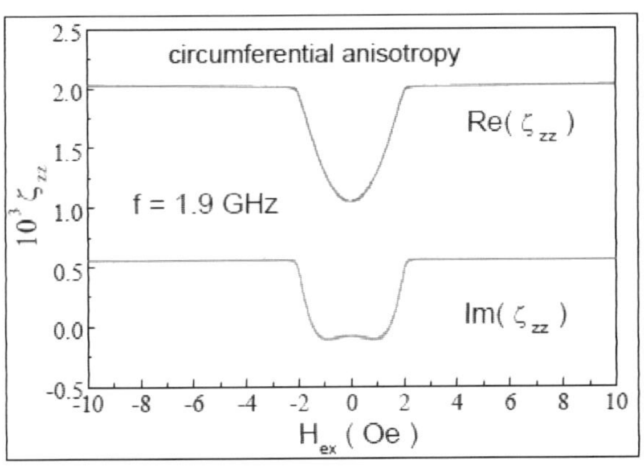

Figure 1: Plots of the surface impedance of amorphous wires with a circumferential anisotropy. The wire has 10 μm diameter, conductivity $\sigma = 7.6 \cdot 10^{15}\ s^{-1}$, anisotropy field H_{ex}=2 Oe, saturation magnetization M_0=500 G, and gyromagnetic constant $\gamma = 2 \cdot 10^7$ (rad/s)/Oe.

The dependence of the impedance on the magnetic properties makes it possible to tune the scattering from a single ferromagnetic wire as follows from Equation (4). Figure 2 presents the current distribution along the wire of finite length near the antenna resonance without a field and in the presence of the field of about the anisotropy field H_K (H_K corresponds to the field needed to magnetize the wire along the axis). The calculation is done neglecting the radiation losses which is justified if the skin effect is not very strong. A very large difference is observed. It should be noted that the antenna resonance frequency defined by the half wavelength condition is chosen such that the skin effect is not very strong ($a\sqrt{\tilde{\mu}}/\delta \sim 1$) and the permeability parameter $\tilde{\mu}$ substantially differs from unity. For a wire with a diameter of 10-20 micron the frequency range is within few GHz. The antenna resonance frequency $f_{res} = c/2l\sqrt{\varepsilon_d}$ can be adjusted by varying the wire length l and the permittivity of the surrounding medium ε_d. If the proper conditions are realised, the induced electric dipole moment of ferromagnetic wire becomes magnetic field dependent as shown in Figure 3.

Scattering from a ferromagnetic wire showing MI effect at GHz frequencies can be used for contactless sensing of external parameters such as magnetic field (or mechanical stress) as was proposed in [9-10]. The procedure is as following.

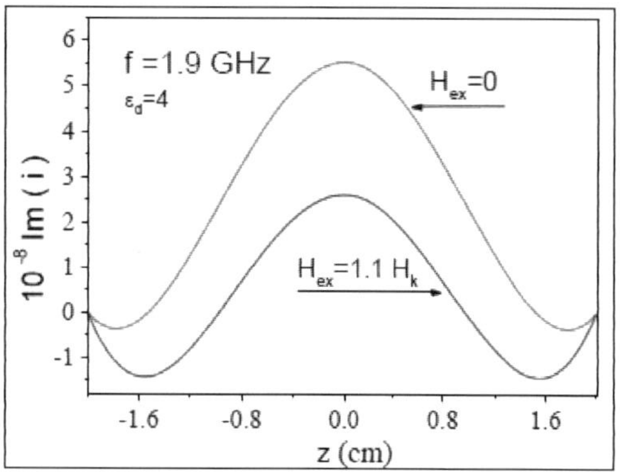

Figure 2: Current distribution along the wire for frequency f=1.9 GHz (near the resonance). Imaginary part is shown. Parameters are the same as for Figure 1.

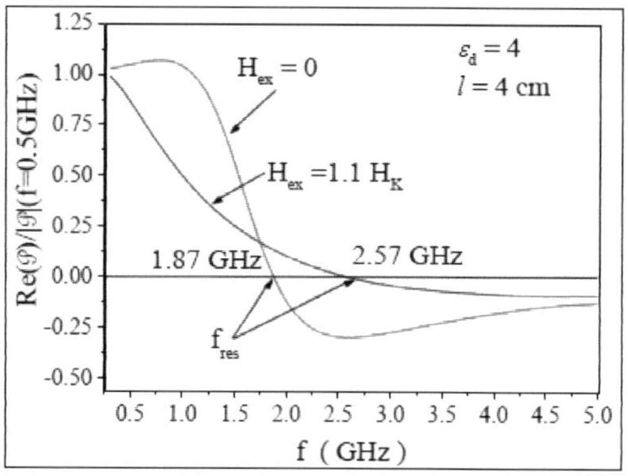

Figure 3: Spectra of real part of dipole moment. Parameters are the same as for Figure 1.

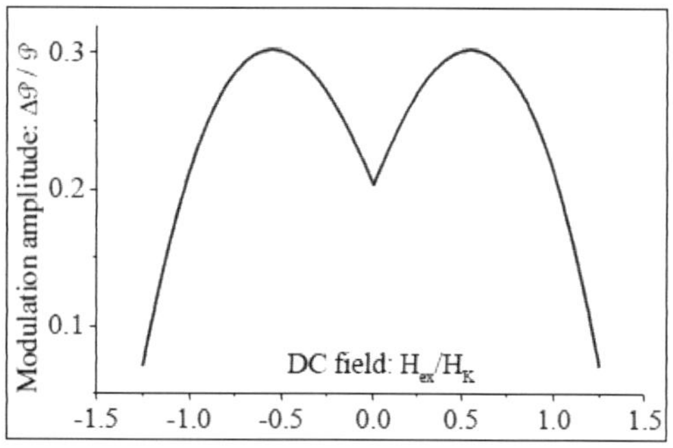

Figure 4: Modulation of the dipole moment with ac bias field as a function of the measured field H_{ex}. This parameter is defined as

$$\Delta\mathcal{P}/\mathcal{P} = Max|\frac{(\mathcal{P}(H_{ex}+H_b)-\mathcal{P}(H_{ex}))}{\mathcal{P}(0)}|.$$

The wire is placed in a microwave field and is subjected to low frequency bias field H_b which modulates the amplitude of the scattered output signal. The magnitude of these modulations will depend on the external dc magnetic field H_{ex}, as shown in Figure 4. This result agrees with the experiment [10]. A similar effect can be expected in the presence of a mechanical stress.

L. V. Panina acknowledges the support for this work under Russian Federation State Contract for organising a scientific work.

References:

[1] Yang S., Lozano K., Lomeli A., Foltz H.D., Jones R. 2005 *Composites A.* V. 36, p. 691.
[2] Qin F.X., Peng H.X., Pankratov N., Phan M.H. et al. 2010 *J. Appl. Phys.* V. 108, ID: 044510.
[3] Yuan B., Yu L., Sheng L., An K., Zhao X.J. 2012 *Phys. D: Appl. Phys.* V. 45, p. 235108.
[4] Panina L.V., Mohri K. 1994 *Appl. Phys. Lett.* V. 65, p. 1189.
[5] Mandal K., Puerta S., Vazquez M., Hernando A. 2002 *IEEE Trans. Magn.* V. 36, p. 3257.
[6] Kraus L., Frait Z., Pirota K.R., Chiriac H. J. 2003 *J. Magn. Magn. Mat.*V. 254-255, p. 399.
[7] Panina L.V., Grigorenko A.N., Makhnovskiy D.P. 2002 *Phys. Rev.*B, V. 66, p. 15411.
[8] Makhnovskiy D.P, Panina L.V. 2003 *J. Appl. Phys.* V. 93, p. 4120.
[9] Herrero-Gomez C., Aragon A.M. et al, 2014 *Appl. Phys. Lett.* V. 105, p. 09240
[10] Herrero-Gomez C., Marın P., Hernando A. (2013) *Appl. Phys. Lett.* V.103, p.144424.
[11] King R. Smith G. Antennas in Matter.The MIT Press, 1981.
[12]Makhnovskiy D.P., Panina L.V., Mapps D.J. 2001 *Phys. Rev.* B, V. 63, p. 14424.

Study of magnetic field distribution in anisotropic single twin-boundary magnetic shape memory (MSM) element in actuators

N Gabdullin and S H Khan

Department of Electrical and Electronic Engineering, School of Mathematics, Computer Science and Engineering, Northampton Square, City, University of London London EC1V 0HB, UK

E-mail: Nikita.Gabdullin.1@city.ac.uk

Abstract. Magnetic shape memory effect exhibited by certain alloys at room temperature is known for almost 20 years. The most studied MSM alloys are Ni-Mn-Ga alloys which exhibit up to 12% magnetic field-induced strain (change in shape) depending on microstructure. A multibillion cycle operation without malfunction along with their "smart" properties make them very promising for application in electromagnetic (EM) actuators and sensors. However, considerable twinning stress of MSM crystals resulting in magneto-mechanical hysteresis decreases the efficiency and output force of MSM actuators. Whereas twinning stress of conventional MSM crystals has been significantly decreased over the years, novel crystals with Type II twin boundaries (TBs) possess even lower twinning stress. Unfortunately, the microstructure of MSM crystals with very low twinning stress tends to be unstable leading to their rapid crack growth. Whilst this phenomenon has been studied experimentally, the magnetic field distribution in anisotropic single twin-boundary MSM elements has not been considered yet. This paper analyses the magnetic field distribution in two-variant single twin-boundary MSM elements and discusses its effects on magnetic field-induced stress acting on the twin boundary.

1. Properties of MSM alloys important for actuation

Properties of MSM alloys are constantly improving. The reversible magnetic field-induced strain as large as 12% reported in [1] doubles the most common 6% maximum strain of first successful Ni-Mn-Ga specimen [2]. Another parameter of immense importance is twinning stress which is related to energy required for propagating twin boundaries changing shape of the MSM element. This parameter is particularly important for MSM actuation due to its relation to internal friction [3]. Pure crystals with low 0.3-0.5MPa twinning stress were reported in [4]. These values are currently attributed to the most common Type I TBs [5]. However, Type II TBs with remarkably low 0.01MPa twinning stress and corresponding 30mT switching field were reported in [6]. These crystals are particularly promising for actuation due to very high efficiency of magneto-mechanical energy conversion. On the other hand, several important aspects of the shape change in Twin II MSM elements reported in [7, 8] question the long lifetime stability of these crystals. Whilst the response of a single twin-boundary MSM specimen to mechanical stresses has been previously studied, the effects related to non-uniform nature of the magnetic field distribution due to non-homogeneity of MSM microstructure and magnetic anisotropy of its twin variants have not been analysed yet.

Content from this work may be used under the terms of the Creative Commons Attribution 3.0 licence. Any further distribution of this work must maintain attribution to the author(s) and the title of the work, journal citation and DOI.
Published under licence by IOP Publishing Ltd

Twin boundaries of both types can exist in the same specimen depending on the condition of their nucleation. However, very low twinning stress is obtained only when a single Type II TB is present. This is distinctly different from the behaviour of Type I TBs which tend to form lamellar structures called fine twins [7, 9]. The concentration of all variants of the same type on one side of a single twin boundary results in a very non-homogeneous magnetic field distribution, especially when anisotropy of MSM permeability is taken into account.

2. Magnetic field distribution in a single twin-boundary MSM element

The analysis summarised in this paper is based on finite element (FE) analysis implemented using commercially available ANSYS Multiphysics software package. Hence, only the magnetic vector potential (MVP) formulation of Maxwell's equations is used in all models. An advantage of such approach is its compatibility with design methodologies available for MSM sensors and actuators [10]. Whereas modelling of variant distribution has been previously discussed [10], magnetic anisotropy of twin variants is usually omitted in existing modelling approaches. However, the overall magnetic anisotropy of MSM element is a consequence of anisotropy of its twin variants. Therefore, each axis in a variant should be assigned with its own magnetisation curve in order to take magnetic anisotropy into account correctly. The modelled MSM element possesses the most studied 5M microstructure [2]. Its saturation magnetisation is 0.69T and initial relative permeability of "easy" and "hard" axes are 50 and 2. The MSM element is blocked in order to prevent variant growth.

Figure 1 shows the non-uniform magnetic field distribution in a single twin boundary MSM element in a strong 0.6T applied magnetic field. A short 2.5mm × 1mm MSM element is used for clarity of the figure. The magnetic field is applied transversely to its longest 2.5mm side resulting in 2.2% elongation. A "hard" variant with a axis along the applied field is allocated above the "easy" variant with c axis along the field. It should be noted that both variants still possess anisotropy which implies that the "hard" variant has its c and the "easy" variant has its a axes normal to the applied field. However, Fig. 1 shows that the field lines change direction inside the MSM element. This happens due to the 45° angle between the TB and the MSM element's sides. This results in both transverse (x) and longitudinal (y) components of the magnetic field in MSM twin variants. However, it is known that longitudinal magnetic fields induce compressive stresses in MSM crystals. A non-uniform magnetic field causing a complex stress distribution in a multi-variant MSM specimen was studied in [11]. Similarly, the non-uniform field distribution in Fig. 1 can be one of the reasons for nucleation of TBs oriented differently than the main (modelled) propagating TB in single twin-boundary MSM elements. This can lead to instability of their microstructure.

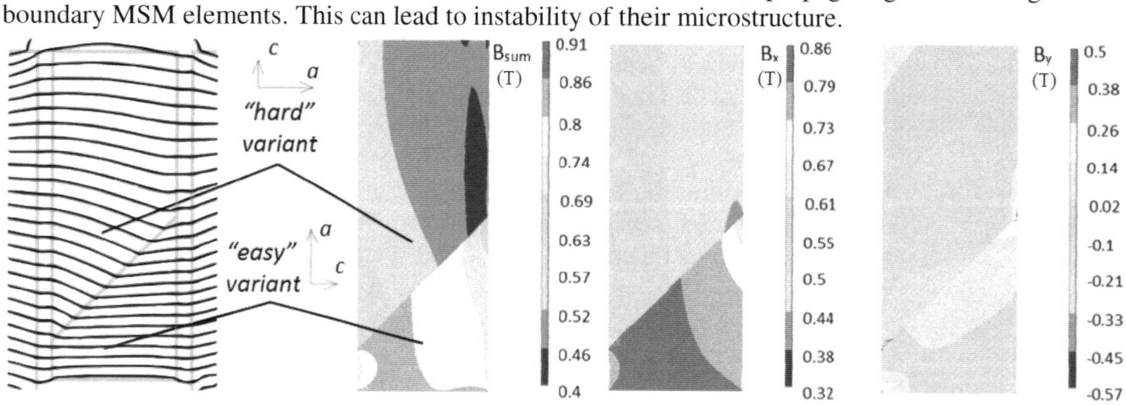

Figure 1. Magnetic field distribution in a single twin-boundary MSM element subjected to a strong applied field.

Figure 2 (a) shows the magnetic field distribution in the same MSM element in a weak applied field not exceeding 0.1T. The effects related to variant anisotropy become even clearer since variants are not saturated. Here, flux lines tend to locally align with the "easy" axis in each variant. This is consistent with the basic theory of the MSM effect. The most interesting point lies on the left side of the TB in Fig. 2 (a), where magnetic flux density inside the MSM element reaches 0.67T due to internal magnetisation of its twin variants. One can observe the general decreasing trend in magnetic flux density along the length of the twin boundary. It is also remarkable that the difference exceeds one order of magnitude, implying greater strains on the left side of the MSM element in comparison with its right side.

It should be also noted that this effect is not as apparent when isotropic twin variants are modelled. Figure 2 (b) illustrates this. Although the general pattern of field distribution remains unchanged, its magnitude is much lower. Therefore, it is significantly important to model anisotropy of MSM twin variants correctly.

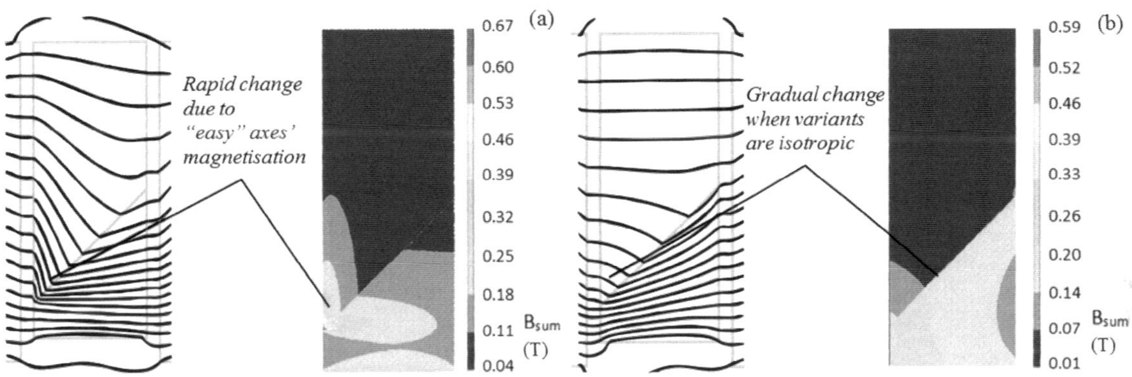

Figure 2. MSM element in a weak applied field with (a) anisotropic and (b) isotropic twin variants.

3. Conclusions

The magnetic field distribution in non-homogeneous two-variant MSM element is studied with FE analysis and its very non-uniform nature is discussed. In addition to non-homogeneity of field distribution which was clearly observed in both weak and strong applied fields, it was also possible to locate a point of particularly high magnetic field energy concentration in a sufficiently weak magnetic field. This result also suggests very non-uniform distribution of mechanical stresses due to magneto-mechanical coupling leading to their concentration in specific areas of the MSM element, supporting findings previously reported by other researchers. The importance of accurate modelling of variant anisotropy is illustrated by comparing the computational results obtained on anisotropic and isotropic twin variants. Further work will include a detailed study of distribution of magnetic field-induced stresses and their effect on the total force acting on the twin boundary, which translates into output force in actuators.

References

[1] Sozinov A, Lanska N, Soroka A and Zou W 2013 *Appl. Phys. Lett.* **102** 021902
[2] Straka L and Heczko O 2005 *J. Magn. Magn. Mater.* **290-291** 829–831
[3] Suorsa I, Pagounis E and Ullakko K 2004 *J. Magn. Magn. Mater.* **272-276** 2029–2030
[4] Pagounis E, Müllner P 2014 *Proc. Actuator* 86
[5] Heczko O, Kopeček J, Straka L, Seiner H 2013 *Mater. Res. Bull.* **48** 5105–5109
[6] Kellis D, Smith A, Ullakko K, Müllner P 2012 *J. Cryst. Growth* **359** 64-68
[7] Straka L, Lanska N, Ullakko K, Sozinov A 2010 *Appl. Phys. Lett.* **96** 131903
[8] Aaltio I, Ge Y, Pulkkinen H, Sjöberg A, Söderberg O, Liu X W and Hannula S-P 2010 *Phys. Procedia* **10** 87-93
[9] Sozinov A, Lanska N, Soroka A, Straka L 2011 *Appl. Phys. Lett.* **99** 124103
[10] Schiepp T, Maier M, Pagounis E, Schlüter A, Laufenberg M 2014 *IEEE Trans. Magn.* **50** 7024504
[11] Haldar K, Chatzigeorgiou G, Lagoudas D C 2015 *Eur. J. Mech. A-Solids* **54** 53-73

IOP Conf. Series: Journal of Physics: Conf. Series **903** (2017) 012013 doi:10.1088/1742-6596/903/1/012013

Change of magnetic domain structure by mechanically induced twin boundary motion in Ni-Mn-Ga single crystal

Vít Kopecký and Oleg Heczko

Dep. of Functional Materials, Institute of Physics ASCR, Na Slovance 2, 18228 Prague 8, Czech Republic

vit.kopecky@fzu.cz

Abstract. The single variant state exhibits usual labyrinth and band magnetic domains depending on orientation of easy magnetization axis. By the passage of single twin boundary induced by mechanical stress the rake and granular domain patterns are formed. These domain patterns are further modified by repeated passage of the twin boundary resulting in similar domain patterns in the sample even though the orientation of the magnetization is different.

1. Introduction

Magnetic shape memory effect or more precisely magnetically induced reorientation, which generates up to 12% deformation in magnetic field [1], is caused by martensitic microstructure reorientation via the movement of highly mobile twin boundaries [2, 3]. There is no phase transformation involved. The field-induced deformation can reach up to 12% in 1 T field but usually the field needed for twin boundary motion is much lower depending on martensite structure [3], crystal quality, twin boundary arrangements [3,4] and particularly on the type of twin boundary [5]. The lowest magnetic field, down to 100 Oe, was observed for single twin boundary of Type II sweeping across the perfect single crystal [5]. This high mobility was confirmed by mechanical stress test showing very low stress, down to 0.05 MPa for boundary motion (twinning stress). This correlates well to observed mobility in magnetic field [6] using the model developed by Likhachev and Ullakko [2].

As the orientation of the crystal and thus the easy magnetic direction changes, the ferroelastic twin boundary is also the magnetic domain wall. In the small field needed for the twin boundary motion the interaction of moving ferroelastic twin boundary with magnetic domains can be important. Magnetic domain structure was investigated mostly in static situation and there are only few reports of the domain evolution with magnetic field [7,8]. In all these investigations the twin microstructure was not controlled and contained many twin boundaries.

Recently we have investigated the behaviour of purposely formed single twin boundary in various single crystals [3, 9] and how the passage of the twin boundary affects the magnetic domain structure [10]. We found that after a single passage of twin boundary, the rake [11] and granular [10] domains were formed which did not bear any similarity of the original domain structure of the crystal with single orientation (single variant) [12]. This proves that using single boundary gives the best opportunity to study the interaction between ferroelastic and ferromagnetic interfaces. Here we present the evolution of the magnetic domains caused by repeated passage of twin boundary induced by mechanical stress.

Content from this work may be used under the terms of the Creative Commons Attribution 3.0 licence. Any further distribution of this work must maintain attribution to the author(s) and the title of the work, journal citation and DOI.
Published under licence by IOP Publishing Ltd

2. Experimental

Domain structure was studied on single crystal $Ni_{50.0}Mn_{28.5}Ga_{21.5}$ (atomic %) from Adaptamat Ltd with dimensions about 8x2x1 mm. The sample was electropolished in single variant state. All domains observations were made at room temperature with the crystal in martensitic 10M (called also 5M) phase. To visualize magnetic domains we used magneto-optical indicator film (MOIF) made from soft ferromagnetic low-anisotropy garnet with in-plane domains. The stray field of bulk magnetic domains in the sample formed the domain pattern in the garnet which was observed by optical microscope Zeiss with linearly polarized light. The whole surface (marked in Fig. 1) was studied but only selected micrographs are presented as the character of the images does not change in whole investigated area.

3. Results and discussion

Fig. 1 shows the experiment. We started with the single variant state in two different orientations magnetized to saturation. Then we formed a single twin boundary manually by repeated small compressive and tensile stress along the long sample dimension. The single twin boundary was then moved across the sample forth and back using mechanical force as shown in the figure. The twin boundary was Type II which has the twinning stress about 0.1 MPa. The strain after passage of single twin boundary is given by the reorientation of the lattice, about 6 %. Despite the fact that the twin boundary is very thin (few nanometers) the single boundary extends across whole sample cross-section as confirmed by optical microscope observation. After repeated passage of the single twin boundary the magnetic domains were observed. The approximate position of domain observation is shown in Fig. 1.

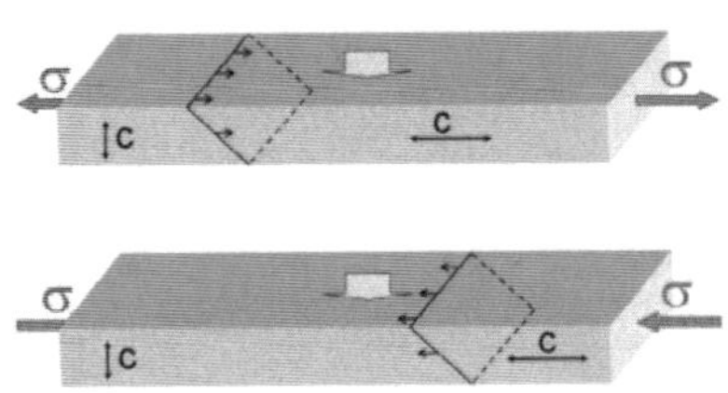

Fig. 1. Schematic drawings of the experiment. The formed twin boundary is marked by black lines in sample cross-section with small arrows. The arrows indicate motion direction induced by mechanical stress (upper – tensile stress, lower – compression). Twin boundary motion results in structure reorientation. The variant orientation is indicated by c-axis. Approximate place of domain observation is pointed by the wide arrow.

Fig. 2 shows the evolution of the magnetic domains starting with the single variant with easy magnetization axis or crystal c-axis along the long sample dimension. It exhibited typical band domains (Fig. 2a) with large domains due to low demagnetization factor and high magnetocrystalline anisotropy. After the first passage of the twin boundary induced by tensile stress (Fig. 1) the rake pattern was formed [10,11]. Despite the similarity to band domains the orientation of magnetization was perpendicular following easy magnetization axis. The second passing of the boundary resulted in crystal orientation identical to initial state. However, the domain pattern was sharply different exhibiting so-called granular domains. After repeated passage of the twin boundary these granular domains became more regular bands inheriting some size features from the rake domains.

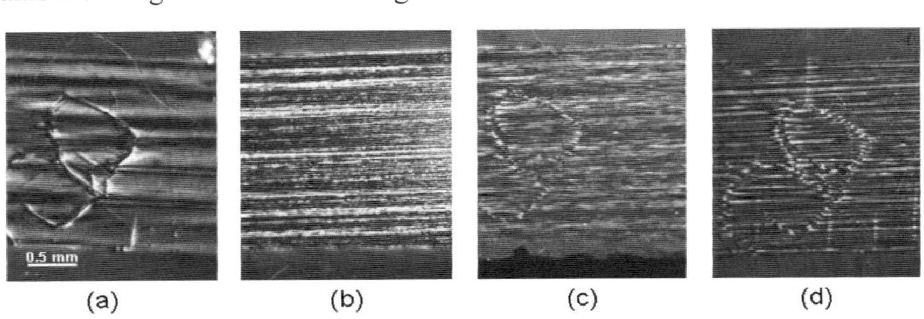

Figure 2. Magnetic domains of single variant state visualized by MOIF. The scratches (irregular round pattern) serve as fiducial marking. (a) Band domains after magnetizing in 1T magnetic field in plane along the long sample axis (horizontally), (b) rake domains after initial passage of twin boundary (c-axis and magnetization perpendicular), (c) granular domain pattern after second passage of twin boundary (c-axis and magnetization in plane), (d) granular domain pattern after 12 passages of twin boundary.

Analogous development is shown in Fig. 3. Here, in contrast to previous case, the initial state was single variant state with c-axis perpendicular resulting in labyrinth or maze domains (or rather bubble-like domains). The initial passage of twin boundary induced by compressive stress, switched the c-axis to in-plane and along the long dimension of sample. However, the band domains were not formed, instead the grainy, irregular low contrast was apparent. We called these granular domains. Previous study indicated that there are no domains walls on the surface and the domains are closing inside [10]. After reverse passage of twin boundary (by tensile stress), a rake pattern was formed similar to the case in Fig. 2. Repeated forth and back passage of twin boundary resulted in well defined rake pattern. Although superficially similar to band domains in general and specifically to granular domains in Fig. 2 after same number of cycles, the vector of magnetization was in this case perpendicular to the surface. It shows that the repeated passage of twin boundary establishes the similar pattern for two different orientation of magnetization inheriting or reflecting the domains in adjacent variant across the twin boundary.

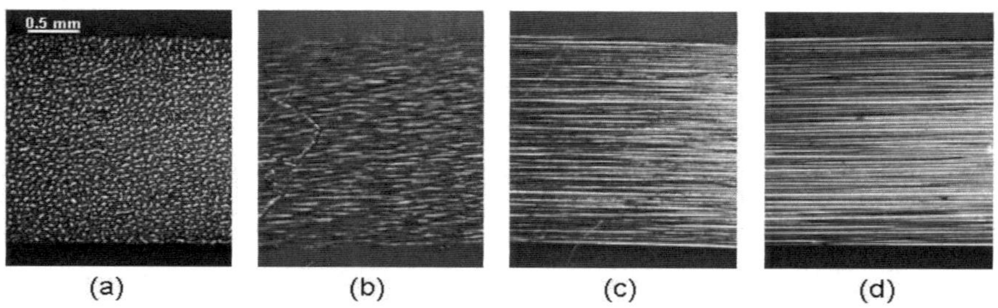

Figure 3. Magnetic domains of single variant state. (a) Labyrinth domains after magnetizing in 1 T magnetic field perpendicular to plane, (b) granular domains after initial passage of twin boundary (c-axis and magnetization in plane), (c) rake domains after second passage of twin boundary (c-axis and magnetization perpendicular), (d) rake domains after 12 passages of twin boundary.

We can conclude that the extraordinary magnetic domain pattern formed by one passage of single twin boundary is further modified by its repeated passage. As such we demonstrated how the magnetic structure can be modified solely by mechanical means. The question is how this newly formed pattern affects the twin boundary mobility and the magnitude of the field needed for the twin boundary motion. This is investigated now and results will be published elsewhere.

Acknowledgment
Support of Czech Scientific Foundation (No. 14-36566G) and Grant Agency of the Czech Technical University in Prague (No. SGS16/245/OHK4/3T/14) are gratefully acknowledged.

References

[1] Sozinov A, Lanska N, Soroka A, and Zou W 2013 *Appl. Phys. Lett.* **102** 021902
[2] Likhachev A A, Ullakko K 2000 *Phys. Lett. A* **275** 142
[3] Heczko O, 2014 *Materials Science and Technology* **30** 1559
[4] Chmielus M et al. 2011 *Acta Mater.* **59** 2948
[5] Heczko O, Straka L and Seiner H, 2013 *Acta Mater.* **61** 622
[6] Kopecký V et al. 2015 *Acta Phys. Pol. A* **128** 754

[7] Lai Y W et al., 2007 *Appl. Phys. Lett.* **90**, 192504
[8] Ge Y, Heczko O, Soderberg O and Hannula S.-P. 2006 *Scripta Mat.* **54**, 2155
[9] Straka L, Hanninen H, Heczko O, 2011 *Appl. Phys. Lett.* **98**, 141902
[10] Kopecký V, Fekete L, Perevertov O and Heczko O 2016 *AIP Advances* **6**, 056208
[11] Lai Y W, Schaefer R, Schultz L and McCord J, 2010 *Appl. Phys. Lett.* **96**, 022507
[12] Heczko O et. al. , 2015 IEEE Trans. Mag. 51, 2505304

Magnetic and dielectric properties of o-$LuFeO_3$/$SrTiO_3$

D V Mamedov[1], T P Gavrilova[1,2], I V Yatsyk[1,2], I F Gilmutdinov[2], Z Y Seidov[3], M A Aljanov[3], M J Najafzade[3], I N Ibrahimov[3], V I Chichkov[4], N V Andreev[4], E Yu Koroleva[5,6] and R M Eremina[1,2]

[1] Kazan E. K. Zavoisky Physical-Technical Institute (KPhTI) of the Kazan Scientific Center of the Russian Academy of Sciences, Sibirsky tract, 10/7, Kazan, 420029,Russia
[2] Kazan (Volga Region) Federal University, Kremlevskaya st., 18, Kazan, 420008 Russia
[3] Institute of Physics, Azerbaijan National Academy of Sciences, H.Cavid ave.131, AZ-1143 Baku, Azerbaijan
[4] National University of Science and Technology MISiS, Leninsky Prospekt, 4, Moscow, 119991 Russia
[5] Ioffe Institute, Politekhnicheskaya ul. 26, St. Petersburg, 194021 Russia; [6] Peter the Great St. Petersburg Polytechnic University, Politekhnicheskaya ul. 29, St. Petersburg, 195251 Russia
e-mail: REremina@yandex.ru

Abstract. Here we present the investigation of the orthorhombic o-$LuFeO_3$ thin film grown on $SrTiO_3$ (001) substrate. Temperature dependencies of magnetization measured in magnetic fields 1000 Oe and 2000 Oe were described in the assumption that ferromagnetic correlated regions are presented in samples together with the paramagnetic phase. Additional argument in favor of the ferromagnetic regions existence in o-$LuFeO_3$ is the magnetic hysteresis curve measured at T=5 and 100K. The temperature dependences of dielectric response were measured and analyzed.

1. Introduction

The $LuFeO_3$ can be crystalized in two modifications: hexagonal and orthorhombic. Thin films of the hexagonal $LuFeO_3$ (h-$LuFeO_3$) deposited on Al_2O_3 and YSZ are a multiferroic material that exhibits spontaneous electric and magnetic polarizations simultaneously [1, 2], which have great importance for future technologies. The orthorhombic crystallographic structure o-$LuFeO_3$ is the thermodynamically stable with standard condition preparation [3], the hexagonal structure is metastable. The free energy of o-$LuFeO_3$ is a few lower than that of the h-$LuFeO_3$. In epitaxial thin films, the film-substrate interfacial energy may favor the h-LuFeO3 structure, if the symmetry of the substrate is triangular or hexagonal. This effect can stabilize the hexagonal structure in epitaxial thin films, to a certain critical thickness [1, 4] Therefore, when $LuFeO_3$ is deposited onto substrate in the form of thin film it can coexist in the form of hexagonal and orthorhombic domains that was established in work [4] where two structural phases of $LuFeO_3$ occurs on a length scale of micrometer, as visualized in real space using X-ray photoemission electron microscopy, and moreover, there is the structural transition at about 1000 °C, from the hexagonal to the orthorhombic phase of $LuFeO_3$ [4] and then it results to interesting physical properties.

The crystal structure of the hexagonal h-$LuFeO_3$ belongs to the space group $P6_3cm$ with lattice parameters a = b = 5.993 Å, c = 11.634 Å and α=β=90°, γ=120°. The crystal structure of the orthorhombic o-$LuFeO_3$ belongs to the space group Pnma with lattice parameters a=5.547 Å; b=7.565 Å; c=5.213 Å. The crystal structure of the substrate material $SrTiO_3$ is *Pm3m*; a = b = c = 3.9046 Å, α

Content from this work may be used under the terms of the Creative Commons Attribution 3.0 licence. Any further distribution of this work must maintain attribution to the author(s) and the title of the work, journal citation and DOI.
Published under licence by IOP Publishing Ltd

IOP Conf. Series: Journal of Physics: Conf. Series **903** (2017) 012014 doi:10.1088/1742-6596/903/1/012014

$= \beta = \gamma = 90°$ (T>105 K) and *I4/mcm*; a = b = 5.510 Å, c = 7.798 Å (T<105 K). Magnetic and dielectric orders in orthorhombic and hexagonal structures $LuFeO_3$ are observed in different temperatures. The strong magnetic order in o-$LuFeO_3$ happens at T_N = 620 K and in h-$LuFeO_3$ at T_N=132K [3]. Ferroelectricity transition in h-$LuFeO_3$ was observed below T_C = 1050 K connecting from a $P6_3/mmc \rightarrow P6_3cm$ structure distortion [3].

2. Samples preparation and experimental details

Single-phase polycrystalline $LuFeO_3$ targets were obtained by solid-state synthesis using stoichiometric quantities of Lu_2O_3 and Fe_2O_3 powders as the starting materials. o- $LuFeO_3$ thin films were deposited from the prepared targets by RF-magnetron sputtering with the ''facing-target'' scheme [5], we used the same technique for the deposition of manganite films in our previous work [6].For the growth of orthorhombic o-$LuFeO_3$ films (001) oriented $SrTiO_3$ (STO) single-crystal substrates were used. A temperature of substrate during the deposition was 700 ^{0}C. A mixture of Ar and O_2 with pressure 1–2 mTorr was used as the working atmosphere. The thickness of the obtained o-$LuFeO_3$ films was 100 nm. The structure of the grown thin films was studied by X-ray diffraction (Bruker D8 DISCOVER, using Cu Kα radiation).

The magnetization was measured on the PPMS-9 device in the temperature range from 4 to 300 K and in magnetic field up to 2T. For simplicity presentation of the magnetization and magnetic susceptibility of o- $LuFeO_3$ we subtracted a diamagnetic contribution of a substrate. In order to separate contributions of the thin film and substrate we measured the mass of a sample and defined the number of moles of $SrTiO_3$ as $\nu=m/\mu$, neglecting the mass of the thin film in comparison with the mass of the substrate. The diamagnetic susceptibility of substrate $SrTiO_3$ is equal $\chi_{diamag}=-6.0\cdot10^{-5}$ emu/mol and can be estimated using diamagnetic susceptibilities of single ions: Sr^{2+} -19; Ti^{4+} -5; O^{2-} -12 given in 10^{-6} emu/mol [7].

The dielectric response was studied at frequency range 0.1 Hz - 10 MHz and temperature region 100-450 K using ultra broadband dielectric spectrometer Novocontrol BDS80 with Quatro-cryosystem. Platinum IDE electrodes were deposited on the thin film by electron beam lithography method.

3. Experimental results

3.1. X-ray diffraction analysis

Figure 1 shows the Θ–2Θ scan of a $LuFeO_3$||STO(001) sample. The peaks correspond to the (001) reflections of the substrate, as well as to the reflections from the (101) plane of the orthorhombic o-$LuFeO_3$. This suggests that the films are textured, with the out-of-plane orientation o-$LuFeO_3$(101)||STO(001).The out-of-plane $d_{(101)}$ = 3.830 Å, somewhat larger than the value for bulk material, i.e. $d^{bulk}_{(101)}$ = 3.799 Å. The strain is $\varepsilon_{(101)}$ = 0.8%.

3.2. Magnetization and magnetic susceptibility

Temperature dependence of magnetization in o-$LuFeO_3$/$SrTiO_3$ was measured at 1000 Oe and 2000 Oe magnetic field in a wide temperature range. The temperature dependence of the magnetic susceptibility obtained as χ=M/H is presented in Figure 2a for both experiments. It is clear visible the difference between absolute values of the magnetic susceptibility measured in different magnetic field, which suggests the existence of the ferromagnetic regions in the sample. In the case of coexistence of the ferromagnetic and paramagnetic phases the magnetic susceptibility can be calculated as:

$$\chi=\chi_{ferr}+\chi_{par}, \quad (1)$$

where χ_{par} is paramagnetic contribution, described by the Curie low χ_{par}=C/T; $\chi_{ferr}=M_{ferr}/H$, where M_{ferr} is the temperature independent ferromagnetic magnetization. The fitting of the temperature dependence of the magnetic susceptibility of o-$LuFeO_3$ thin film was performed using the equation (1) and fit parameters $C=4\cdot10^{-8}$emu·K, M_{ferr}(1000 Oe) $=1.15\cdot10^{-5}$ emu·Oe; M_{ferr}(2000 Oe)$=1.9\cdot10^{-5}$ emu·Oe; fit curves are given in Figure 2a.

To get the additional argument of the ferromagnetic phase existence in o-$LuFeO_3$ we have performed the measurements of the field dependence of the magnetization at T=5 and 100K (see Figure 2b). The hysteresis loop observed in the magnetic field range from -0.6 kOe to 0.6 kOe for both temperatures 5K and 100 K. Therefore ones again we conclude that ferromagnetic regions are present in the sample together with the paramagnetic phase in o-$LuFeO_3$ thin films deposited on $SrTiO_3$. Previously weak ferromagnetism and hysteresis loops were observed in the bulk o-$LuFeO_3$ [8], assuming the preserving of the magnetic properties during the transition from bulk samples to thin films.

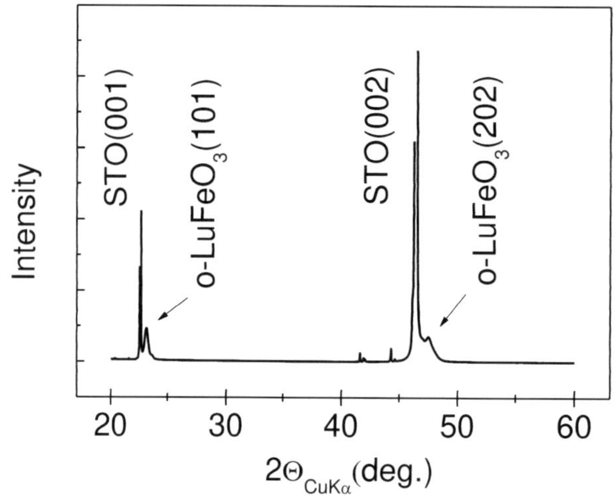

Figure 1. θ–2θ XRD scan of a o-$LuFeO_3$ film on the STO(001) substrate.

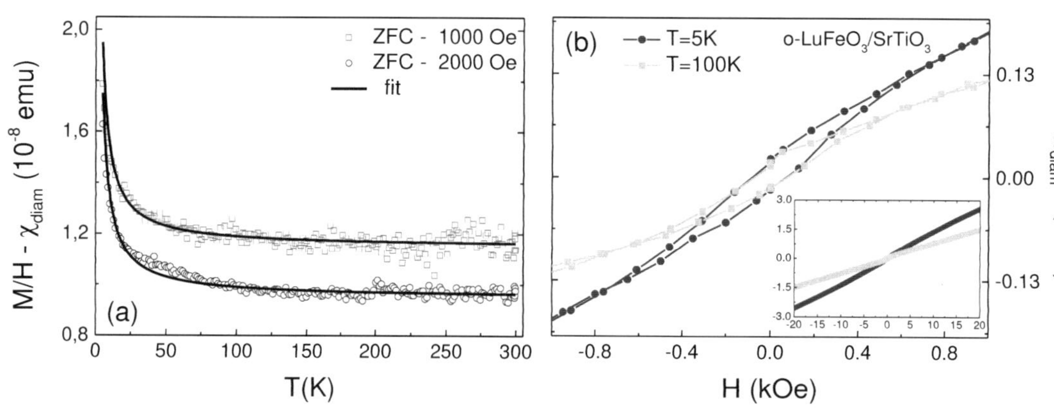

Figure 2. a) Temperature dependencies of M/H- χ_{diam} for $LuFeO_3/SrTiO_3$ in magnetic fields 1000 Oe and 2000 Oe; fitting was performed using the equation (1) (see the text); b) hysteresis curve measured at a temperature of 5 and 100K for $LuFeO_3/SrTiO_3$.

3.3. Dielectric behavior

Temperature dependence of complex dielectric response of o-$LuFeO_3/SrTiO_3$ was measured at the frequency range 0.1 Hz - 10 MHz and at temperatures 100-450 K. In order to separate a contribution from the thin film and the substrate we measured dielectric properties of $SrTiO_3$ substrate with the same electrodes and in the same conditions. Calculated temperature dependencies of dielectric permittivity ε' of the thin film o-$LuFeO_3$ for several measuring frequencies are presented in Figure 3a. The schematic view and optic images of the sample with electrodes are given in Figure 3b and 3c, respectively.

The anomaly of the intrinsic dielectric response was observed in [3] near the magnetic transition point T_N~600 K that clearly told about a coupling between dielectric and magnetic properties in o-$LuFeO_3$. In our case this anomaly has corresponded to the observed increase of ε' with increasing temperature up to 450 K. We also observed a frequency dependent maximum of permittivity near 130 K. Similar low temperature anomaly, indicating possible charge ordering in the temperature range of 150-300 K, was observed in o-$LuFeO_3$ ceramics in [9], assuming the preserving of the dielectric properties during the transition from bulk samples to thin films how it was in the case of magnetic properties.

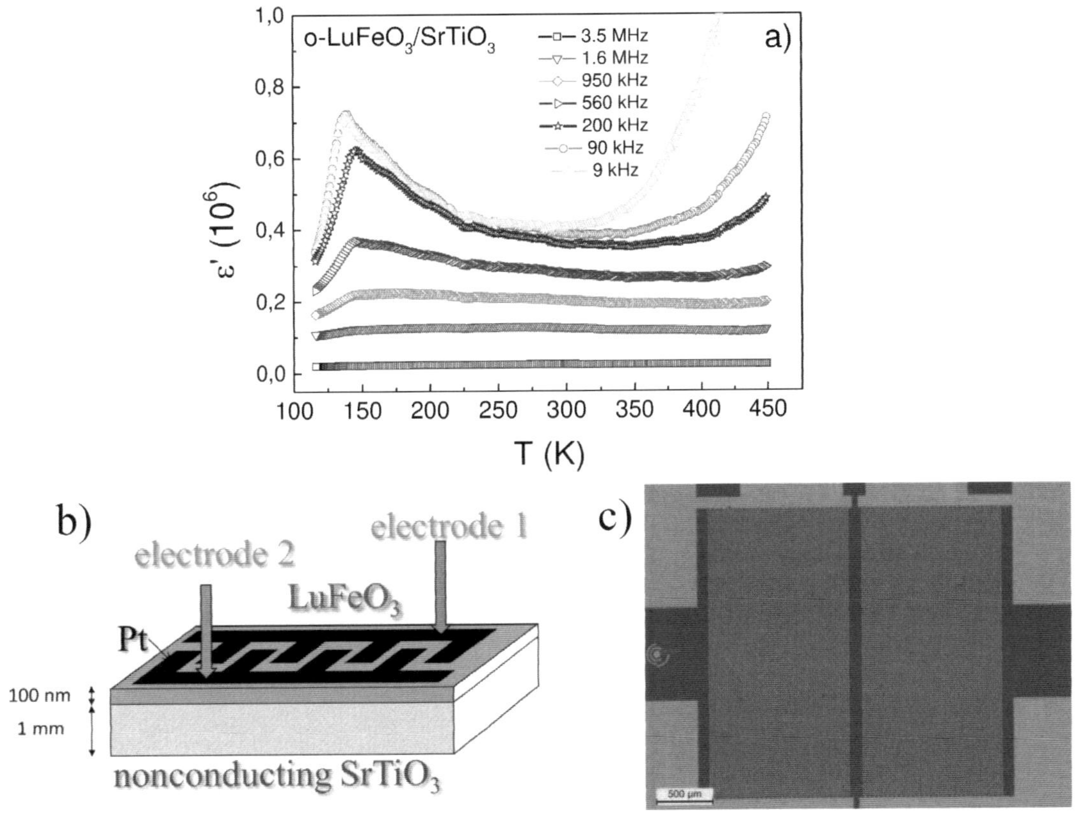

Figure 3. a) Temperature dependences of the real ε' part of the permittivity at different frequencies for the thin film $LuFeO_3$, b) schematic and c) optic images of the sample with electrodes patterned into interdigitated structure.

4. Conclusions

The magnetic and dielectric properties of the radio frequency magnetron sputtered orthorhombic $LuFeO_3$ thin films were investigated. Magnetic properties of o-$LuFeO_3/SrTiO_3$ thin film were described in the assumption that ferromagnetic correlated regions are presented in samples together with the paramagnetic phase. We also observed a frequency dependent maximum of permittivity near 130 K. indicating possible charge ordering in the temperature range of 150-300 K

4. Acknowledgments

This work was supported by the Science Development Foundation under the President of the Republic of Azerbaijan Grant № EIF-2013-9(15)-46/09/1.

References

[1] Wang W, Zhao J, Wang W, Gai Z, Balke N, Chi M, Lee H N, Tian W, Zhu L, Cheng X, Keavney D J, Yi J, Ward T Z, Snijders P C, Christen H M, Wu W, Shen J, Xu X 2013 *Phys. Rev. Lett.* **110** 237601

[2] Disseler S M, Borchers J A, Brooks C M, Mundy J A, Moyer J A, Hillsberry D A, Thies E L, Tenne D A, Heron J, Holtz M E, Clarkson J D, Stiehl G M, Schiffer P, Muller D A, Schlom D G and Ratcliff W D 2015 *Phys. Rev .Lett.* **114** 217602

[3] Chowdhury U, Goswami S, Bhattacharya D, Ghosh J, Basu S and Neogi S 2014 *App. Phys. Lett.* **105** 052911

[4] Cao S, Zhang X, Sinha K, Wang W, Wang J, Dowben P A and Xu X 2016 *Appl. Phys. Lett.* **108** 202903

[5] Antonova E A, Ruzinov V L, Stark S Yu, Tchitchkov V I 1991 *Supercond. Phys. Chem. Technol.* **4** 1624

[6] Andreev N V, Sviridova T A, Chichkov V I, Volodin A P, Van Haesendonck C, Mukovskii Ya M 2014 *Journal of Alloys and Compounds* **586** S343

[7] Bain A G and Berry J F 2008 *J of Chem.Educ.* **85** 532

[8] Zhu W, Pi L, Tan S and Zhang Y 2012 *App. Phys. Lett.* **100** 052407

[9] Zhang L, Chen X M 2009 *Solid State Comm.* **149** 1317

IOP Conf. Series: Journal of Physics: Conf. Series **903** (2017) 012015 doi:10.1088/1742-6596/903/1/012015

Anisotropic magnetic property of nanocomposite $Nd_2Fe_{14}B/Mo/\alpha$-Fe multilayer films

K Kobayashi[1], D Ogawa[1], K Koike[1*], H Kato[1], M Oogane[2], T Miyazaki[2], Y Ando[2] and M Itakura[3]

[1]Graduate school of Science and Engineering, Yamagata University, Yonezawa 992-8510, Japan
[2]Graduate school of Engineering, Tohoku University, Sendai 980-8579, Japan
[3]Interdisciplinary Graduate School of Engineering Sciences, Kyushu University, Kasuga 816-8580 Japan

e-mail: kkoike@yz.yamagata-u.ac.jp

Abstract. We fabricated the MgO(100)/Mo/[$Nd_2Fe_{14}B$/Mo/Fe/Mo]$_5$/Mo multilayer films, in which we tried to avoid the negative J_{ex} interfaces and thermal diffusion between $Nd_2Fe_{14}B$ and α–Fe layers by forming the stacked structure with ultra thin Mo interlayer. The films without Mo interlayer showed isotropic magnetic properties, while films with Mo interlayer thickness t_{Mo} = 1 nm exhibited the perpendicular anisotropy with the coercivity of 7 kOe. Shapes for the major loop and the first order reversal curves (FORCs) suggested an existence of positive exchange coupling between the Nd-Fe-B and the α–Fe layers via the 1 nm thick Mo interlayer.

1. Introduction

Since an idea of exchange-coupled hard/soft nanocomposite magnets was proposed, predicting a huge maximum-energy-product $(BH)_{max}$ value [1], [2], much experimental effort was made so far. For instance, Sm-Co/α–Fe nanocomposite films showed excellent properties with a high $(BH)_{max}$ [3], [4]. On the other hand, magnetization value of the Nd-Fe-B system has a larger than that of Sm-Co system. Therefore, many researchers are tried to make the Nd-Fe-B/α–Fe or Nd-Fe-B/Fe-Co nanocomposite films [5], [6]. [7], [8]. However, reported experimental $(BH)_{max}$ values in these Nd-Fe-B/α–Fe systems are still much smaller than the predicted one. The reason for the gap would be related mainly to a difficulty in controlling the size and alignment of magnetic grains. Moreover, Toga *et al.* suggested, according to the first-principle calculations [9] the magnitude of exchange-coupling constant J_{ex} in the $Nd_2Fe_{14}B$ /α–Fe system depends on the crystal orientation at the interface, even taking a negative J_{ex} value for a particular combination. We recently confirmed experimentally that positive and negative J_{ex} values were realized in $Nd_2Fe_{14}B$(001)/α–Fe(100) and the $Nd_2Fe_{14}B$(100)/α–Fe interfaces, respectively [10], [11]. These results thus suggest that we need to control the crystallographic orientation at the interfaces in order to avoid interfaces with a negative J_{ex} value.

In this work, we fabricated the $Nd_2Fe_{14}B$/Mo/Fe/Mo multilayer films, in which we tried to avoid the negative J_{ex} interfaces by keeping the layer-by-layer structure and the perpendicular alignment of the $Nd_2Fe_{14}B$ grains.

2. Experimental procedure

Nd-Fe-B/Mo/Fe/Mo multilayer films were prepared by UHV sputtering system onto MgO(100) substrate. The films have the stacking sequence of MgO(100)/Mo(20 nm)/[Nd-Fe-B(30 nm)/Mo(0, 1 nm)/Fe(5 nm)/Mo(0, 1 nm)]$_5$/Mo(10 nm). The $Nd_7Fe_{80}B_{13}$ alloy and Nd metal with 3N-purity and the Fe and Mo metals with 4N-purity were used for the sputtering targets. The Ar gas pressure P_{Ar} during

Content from this work may be used under the terms of the Creative Commons Attribution 3.0 licence. Any further distribution of this work must maintain attribution to the author(s) and the title of the work, journal citation and DOI.
Published under licence by IOP Publishing Ltd

IOP Conf. Series: Journal of Physics: Conf. Series **903** (2017) 012015 doi:10.1088/1742-6596/903/1/012015

the deposition of the Mo buffer- and capping layers and the Nd-Fe-B/Mo/Fe/Mo multilayers were 0.07 Pa and 0.2 Pa, respectively. The Nd-Fe-B layer was deposited by co-sputtering the $Nd_7Fe_{80}B_{13}$ and Nd targets due to control the film composition. The MgO(100)/Mo(100) epitaxial growth was induced depositing the Mo buffer layer onto heated substrate at 300°C in two-step process, which is crystallized during an annealed at 670°C [12]. During the deposition of multilayer parts, the substrate was heated at 300°C. Then, the films were annealed successively in the range from T_a = 400°C to 700°C in a vacuum under 10^{-7} Pa. Magnetization curve was measured at 20°C using Vibrating sample magnetometer (VSM) with magnetic field up to 12 kOe. Crystal structure of the films was characterized by $\theta/2\theta$ scan of X-ray diffraction (XRD) techniques with Cu-Kα radiation. Film thickness was evaluated by X-ray reflectometry (XRR). Surface morphology of the films was observed by atomic force microscopy (AFM).

3. Result and Discussion

Fig. 1 shows XRD patterns for the Mo(20 nm)/[Nd-Fe-B(30 nm) /Fe(5 nm)]$_5$/Mo(10 nm) and the Mo(20 nm)/[Nd-Fe-B(30 nm)/Mo(1 nm)/Fe(5 nm)/Mo(1 nm)]$_5$/Mo(10 nm) multilayer films, respectively. When the films without the Mo interlayer were annealed above T_a = 500°C, Nd_2Fe_{17} phase was appeared due to thermal diffusion between $Nd_2Fe_{14}B$ and α–Fe layers, while films with the Mo interlayer was not shown Nd_2Fe_{17} phase in the XRD patterns. Also, films with the Mo interlayer were appeared α–Fe phase. This result suggests that Mo interlayer avoid inter diffusion between $Nd_2Fe_{14}B$ and α–Fe layers. However, no peaks from $Nd_2Fe_{14}B$ phase were shown in these XRD patterns.

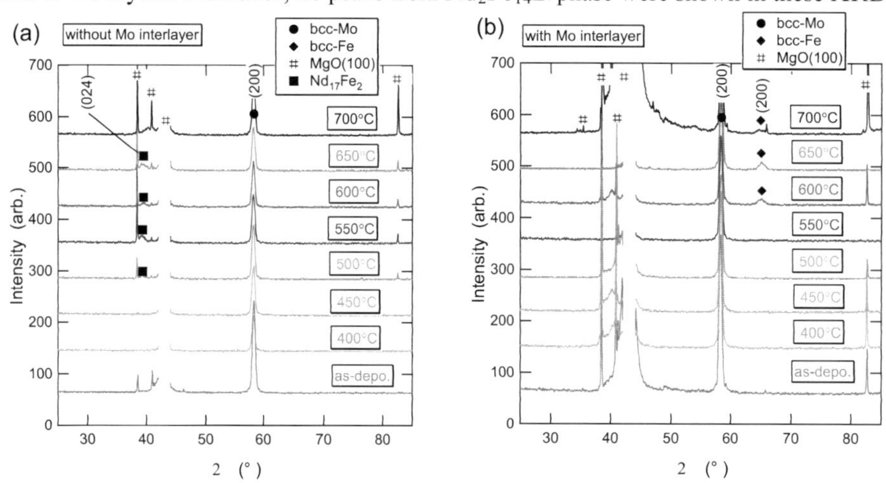

Fig. 1 XRD patterns for (a) Mo(20 nm)/[Nd-Fe-B(30 nm) /Fe(5 nm)]$_5$/Mo(10 nm) and (b) Mo(20 nm)/[Nd-Fe-B(30 nm)/Mo(1 nm)/Fe(5 nm)/Mo(1 nm)]$_5$/Mo(10 nm) multilayer films.

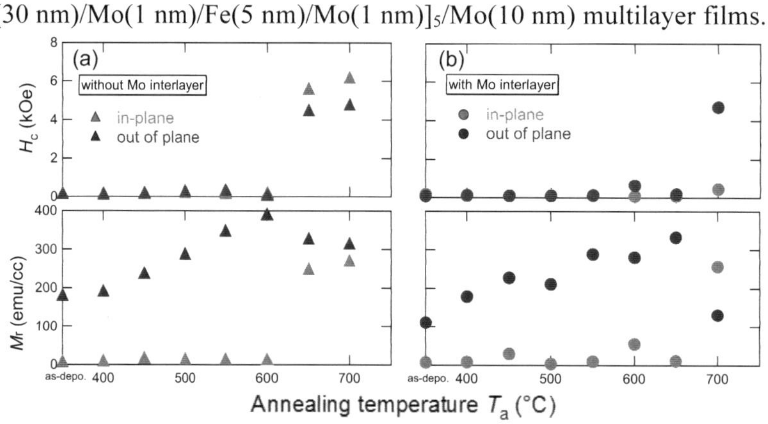

Fig. 2 Coercivity and remanence for the multilayer films (a) without Mo interlayer and (b) with Mo interlayer as a function of annealing temperature.

Fig. 2 shows the coercivity H_c and the remanence M_r of films in Fig. 1 as a function of annealing temperature. Hard magnetic properties of the films appeared to originate from crystallization behaviour of amorphous Nd-Fe-B phase above T_a = 650˚C. The films with Mo interlayer above T_a = 650˚C showed a significant increase in the H_c of about 7 kOe. The films without Mo interlayer showed isotropic magnetic properties, while films with Mo interlayer exhibited the perpendicular magnetic anisotropy. Shapes for the FORCs of film with Mo interlayer which had steep slope suggested an existence of positive exchange coupling between $Nd_2Fe_{14}B$ and α–Fe layers via the 1 nm thick Mo interlayer. (see Fig. 3). HR-TEM observation of the multilayer films is now in progress in order to deeply understand the effect of Mo interlayer on above mentioned notable phenomena.

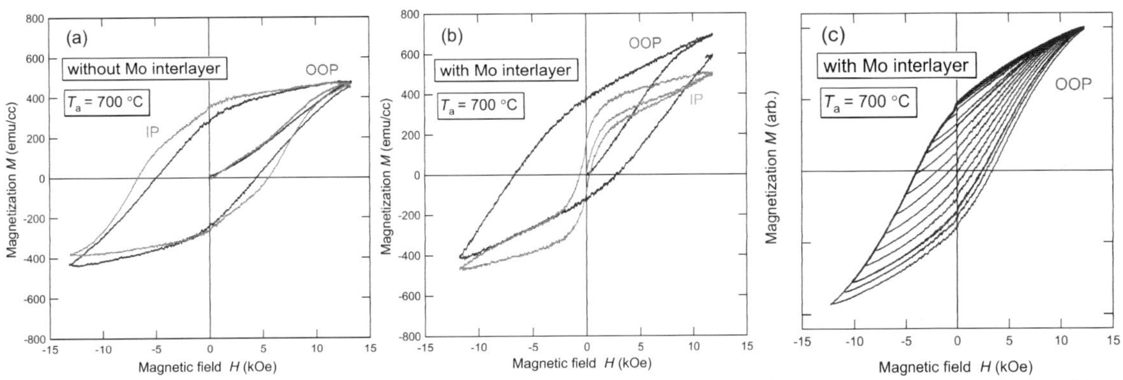

Fig. 3 Out of plane (OOP) and in-plane (IP) of magnetization curves for the annealed multilayer films (a) without Mo interlayer and (b) with Mo interlayer. (c) FORCs of sample of (b).

4. Conclusion

The insertion of ultra-thin Mo interlayer (thickness t_{Mo} = 1 nm) is effective method to improve the magnetic anisotropy of the nanocomposite Nd-Fe-B/Mo/Fe/Mo multilayer films. By optimizing the annealing temperature, the anisotropic hard magnetic properties of the films increases by more than that of the film without Mo interlayer.

5. References

[1] Skomski R and Coey J M D, 1993 *Phys. Rev.* B **48** 15812.
[2] Kneller E and Hawig R, 1991 *IEEE Trans. Magn.* **27** 3588.
[3] Zhang J, Takahashi Y K, Gopalan, and R, Hono K, 2005 *Appl. Phys. Lett.* **86** 122509.
[4] Neu V, Sawatzki S, Kopte M, Mickel C, and Schultz L, 2012 *IEEE Trans. Magn.* **48** 3599.
[5] Parhofer S M, Wecker J, Kuhrt C, Gieres G, and Schultz L, 1996 *IEEE Trans. Magn.* **32** 4437.
[6] Shindo M, Ishizone M, Sakuma A, Kato H, and Miyazaki T, 1997 *J. Appl. Phys.* **81** 4444.
[7] Kato H, Ishizone M, Koyama K, and Miyazaki T, 2005 *J. Magn. Magn. Mater.* **290-291** 1221.
[8] Cui W, Takahashi K Y, Hono K, 2012 *Adv. Mater.* **24** 6530.
[9] Toga Y, Moriya H, Tsuchiura H, and Sakuma A, 2011 *J. of Phys: Conf. Ser.* **266** 012046.
[10] Ogawa D, Koike K, Kato H, Mizukami S, Miyazaki T, Oogane M, and Ando Y, 2013 *J. Korean Phys. Soc.* **63** 489.
[11] Ogawa D, Koike K, Mizukami S, Miyazaki T, Oogane M, Ando Y, and Kato H, 2015 *Appl. Phys. Lett.* **84** 4230.
[12] Koikc K, Kusano T, Ogawa D, Kobayashi K, Kato H, Oogane M, Miyazaki T, Ando Y, and Itakura M, 2016 *Nanoscale Res. Lett.* **11** 33.

Acknowledgments
This work was supported partly by the Industry-Academia Collaborative R&D Programs, Science and Technology Agency (JST), Japan; and by Grant-in-Aid for Scientific Research KAKENHI (grant No. 16H04488) from JSPS, Japan.

First principles and metadynamics study of the spin-reorientation transition in Fe/Au(001) films

B. Nagyfalusi[1], **L. Udvardi**[1,2] **and L. Szunyogh**[1,2]

[1] Department of Theoretical Physics, Budapest University of Technology and Economics, Budafoki út 8, H-1111 Budapest, Hungary
[2]MTA-BME Condensed Matter Research Group, Budapest University of Technology and Economics, Budafoki út 8, H-1111 Budapest, Hungary

E-mail: nagyfalusi@phy.bme.hu

Abstract. Based on first principles calculations, we investigate the magnetic anisotropy and spin reorientation transition (SRT) for Fe_n/Au(001) (n=2, 3) films. The SRT occurs at three atomic layer of Fe in agreement with experiments due to competing on-site and two-site anisotropy. We also study the temperature dependence of the magnetic anisotropy energy (MAE) by means of metadynamics Monte Carlo simulations.

1. Introduction

Magnetic anisotropy in multilayers and nanoparticles is of high technological importance in magnetic data storage [1, 2]. In general, thickness dependent SRT is explained as the result of competing volume and surface anisotropies. The most important contribution to the volume part is the shape anisotropy related to the magnetic dipole-dipole interaction and prefers the in-plane orientation of the magnetization, whereas the surface anisotropy due to the spin-orbit coupling favours in most case a normal-to-plane configuration. In the present paper we study the magnetic anisotropy of thin Fe films on Au(001) surface from first principles. The Fe/Au(001) system has been a subject of numerous investigations concerning its structural and magnetic properties, see eg. Ref. [3] and the references therein. Fe monolayers grown on Au(001) have also been referred as a model two-dimensional ferromagnet [4]. The SRT for Fe layers on Au(001) occurs between two and three atomic layers [3]. Based on our computational results we will argue that, in this particular case, the SRT is the consequence of the competition between the on-site uniaxial anisotropy and the anisotropy of the exchange couplings.

2. Methods of calculations

The electronic structure of the overlayer was calculated in terms of the fully relativistic screened Korringa-Kohn-Rostoker (KKR) method applying the surface Green's function technique to treat the semi-infinite Au system [5]. Since the lattice mismatch between the bcc Fe(001) and the fcc Au(001) is 0.6 % we supposed a perfect 2D translational symmetry for the whole system using the lattice constant of Au (2.87 Å). The Fe-Fe interlayer distance has been chosen to be the same as the bulk value (1.44 Å). Due to the fact that the magnetic anisotropy energy is sensitive to the lattice relaxation the electronic structure has been determined for different values of the

Content from this work may be used under the terms of the Creative Commons Attribution 3.0 licence. Any further distribution of this work must maintain attribution to the author(s) and the title of the work, journal citation and DOI.
Published under licence by IOP Publishing Ltd

Fe-Au interlayer distance changing from 1.6 Å to 1.7 Å. The range of the interlayer distance is in agreement with measurements [3] and previous ab-initio calculations [6].

We describe the magnetic system in terms of an extended Heisenberg model:

$$H = -\frac{1}{2}\sum_{pi,qj} \mathbf{s}_{pi}^{\mathrm{T}} \cdot \mathbf{J}_{pi,qj} \cdot \mathbf{s}_{qj} + \sum_{pi} K_p (\mathbf{s}_{pi} \cdot \hat{\mathbf{z}})^2 \,, \quad (1)$$

where $\mathbf{s}_{pi}$ is a unit vector parallel to the atomic magnetization at site i in layer p, T denotes the transpose of the vector, $\mathbf{J}_{pi,qj}$ is a 3×3 exchange tensor, K_p is the uniaxial anisotropy constant at layer p and $\hat{\mathbf{z}}$ denotes a unit vector normal to the surface of the magnetic layer. The exchange tensor and the uniaxial anisotropy constant have been determined by using the relativistic torque method [7]. The thermodynamic quantities of the system such as the magnetization, the magnetic susceptibility and the specific heat, have been calculated by means of Monte Carlo (MC) simulations. The temperature dependence of the magnetic anisotropy has been calculated by applying metadynamics [8] implemented in the Metropolis MC simulations. The free energy has been sampled along a collective variable (CV) defined as the z component of the normalized magnetization: $\eta = M_z/M$, where $M_z = \hat{\mathbf{z}}\mathbf{M}$, and $M = |\mathbf{M}|$ with the total magnetization $\mathbf{M} = \sum_{p,i} \mathbf{s}_{pi}$.

3. Results and discussion

The calculated magnetic moments for $Fe_2Au(001)$ and $Fe_3Au(001)$ and the corresponding Curie temperatures are summarized in Table 1. The values changed within a few percent in the whole relaxation range. The Curie temperatures are in relatively good agreement with the experiments in Ref. [9].

Table 1. Calculated magnetic moments of the Fe layers and simulated Curie temperatures for $Fe_2/Au(001)$ and $Fe_3/Au(001)$. The first layer is on the top of the Au(001) surface.

	1	2	3	T_C^{calc}	T_C^{exp}[9]
$Fe_2/Au(001)$	$2.7\,\mu_B$	$3.0\,\mu_B$		440 K	430 K
$Fe_3/Au(001)$	$3.0\,\mu_B$	$2.3\,\mu_B$	$3.0\,\mu_B$	500 K	550 K

In our model the exchange interaction between two spins is described by a 3×3 matrix. One third of the trace of this matrix is associated with the usual exchange coupling appearing in a conventional Heisenberg model, the anti-symmetric part can be identified as the Dzyaloshinsky-Moriya interaction and the symmetric part is the pseudo-dipolar interaction. Beyond the on-site anisotropy, this two-site term contributes to the MAE. Both terms are the consequence of the spin orbit coupling. The calculated total MAE and the on-site anisotropy contributions of the Fe layers are given in Table 2. For $Fe_2/Au(001)$ both the on-site and two-site terms prefer normal-to-plane magnetization in the relaxation range considered. In the case of $Fe_3/Au(001)$ the on-site anisotropy still prefers normal to plane configuration, but it is counterbalanced by the two-site anisotropy resulting in to an in-plane magnetization in the whole relaxation range.

Crespo *et al.* [10] have shown that the magnetization as a collectice variable can be used to explore the free energy of the Ising model. In the left panel of Fig. 1 the parabolic behavior of the biasing potential is depicted as a function of the CV. Obviously, $\eta = \pm 1$ belongs to the normal-to-plane, while $\eta = 0$ to the in-plane magnetic configuration. The curvature of the parabolas is decreasing as the temperature is increasing and the biasing potential will be flat when the system reaches the paramagnetic phase. In the right panel of Fig. 1 the MAE extracted from the curvature of the biasing potential is shown as a function of the magnetization. According to the Callen-Callen's law [11], at low temperature the MAE curve behaves as $(M/M_0)^3$. At

Table 2. Calculated total MAE and sum of the layer-dependent on-site anisotropy constants for Fe_2/Au(001) and Fe_3/Au(001). All energies are given in meV.

	Fe_2/Au(001)		Fe_3/Au(001)	
$d_{\mathrm{Fe-Au}}$	$E_z - E_x$	$K_1 + K_2$	$E_z - E_x$	$K_1 + K_2 + K_3$
1.58 Å	−0.669	−0.263	0.073	−0.072
1.61 Å	−0.649	−0.223	0.088	−0.076
1.64 Å	−0.605	−0.184	0.118	−0.086
1.67 Å	−0.565	−0.150	0.142	−0.095

higher temperature it is close to $(M/M_0)^2$ due to the two-site anisotropy present in the model, similar to the case of bulk FePt [12].

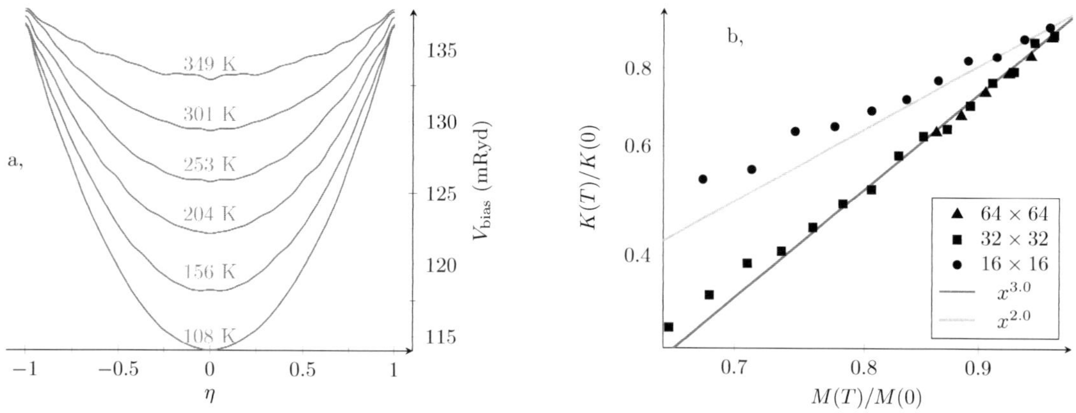

Figure 1. Biasing potential at different temperatures in the metadynamics simulations of Fe_2/Au(001) with the interlayer distance, $d_{\mathrm{Fe-Au}} = 1.58$ Å (left panel). The normalized MAE as a function of the normalized magnetization is shown in the right panel on a log-log scale. Different symbols refer to different sizes of the 2D lattice used in the simulations.

4. Conclusions

In the present paper we used metadynamics to study the temperature dependence of the magnetic anisotropy for Fe_2/Au(001) system. From ab-initio calculation we concluded that the thickness driven SRT is the consequence of the competing on-site anisotropy and the anisotropy of the exchange interactions due to the spin-orbit coupling.

This work was supported by the National Research, Development and Innovation Office of Hungary under project No. K115575.

[1] Falicov L, Pierce D, Bader S, Gronsky R, Hathaway K, Hopster H, Lambeth D, Parkin S, Prinz G, Salamon M, Schuller I and Victora R 1990 *Journal of Materials Research* **5** 1299
[2] Weller D and Moser A 1999 *IEEE Transactions on Magnetics* **35** 4423
[3] Wilgocka-Ślezak D, Freindl K, Kozio A, Matlak K, Rams M, Spiridis N, Ślezak M, Ślezak T, Zajac M and Korecki J 2010 *Phys. Rev. B* **81**(6) 064421
[4] Himpsel F J, Ortega J E, Mankey G J and Willis R F 1998 *Advances in Physics* **47** 511
[5] Szunyogh L, Újfalussy B and Weinberger P 1995 *Phys. Rev. B* **51**(15) 9552
[6] Kellar S A, Chen Y, Huff W R A, Moler E J, Hussain Z and Shirley D A 1998 *Phys. Rev. B* **57**(3) 1890
[7] Udvardi L, Szunyogh L, Palotás K and Weinberger P 2003 *Phys. Rev.* B **68** 104436
[8] Laio A and Parrinello M *Proceedings of the National Academy of Science*
[9] Dürr W, Taborelli M, Paul O, Germar R, Gudat W, Pescia D and Landolt M 1989 *Phys. Rev. Lett.* **62** 206
[10] Crespo Y, Marinelli F, Pietrucci F and Laio A 2010 *Phys. Rev. E* **81** 055701
[11] Callen E R and Callen H B 1963 *Phys. Rev.* **129**(2) 578
[12] Mryasov O N, Nowak U, Guslienko K Y and Chantrell R W *Europhysics Letters* **69** 805

IOP Conf. Series: Journal of Physics: Conf. Series **903** (2017) 012017 doi:10.1088/1742-6596/903/1/012017

Charge transfer and magnetization of a MoS_2 monolayer at the Co(0001)/MoS_2 interface

T. Garandel[1,2], R. Arras[1], X. Marie[2], P. Renucci[2], and L. Calmels[1]

[1]CEMES-CNRS and Université de Toulouse, 29 rue Jeanne Marvig, BP 94347, 31055 Toulouse, France

[2]Université de Toulouse, INSA-CNRS-UPS, LPCNO, 135 Avenue de Rangueil, 31077, Toulouse, France

garandel@cemes.fr

Abstract. The Co/MoS_2 system may constitute a fundamental building block for future spintronic devices based on a single MoS_2 transition metal dichalcogenide monolayer. Here, the *hcp* Co(0001)/MoS_2 interface electronic structure as well as magnetic properties are investigated by first principles calculations based on the density functional theory. The charge transfer due to covalent bonding between S and Co atoms at the interface has been calculated for the lowest energy configuration obtained after optimization of the atomic coordinates. This charge transfer is different for majority and minority spin electrons, which induces a magnetization of the MoS_2 layer bellow the Cobalt contact. The connection between the charge transfers at the interface and the modification of the magnetic properties is discussed.

1. Introduction

In the past five years, two-dimensionnal (2D) semiconductor materials based on transition metal dichalcogenide monolayers (MoS_2, WSe_2…) have encountered a large interest [1], due to their original properties [2] that could be used in nano-electronics, optoelectronics and spintronics. In view of future devices based on spintronics, electrical spin-injection into these 2D semiconducting single layers is a strong requirement. The more efficient way to realize such an injection at room temperature is to use a ferromagnetic metallic electrode, provided that the possible conductivity mismatch problem has been circumvented [3]. The understanding of the electronic and magnetic properties of MoS_2 at the interface is thus a key step to optimize such an injection. Here we focus on the Co(0001)/MoS_2 interface, and we analyze by first principles calculations based on the density functional theory the magnetic properties induced in the MoS_2 single layer at this interface. For realistic calculations, we considered supercells built from 5x5 Co(0001) atomic layers with the *hcp* stacking and a 4x4 MoS_2 single layer. These stacking would correspond to the very small lattice mismatch of 0.8%.

2. First principles methods and atomic structure of the Co(0001)/MoS_2 interface

The atomic and electronic structures of the interface have been calculated from first principles, using the code Wien2k [4] based on the density functional theory. This code uses a basis of Linearized Augmented Plane Waves+local orbitals (LAPW+lo) and the Perdew, Burke and Ernzerhof exchange-correlation potential to solve the Kohn-Sham equation, atomic sphere radii of 1.8, 1.8 and 2.0 atomic units, respectively for Co, S and Mo atoms, 24 different Bloch vectors for sampling the irreducible wedge of the two-dimensional Brillouin zone and a basis size corresponding to an energy cut-off of 151

Content from this work may be used under the terms of the Creative Commons Attribution 3.0 licence. Any further distribution of this work must maintain attribution to the author(s) and the title of the work, journal citation and DOI.
Published under licence by IOP Publishing Ltd

eV ($R_{min}K_{max}$=6.0, with R_{min}=1.8 atomic units). The lowest energy atomic configuration has been selected from several symmetric slabs, which correspond to different possible interface atomic structures and different ways of superimposing the 4x4 MoS_2 single layer on the top of the 5x5 Co(0001) surface. Each slab contains a Co layer with a thickness of 5 Co(0001) monolayers, surrounded on each of its sides by a MoS_2 monolayer. The vacuum between periodically repeated MoS_2 monolayers has a thickness larger than 1 nm.

The lowest energy configuration, obtained after minimizing atomic forces, corresponds to the case where one of the S atoms of the 4x4 MoS_2 cell is located exactly in front of one of the hollow sites of the Co surface. The average distance between the Co and S atomic layers at the interface is of 0.205 nm, which means that these two layers are covalently bonded. The resulting Co(0001)/MoS_2 interface contains 25 Co and 16 S atoms. Its atomic structure can be described as follows: Each interface Co atom is bonded to a single interface S atom. One of the16 S atoms (labelled S_3) is covalently bonded to 3 different equivalent Co atoms with an interatomic distance d_{CoS}=0.236 nm; 3 of the interface S atoms (labelled S_2) are each bounded to 2 equivalent Co atoms with an interatomic distance d_{CoS}=0.234 nm; all the other interface S atoms are only bonded to one interface Co atom (d_{CoS} between 0.221 and 0.222 nm). The bonded MoS_2 layer is slightly corrugated.

3. Spin magnetic moments at the Co(0001)/MoS_2 interface

The spin magnetic moment of Co atoms is on average 8% lower at the Co(0001)/MoS_2 interface than in bulk *hcp* Co, with values that depend on the kind of interface S atom to which they are covalently bonded: The interface Co atoms that show the highest spin magnetic moment (1.62 μ_B) are those bonded to the S_3 atom, followed by the Co atoms which are bonded to the S_2 atoms (1.57 μ_B). All the other interface Co atoms have a spin-magnetic moment between 1.48 and 1.50 μ_B. The lowering of the interface Co atom spin magnetic moment is consequently more important when Co atoms exhibit a stronger covalent bond with S atoms, with a shorter Co-S bond length. All the interface S atoms have a small spin magnetic moment which has the same sign as the Co atom magnetic moments, and take values between 0.012 and 0.016 μ_B without a clear correlation with the length of the corresponding Co-S bond. The spin magnetic moment of S atoms in the external S layer is even smaller (between 0.003 and 0.004 μ_B). The spin magnetic moment of Mo atoms (between -0.029 and -0.024 μ_B) is antiferromagnetically coupled to the Co and S magnetic moments, except when these Mo atoms are bonded to one or to two S_2 atoms (Mo spin magnetic moments are in this case positive and respectively take the values of 0.008 and 0.050 μ_B).

4. Charge transfers at the Co(0001)/MoS_2 interface

To calculate the charge transfer between atoms induced by covalent bonding at the Co(0001)/MoS_2 interface, we proceeded in the following way: in a first step, we calculated the majority spin electron density $n_\uparrow(\mathbf{r})$ for the lowest energy atomic structure of the whole slab. In a second step, we calculated the majority spin density $n_{\uparrow,Co}(\mathbf{r})$ when all the Co atoms keep exactly the same positions, but Mo and S atoms are removed. In a third step, we obtainted the majority spin density $n_{\uparrow,MoS2}(\mathbf{r})$, calculated when Mo and S atoms keep the same positions, but Co atoms are removed. The majority spin space-dependent charge transfer can finally be obtained from $\Delta n_\uparrow(\mathbf{r})=n_\uparrow(\mathbf{r})-\{n_{\uparrow,Co}(\mathbf{r})+n_{\uparrow,MoS2}(\mathbf{r})\}$. The minority spin charge transfer $\Delta n_\downarrow(\mathbf{r})$ can be obtained in exactly the same way. Fig.1 shows a side view of the unit cell (we can see that the interface atoms of the same chemical species are not all equivalent) and of the calculated three-dimensional minority spin charge transfer $\Delta n_\downarrow(\mathbf{r})$. Red and green areas respectively correspond to a local excess and a local lack of minority spin electrons. We see on this figure that charge transfer between Co and S atoms at the interface are exactly along the Co-S bonds. A similar figure (not shown here) has been drawn for majority spin electrons. These figures show that the reduction of the Co magnetic moment at the interface is due to an excess of minority spin electrons concomitant with a lack of majority spin electrons on this layer. Similarly, this figure shows that the magnetic moment of Mo atoms is due to a lack of minority spin (and to an excess of majority spin) electrons for the Mo atoms which have a positive spin magnetic moment (right hand side of Fig.1b), and mostly to an excess of minority spin electrons for Mo atoms which have a negative magnetic moment (left hand side of Fig.1b).

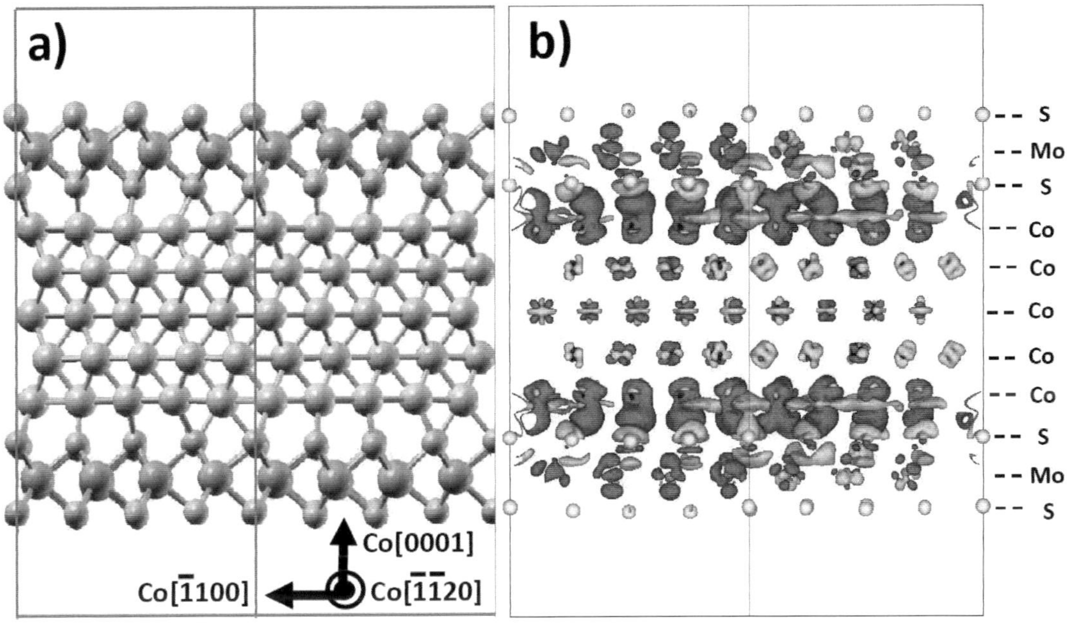

Fig1. a) Side view of the atomic structure of the unit cell used in the calculation; Co, Mo and S atoms are respectively shown with blue, green and yellow spheres. b) Side view of the calculated three-dimensional minority spin charge transfer. Red and green areas respectively correspond to a local excess and a local lack of minority spin electrons.

5. Conclusion

We interpreted the magnetic properties induced in a MoS_2 layer by a Co(0001) contact in terms of charge transfer. A decrease of the interface Co atom spin magnetic moment is observed, together with a small magnetization of interface S atoms. Mo atoms also hold small magnetic moments which can take positive or negative values, depending on the atomic environment of the S atoms to which they are bonded. Such Co(0001)/MoS_2 interface constitute an elementary block of a full spintronic device. Results on the semiconductor or metallic nature of MoS_2 below the Cobalt contact, as well as on the electron spin polarization at the Fermi level in the MoS_2 layer bonded to Co will be published elswhere.

The authors acknowledge the Université Fédérale de Toulouse- Midi-Pyrénées and Région Midi-Pyrénées for the PhD grant SEISMES as well as the grant NEXT n° ANR-10-LABX-0037 in the framework of the « Programme des Investissements d'Avenir". This work was granted access to the HPC resources of CALMIP supercomputing center under the allocation p1446 (2014-2016).

References

[1] B. Radisavljevic, A. Radenovic, J. Brivio, V. Giacometti, A. Kis, Nat.Nanotechnol. **6**,147-150 (2011).
[2] D. Xiao, G. Liu, W. Feng, X. Xu, and W. Yao, Phys. Rev. Lett. **108**, 196802 (2012).
[3] A. Fert and H. Jaffrès, Phys. Rev. B 64, 184420 (2001)
[4] P. Blaha, K. Schware, G. K. H. Madsen, D. Kvasnicka, and J. Luitz, WIEN2k, an augmented plane wave+local orbitals program for calculating crystal properties, Vienna Univ. of Technology, Vienna, 2001.

Magnetic properties of multilayers based on anisotropic Heisenberg films with dipolar long-range interaction

P V Prudnikov[†], **V V Prudnikov**[‡], **A P Soldusova**[§] **and A N Purtov**[*]
Omsk State University, Pr. Mira 55A, Omsk 644077, Russia
E-mail: [†]prudnikp@univer.omsk.su, [‡]prudnikp@univer.omsk.su, [§]anka_omsk@mail.ru, [*]iredlle@gmail.com

Abstract. The Monte Carlo simulation of the critical behavior of multilayer structures based on anisotropic Heisenberg mode are presented. The non-equilibrium critical relaxation of structures consisting of two magnetic layers with dipolar interlayer interaction is studied. The investigation of non-eqilibrium critical behavior of multilayer structure wich correspond to the nanoscale superlattice Co/Cr demonstrates that the aging effects can be observed in a wider temperature range than for bulk magnetic systems.

Magnetic order in the multilayers is complex due to a strong influence of the shape and the magnetocrystalline anisotropies of the sample. It is well known now the fundamental role of competing interactions in the emerging features of low-dimensional systems. Among a wide number of numerical and theoretical investigations on equilibrium and dynamical properties of several model Hamiltonian of low-dimensional magnets, Heisenberg-like models are one of the most widely used to approach real magnetic materials [1].

This study includes the Monte Carlo simulation of the critical behavior of multilayers based on anisotropic Heisenberg films with dipolar interaction. Competition between short-range exchange interaction, uniaxial anisotropy and long-range dipolar interaction results in appearance of different domain structures and in occurrence of the phase transitions.

Magnetic film is constructed as a spin lattice with Heisenberg Hamiltonian

$$H = -J\sum_{i,j}\left[\mathbf{S}_i\mathbf{S}_j - \Delta(N)S_i^zS_j^z\right] - A\sum_i {S_i^z}^2 - \mathbf{h}\sum_i \mathbf{S}_i + D\sum_{i\neq j}\left(\frac{\mathbf{S}_i\mathbf{S}_j}{r_{ij}^3} - 3\frac{(\mathbf{S}_i\mathbf{r}_{ij})(\mathbf{S}_j\mathbf{r}_{ij})}{r_{ij}^5}\right), \quad (1)$$

where $\mathbf{S}_i$ is a three-dimensional spin in the lattice cite i, J is exchange constant, A characterizes uniaxial anisotropy in out-of-plane direction, $\mathbf{h}$ is applied magnetic field, D characterizes the strength of dipolar interaction and $r_{ij} = r_i - r_j$ is the distance between cites i and j.

Anisotropy parameter Δ characterizes the amount of anisotropy: $\Delta = 0$ corresponds to the isotropic Heisenberg case, $\Delta = 1$ – the case of XY-model. Microscopic nature of anisotropy in films of $\mathrm{Fe}, \mathrm{Co}, \mathrm{Ni}$ and it's dependence from film thicknesses N is determined by influence of crystalline field of substrate surface, magnetic single-ion anisotropy and dipole-dipole interaction of magnetic moments of atoms in film and their concurrence [2].

We measured the magnetization the total magnetization $\boldsymbol{m} = \boldsymbol{m}_1 + \boldsymbol{m}_2$ and the staggered magnetization $\boldsymbol{m} = \boldsymbol{m}_1 - \boldsymbol{m}_2$, where $\boldsymbol{m}_1, \boldsymbol{m}_2$ are magnetizations of adjacent magnetic layers. The modeling of the trilayer structure consisting of two ferromagnetic films with dipolar interlayer interaction separated by nonmagnetic metal film demonstrates the effective antiferromagnetic character of this interaction. The simulations were carried out for layers with

Content from this work may be used under the terms of the Creative Commons Attribution 3.0 licence. Any further distribution of this work must maintain attribution to the author(s) and the title of the work, journal citation and DOI.
Published under licence by IOP Publishing Ltd

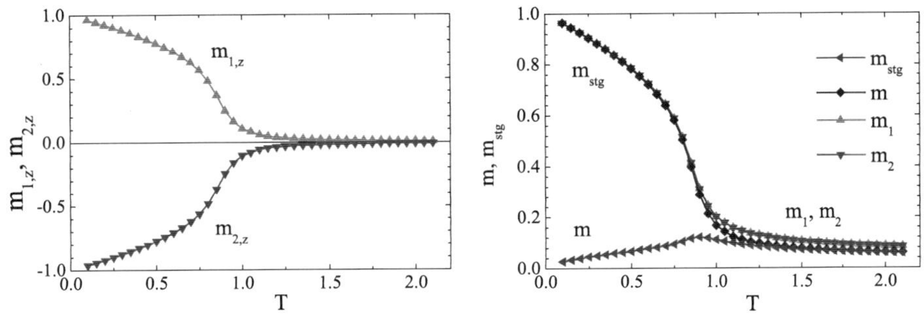

Figure 1. Temperature dependencies of the adjacent layer magnetization $m_{i,z}$ (left), "staggered" m_{stg} and total m magnetization (right) of multilayer structure with dipole interaction. The temperature is measured in units J/k_b

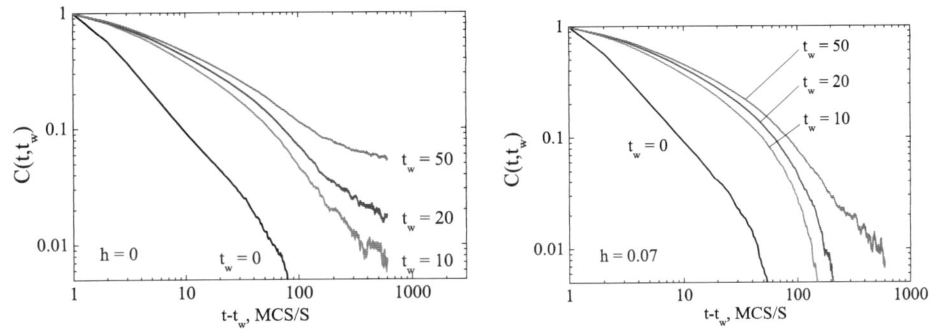

Figure 2. Ageing effects in non-equilibrium relaxation of multilayers from high-temperature initial state for $h = 0$ (left) and 0.07 (right)

sizes $L \times L \times N$ with the use of periodic and free boundary conditions for the in-plane and out-plane directions, respectively. N is the thickness of ferromagnetic layers. The Metropolis algorithm was used for Monte Carlo simulations. Magnetic properties are calculated for this system with $A = 0.1J$, $D = 0.01J$, $\Delta = 0$ (Figure 1). Dipolar interaction is resulted in antiparallel orientations of adjacent ferromagnetic layers.

In investigations of influence of the initial states on characteristics of non-equilibrium critical behaviour we computed the magnetization

$$\boldsymbol{m}(t) = \left\langle \frac{1}{NL^2} \sum_{i=1}^{NL^2} \boldsymbol{S}_i(t) \right\rangle, \tag{2}$$

and the time-dependent autocorrelation function

$$C(t, t_w) = \left\langle \frac{1}{NL^2} \sum_{i=1}^{NL^2} \boldsymbol{S}_i(t)\boldsymbol{S}_i(t_w) \right\rangle - \boldsymbol{m}(t)\boldsymbol{m}(t_w), \tag{3}$$

The time variable t_w characterizes the age of a sample and is called waiting time. Autocorrelation function in nonequilibrium processes depend not only on the difference $t - t_w$ but also on each time variable individually at $t - t_w$ and t_w much smaller than the relaxation time of the system t_{rel}. The non-equilibrium evolution starts when the system after initial preparation is placed to the thermostat with $T = T_c$. At the waiting time t_w, we have begun to measure two-time correlation function within the $1 \ll t, t_w \ll t_{\mathrm{rel}}$ range. The aging of a system

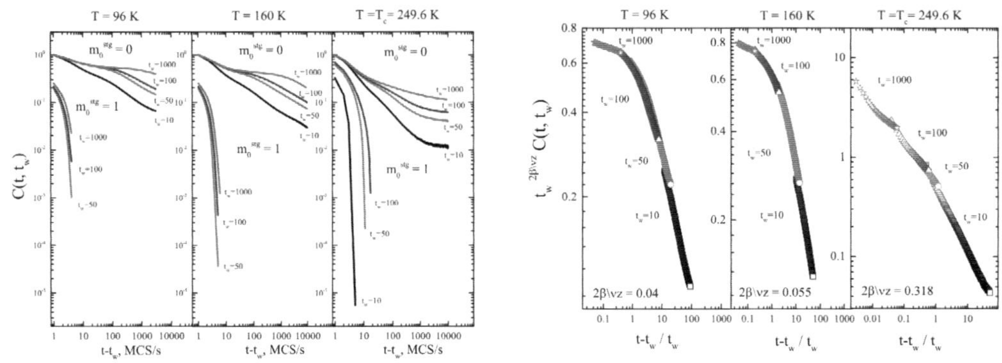

Figure 3. Aging effects of $C(t, t_w)$ for evolution from different initial states $m_0^{\rm stg} = 0$, $m_0^{\rm stg} = 1$ (left) and scaling "collaps" of $F(t/t_w) = t_w^{2\beta/\nu z} C(t, t_w)$ for initial state $m_0^{\rm stg} = 0$ (right)

is manifested in slowing down relaxation processes with the time passing from the preparation of a sample. The trilayer structure with dipolar interaction demonstrates the aging effects (Figure 2). The presence of external magnetic field h leads to increasing of relaxational processes.

The features of non-equilibrium critical dynamics [3] can be considered as basis for understanding and interpretation of the experimental data obtained for Co/Cr [4]. The next values of modelling parameters $\Delta(N = 3) = 0.7$, $A = 0$, $D = 0$ and $J_1 \approx 4{\cdot}10^{-21}$J are used for the multilayer structure, which correspond to the nanoscale superlattice [Co(0.6 nm)/Cr(0.78 nm)] [4]. The investigation of the magnetization relaxation in the multilayer superlattice has revealed the magnetic aging phenomena (Figure 3.left). The realization of dynamical scaling which demonstrated in Figure 3 gives the possibility to determine the critical exponent $\beta/\nu z = 0.159$. The nanoscale periodicity creates in these magnetic multilayer structures mesoscopic spatial magnetic correlations with slow relaxation dynamics when quenching of the system into a non-equilibrium state is realized. In comparison with bulk magnetic systems where the slow dynamics and aging phenomenon are displayed close to critical point, the magnetic superlattices, structured on the nanoscale, give possibility to increase relaxation time due to the increasing in the characteristic spin-spin correlation length. The aging can be observed in the multilayer structures in a wider temperature range $T < Tc = 249.6$K than for bulk magnetic systems.

Therefore we have studied the magnetic properties of the trilayer structure consisting of two ferromagnetic films separated by nonmagnetic metal film. It were considered long-range dipolar interlayer interaction or RKKI ineraction, which lead to the effective antifferomanetic interaction for specified thickness of nonmagnetic layer. The study of aging effects in multilayered structures shows that these effects brightly appear in non-equilibrium relaxation from initial states.

Acknowledgments

This work was supported by Grant No. MD-6024.2016.2 of Russian Federation President and by Project of the Ministry of Education and Science of the Russian Federation. The simulations were supported by the Supercomputing Center of Lomonosov Moscow State University, Moscow and Saint Petersburg Joint Supercomputer Center of the Russian Academy of Sciences.

References

[1] Vaz C A F, Bland J A C, Lauhoff G 2008 Rep. Prog. Phys. **71** 056501
[2] Prudnikov P, Prudnikov V, Menshikova M and Piskunova N 2015 J. Magn. Magn. Mater. **387** 77
[3] Prudnikov V, Prudnikov P, Pospelov E 2016 J. Stat. Mech. **043303**
[4] Mukherjee T, Pleimling M, Binek Ch. 2010 Phys. Rev. B **82** 134425

IOP Conf. Series: Journal of Physics: Conf. Series **903** (2017) 012019 doi:10.1088/1742-6596/903/1/012019

Dependence of magnetic anisotropy on MgO sputtering pressure in $Co_{20}Fe_{60}B_{20}$/MgO stacks

A Kaidatzis, C Serletis and D Niarchos

Institute of Nanoscience and Nanotechnology, NCSR Demokritos, 15310 Aghia Paraskevi, Athens, Greece

a.kaidatzis@inn.demokritos.gr

Abstract. We investigated the dependence of magnetic anisotropy of Ta/$Co_{20}Fe_{60}B_{20}$/MgO stacks on the Ar partial pressure during MgO deposition, in the range between 0.5 and 15 mTorr. The stacks are studied before and after annealing at 300°C and it is shown that magnetic anisotropy significantly depends on Ar partial pressure. High pressure results in stacks with very low perpendicular magnetic anisotropy even after annealing, while low pressure results in stacks with perpendicular anisotropy even at the as-deposited state. A monotonic increase of magnetic anisotropy energy is observed as Ar partial pressure is decreased.

1. Introduction

Magnetic tunnel junctions (MTJs) with perpendicular magnetic anisotropy (PMA) based on CoFeB/MgO/CoFeB trilayers are being studied for several years for non-volatile and high density data storage applications [1]. The magnetic anisotropy of such multilayers strongly depends on composition, layer thickness, deposition conditions, and annealing temperature; precise tailoring of these parameters must be performed in order to obtain PMA, which originates from the CoFeB/MgO interface [1]. Furthermore, increasing PMA energy is of crucial importance for increasing the thermal stability of MTJ cells and allowing the miniaturization of data storage devices.

In this work, we study the dependence of the magnetic anisotropy of Ta/CoFeB/MgO half-MTJs on the sputtering pressure during MgO deposition. Previous studies of the effect of the sputtering conditions on the properties of CoFeB/MgO-based tunnel junction with in-plane anisotropy (IPA), have shown the importance of Ar pressure during MgO sputtering; it was shown that the smoothness of the CoFeB/MgO interface crucially depends on Ar sputtering pressure, affecting the MTJs' tunnel magneto-resistance [2, 3, 4]. Here we show that the magnetic anisotropy axis is greatly affected by the Ar partial pressure during MgO sputtering and we determine the pressure range for obtaining PMA.

2. Experimental details

Ta(6 nm)/$Co_{20}Fe_{60}B_{20}$(1 nm)/MgO(2 nm) multilayers were deposited on Si(100)/SiO_2(500 nm) substrates, using ultra-high vacuum magnetron sputtering. The substrates were always kept at room-temperature and were rotating. Direct current power was used for metal deposition (1.5 W/cm^2 power density, 3 mTorr Ar working pressure, 0.018 nm/s deposition rate). Radio-frequency power was used for MgO deposition (power density was 10 W/cm^2 for all samples deposited). A ceramic MgO target and metallic Ta and $Co_{20}Fe_{60}B_{20}$ targets were used as source materials. During MgO deposition, the Ar working pressure was between 0.5 and 15 mTorr. All samples were annealed in high-vacuum.

Content from this work may be used under the terms of the Creative Commons Attribution 3.0 licence. Any further distribution of this work must maintain attribution to the author(s) and the title of the work, journal citation and DOI.
Published under licence by IOP Publishing Ltd

The deposition rate was monitored in-situ by means of a quartz crystal microbalance, calibrated against x-ray reflectivity measurements of test samples. Atomic force microscopy (AFM) imaging was employed for studying the samples' surface morphology. AFM measurements were performed using a NT-MDT Smena microscope and commercial AFM probes (Bruker PPP-FMR). The magnetostatic characterization was performed using a Quantum Design MPMS SQUID magnetometer.

3. Results and discussion

The multilayer surface morphology was probed by AFM; because of the very low thickness of the MgO layer, the multilayer surface morphology follows the morphology of the CoFeB/MgO interface, which is the focus of this study. The samples are measured before and after annealing at 300°C. The results of the AFM imaging are shown in Figure 1. These results show a clear correlation between the surface morphology and the Ar partial pressure: the root mean square surface roughness increases almost monotonically from 0.40 to 0.57 nm, as Ar pressure increases. This trend is in agreement to previous studies [3] and the effect can be explained by taking into account the higher kinetic energy of the deposited atoms as sputtering pressure decreases. The effect of sputtering pressure is also significant on the deposition rate: higher pressure results in lower deposition rate as sputtered atoms collide with Ar species and are deflected away from the substrate. After annealing, there is an overall increase of surface roughness, although the same trend as before is observed: surface roughness increase from 0.57 to 0.89 nm as Ar pressure increases.

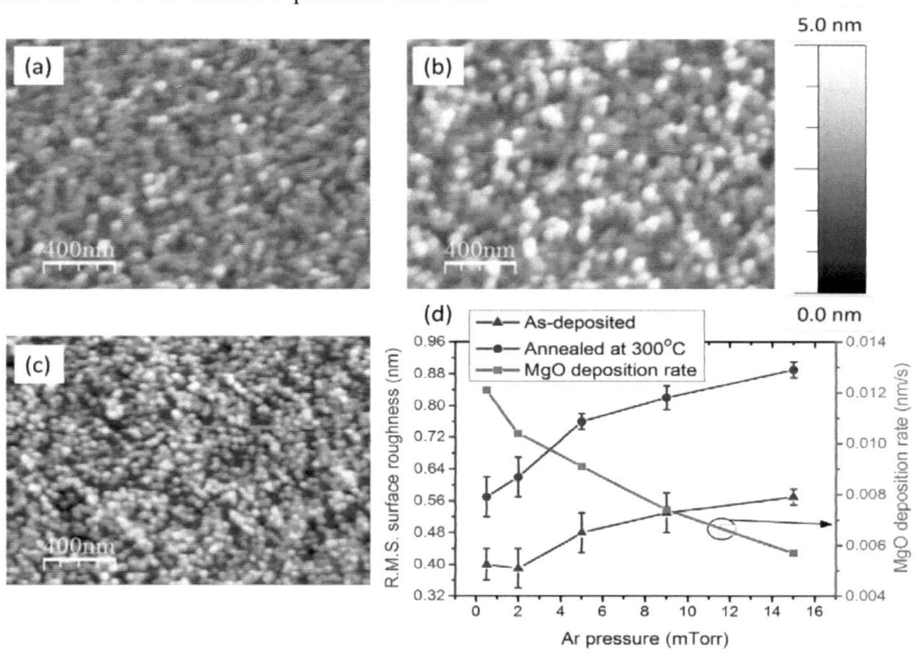

Figure 1. Representative AFM images of the multilayer surface morphology after annealing at 300°C. Deposition pressure is (a) 0.5 mTorr, (b) 5 mTorr, and (c) 15 mTorr. (d) Root mean-square surface roughness and deposition rate as a function of Ar partial pressure.

Hysteresis loops have been obtained with the magnetic field applied both perpendicular and in the plane of the samples' surface. As seen in Figure 2, all annealed samples have PMA, but the anisotropy energy decreases as Ar pressure increases. Figure 2(d) shows the effective anisotropy-CoFeB thickness product variation as a function to the Ar pressure. The effective anisotropy k_{eff} is experimentally determined from the perpendicular and in-plane hysteresis loops, by comparing the energy difference required to saturate the film along the two directions [5]. As-deposited samples have in-plane anisotropy for Ar pressure higher than 5 mTorr, while PMA with low effective anisotropy energy is obtained for lower sputtering pressure. After annealing, all samples have PMA, with the effective anisotropy energy increasing as Ar pressure decreases. It should be noted that volume saturation magnetization does not significantly depends on Ar pressure and it is 1020±47 emu/cm^3 and 850±39 emu/cm^3 before and after annealing, respectively. This decrease is attributed to the interdiffusion of CoFeB and Ta upon annealing, which increases the thickness of the magnetic dead layer present at the Ta/CoFeB interface [5, 6].

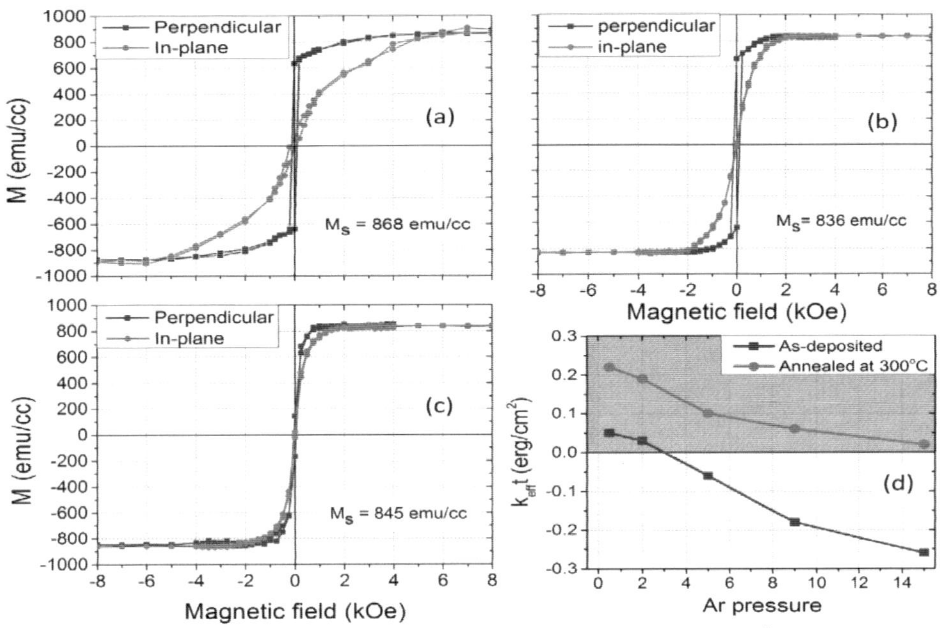

Figure 2. SQUID magnetometry hysteresis cycles after annealing at 300°C. MgO films deposited at (a) 0.5 mTorr, (b) 5 mTorr, (c) 15 mTorr. (d) Variation of the effective anisotropy energy-CoFeB thickness product with Ar pressure. The shaded area indicates PMA.

4. Conclusions

We have shown that the magnetic anisotropy axis and the effective anisotropy energy of Ta/CoFeB/MgO half-MTJs significantly depend on MgO sputtering pressure. Low pressure results in multilayers with PMA, even at the as-deposited state. As pressure increases, the effective anisotropy energy decreases and an in-plane easy magnetization axis is favoured. This trend coincides to the increasing MgO/CoFeB interface roughness as the sputtering pressure increases.

Acknowledgements

Funding from the E.C. through a FP7-ICT project (Grant No. 318144) is acknowledged.

References

[1] Ikeda S, Miura K, Yamamoto H, Mizunuma K, Gan H D, Endo M, Kanai S, Hayakawa J, Matsukura F and Ohno H, 2010 *Nat. Mat.* **9** 721.
[2] Shen W, Mazumdar D, Zou X, Liu X, Schrag B and Xiao G, 2006 *Appl. Phys. Lett.* **88** 182508.
[3] Wrona J, Langer J, Ocker B, Maass W, Kanak J, Stobiecki T and Powroznik W, 2010 *J. Phys.: Conf. Ser.* **200** 052032
[4] Harnchana V, Hindmarch A T, Brown A P, Brydson R M and Marrows C H, 2010 *J. Phys.: Conf. Ser.* **241** 012039
[5] Sokalski V, Moneck M T, Yang E and Zhu J-G, 2012 *Appl. Phys. Lett.* **101** 072411
[6] Kaidatzis A, Bran C, Psycharis V, Vázquez M, García-Martín J M and Niarchos D, 2015 *Appl. Phys. Lett.* **106** 262401

Modification of magnetization orientation in Pt/Co/Pt ultrathin films by femtosecond laser pulses

J. Kisielewski[1], Z. Kurant[1], I. Sveklo[1], M. Tekielak[1], E. Dynowska[2], M. Jakubowski[2], A. Wawro[2] and A. Maziewski[1]

[1] Department of Physics of Magnetism, University of Białystok, Ciołkowskiego 1L, 15-245 Białystok, Poland

[2] Institute of Physics, Polish Academy of Sciences, al. Lotników 32/46, 02-668 Warsaw, Poland

E-mail: `jankis@uwb.edu.pl`

Abstract. Ultrathin Pt/Co/Pt films were irradiated with femtosecond laser pulses of sufficiently high energy to create irreversible structural and in consequence magnetic modifications. The effect was studied as a function of an initial structure of the Co film, determined by different temperatures during preparation of the Pt buffer layer, and a Co film thickness, which determines the initial out-of-plane or in-plane magnetization orientation. A creation of many magnetic phases is shown. The out-of-plane magnetization state is found to be created for significantly broader fluence range for the Co film deposited on the Pt buffer layer grown at 750°C in comparison to that grown at room temperature. Reversible magnetization dynamics was also studied in the low fluence regime, using a pump-probe setup. The results were discussed using simulations of temperature dynamics.

Laser annealing is an effective method of tailoring magnetic properties of ultrathin magnetic films. Particularly it enables modification of magnetic anisotropy in Pt/Co/Pt [1, 2]. In this work we show the influence of the initial structure of the Co film, determined by the temperature during the growth of the Pt buffer layer, on an appearance of the laser-induced magnetic phases. Laser pulse induced magnetization dynamics was also studied with a support of modeling of pulse-driven temperature changes.

The samples $Al_2O_3(001)/Pt(5nm)/Co(d_{Co})/Pt(5nm)$ were produced by the MBE technique. The Pt buffers were grown at room temperature (RT) and at 750°C (high temperature, HT). The other films of the stacks were deposited at RT. The discussed Co thickness d_{Co}=1.5 and 3.5 nm corresponds to the out-of-plane and in-plane magnetization orientation, respectively. All four combinations of the growth temperature and d_{Co} were used. Laser annealing was performed using a Ti:sapphire oscillator with a regenerative amplifier. Single laser pulses of duration of 40 fs and wavelength of 800 nm were released manually and focused onto the sample by a lens. Arrays of spots were created at each sample, with pulse energies between 27 and 141 μJ. The laser-annealed samples were studied using a magneto-optical Kerr microscope in polar (PMOKE) configuration. The optical and PMOKE remanence images of the spots revealed a complicated structure of ring-shaped areas of the modified magnetic and optical properties, similar to ones reported before [1, 2]. Similar as before, we assume that any laser-induced irreversible modification depends on the local pulse fluence only. Using a typical analysis method [3] we obtained the fluence values for all boundaries between the regions observed within the spots, as presented in Fig. 1.

Content from this work may be used under the terms of the Creative Commons Attribution 3.0 licence. Any further distribution of this work must maintain attribution to the author(s) and the title of the work, journal citation and DOI.
Published under licence by IOP Publishing Ltd

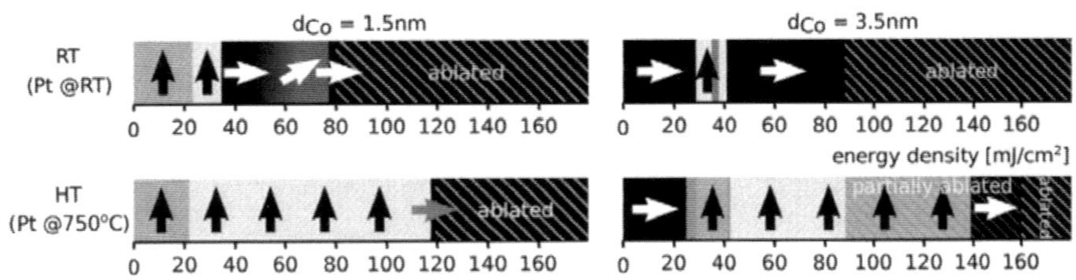

Figure 1. Laser-induced magnetization direction, as a function of the pulse fluence, for two values of d_{Co} and two types of the Pt buffer layer. At high fluences the sample is ablated.

Fluences up to about 20 mJ/cm^2 do not affect the samples, thus the magnetization direction is unchanged. Above 20 mJ/cm^2 a modification starts to occur, and the out-of-plane magnetization state appears also for thicker Co, while for the thinner Co the magnetooptical signal is enhanced (brighter color in Fig. 1). Comparing RT and HT samples, a huge difference can be found in the fluence range responsible for the out-of-plane state. An ablation threshold is much higher for the HT samples. A partial ablation occurs for the thicker Co, but the out-of-plane state with a square hysteresis loop is still present there.

An explanation of some RT/HT differences may be provided by the XRD studies of the RT and HT buffers. The latter one shows a higher crystalline quality and much smoother surface. This means a better growth of following layers. Lower number of defects results in more tight atomic bonds in a crystal, what makes it more difficult to be destroyed e.g. by thermal motion.

To have a clearer picture of the laser annealing process, a temperature dynamics in the pulse laser-heated sample was modeled. A two-temperature model (2TM), involving two interacting baths of electrons and a lattice, was used [4]. Additionally, phase transitions in contributing metals were involved, using standard enthalpy functions, and the sapphire substrate was included, as a heat sink. The bulk thermodynamic parameters of the materials were used in the model, what may cause some errors, however the real "ultrathin" values, which may differ due to e.g. reduced mean free path of the carriers, are in principle unknown. As the model assumes an ideal Pt/Co/Pt structure, it does not distinguish the RT and HT cases. The calculations are demonstrated in Fig. 2(a) for d_{Co}=1.5 nm and for the fluence of 30 and 90 mJ/cm^2. The electrons absorb the energy from the laser pulse, reaching extreme temperatures, and next they thermalize with the lattice. Structural modifications depend on the increase of the lattice temperature. A cooling rate, determined by a heat transfer to the substrate is also important, as it defines how long the sample stays hot and an effective mixing can take place. According to the results of the calculations, after a 30 mJ/cm^2 pulse the lattice temperature is still below the melting point, however atoms can mix intensively [5]. On the contrary, a 90 mJ/cm^2 pulse is capable to melt Pt and even boil Co. The calculations for d_{Co}=3.5 nm look very similar, however the reached temperatures are a few percent lower, as given energy is distributed over thicker material. In contrast to the qualitative simulation, a real process of melting and boiling within a few ps looks like an explosion, so one can expect ablation, what corresponds well to the experiment.

To cast more light onto the process of laser annealing, the magnetization dynamics was measured using a pump-probe experimental setup. The sample of d_{Co}=1.5 nm on the RT Pt buffer was excited by the pump beam of a wavelength of 488nm (from optical parametric amplifier) with a repetition rate of 1 kHz. The magnetization, preferably aligned out-of-plane, was saturated with a perpendicular external magnetic field ±1.6 kOe. The second less intense beam of a wavelength of 800nm was used to probe the magnetization state in the sample, using a Wollaston prism and a double balanced detector. In Fig. 2(b) the measured

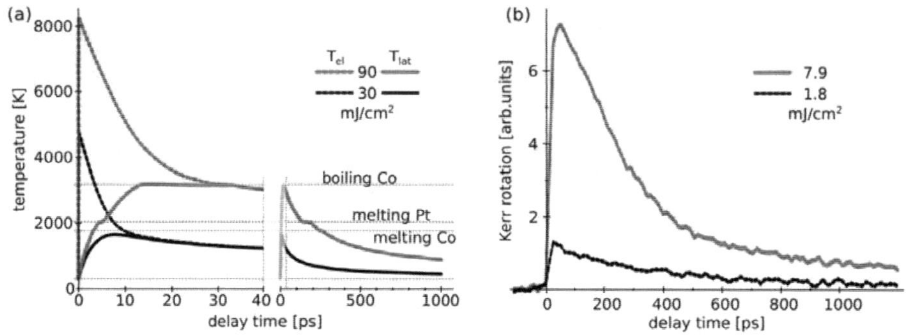

Figure 2. (a) 2TM simulations of in sapphire/Pt(5nm)/Co(1.5nm)/Pt(5nm) after heating by a laser pulse; (b) pump-probe measurements of the laser-induced reversible change of Kerr rotation in Co(1.5nm) on RT Pt buffer.

magnetization dynamics is presented for the pulse fluence of 1.8 and 7.9 mJ/cm^2, which is far below any threshold of single-pulse modification, presented in Fig. 1. An ultrafast change of the magnetic signal is followed by a slow relaxation. The amplitude is fluence-dependent and the direction of changes was found to depend on a direction of the applied magnetic field, what allows us to identify this process as a thermal-origin laser-induced reduction of magnetization (demagnetization) [6].

Experimental value of the long-term decay time is found to be about 300 ps and it corresponds well to the value derived from the simulated curves of the temperature dynamics for the respective fluences, where it is related to a heat transfer into the substrate. With the 7.9 mJ/cm^2 curve we noticed, that repeating the measurement many times (on a time scale of minutes) the amplitude of demagnetization becomes gradually reduced. This non-repeatability suggests an onset of irreversible modifications in the sample, whose thresholds in the multi-pulse irradiation regime can be significantly lower as compared to the single-pulse case [2]. The results of the simulated temperature dynamics support this suggestion, as the lattice temperature for the 7.9 mJ/cm^2 pulse reaches about 700 K, which temperature in the classic annealing of Pt/Co systems is reported to bring intermixing of atoms [5]. Obviously, during several tens of picoseconds under 700 K some minor sample variations may happen only, however after thousands of pulses accumulating modifications may finally become significant.

Modifications of magnetization orientation in Pt/Co/Pt ultrathin films by fs laser pulses were found to be strongly dependent on buffer creation conditions. Some results are consistent with the temperature dynamics simulations, however deeper understanding of these effects requires more detailed studies.

Acknowledgments
The work was supported by the National Science Centre (Poland), DEC-2012/06/M/ST3/00475.

References
[1] Kisielewski J, Dobrogowski W, Kurant Z, Stupakiewicz A, Tekielak M, Kirilyuk A, Kimel A, Rasing T, Baczewski L T, Wawro A, Balin K, Szade J and Maziewski A 2014 *J. Appl. Phys.* **115** 053906
[2] Kisielewski J, Kurant Z, Sveklo I, Tekielak M, Wawro A and Maziewski A 2016 *J. Appl. Phys.* **119** 193901
[3] Chalupský J, Burian T, Hájková V, Juha L, Polcar T, Gaudin J, Nagasono M, Sobierajski R, Yabashi M and Krzywinski J 2013 *Opt. Express* **21** 26363–26375
[4] Anisimov S, Kapeliovich B and Perelman T 1974 *Sov. Phys. JETP* **39** 375
[5] Galeotti M, Atrei A, Bardi U, Cortigiani B, Rovida G and Torrini M 1993 *Surf. Sci.* **297** 202 – 208
[6] Beaurepaire E, Merle J C, Daunois A and Bigot J Y 1996 *Phys. Rev. Lett.* **76**(22) 4250–4253

IOP Conf. Series: Journal of Physics: Conf. Series **903** (2017) 012021 doi:10.1088/1742-6596/903/1/012021

Thickness and angular dependent magnetic anisotropy of $La_{0.67}Sr_{0.33}MnO_3$ thin films by Vectorial Magneto Optical Kerr Magnetometry

S K Chaluvadi[1], P Perna[2], F Ajejas[2], J Camarero[2, 3], A Pautrat[4], S Flament[1] and L Méchin[1]
[1] *GREYC (UMR 6072), CNRS - ENSICAEN – UCBN, 6 bvd Maréchal Juin, 14050 Caen Cedex, France*
[2]*IMDEA-Nanociencia, Campus de Cantoblanco, 28049 Madrid, Spain*
[3]*Universidad Autonoma de Madrid, Campus de Cantoblanco, 28049 Madrid, Spain*
[4] *CRISMAT (UMR 6508), CNRS - ENSICAEN – UCBN, 6 bvd Maréchal Juin, 14050 Caen Cedex, France*
E-mail : sandeep-kumar.chaluvadi@unicaen.fr; laurence.mechin@ensicaen.fr

Abstract. We investigate the in-plane magnetic anisotropy in $La_{0.67}Sr_{0.33}MnO_3$ thin films grown on $SrTiO_3$ (001) substrate using angular dependent room temperature Vectorial Magneto-Optical Kerr Magnetometry. The experimental data reveals that the magnetic anisotropy symmetry landscape significantly changes depending upon the strain and thickness. At low film thickness (12 and 25 nm) the dominant uniaxial anisotropy is due to interface effects, step edges due to mis-cut angle of $SrTiO_3$ substrate. At intermediate thickness, the magnetic anisotropy presents a competition between magnetocrystalline (biaxial) and substrate step induced (uniaxial) anisotropy. Depending upon their relative strengths, a profound biaxial or uniaxial or mixed anisotropy is favoured. Above the critical thickness, magnetocrystalline anisotropy dominates all other effects and shows a biaxial anisotropy.

1. Introduction

$La_{0.67}Sr_{0.33}MnO_3$ (LSMO) is considered to be a promising candidate for spintronic devices. It exhibits nearly 100% spin polarization and room temperature ferromagnetism (T_c~370K). LSMO thin film properties are very sensitive to thickness, strain, temperature etc. Due to the different lattice parameters between LSMO and $SrTiO_3$ (STO) (001), rhombohedral LSMO undergoes an *in-plane* biaxial tensile strain when grown on STO (001) and imposes a tetragonal distortion in to film. As STO (001) is cubic, the *in-plane* biaxial tensile strain induced into the film is isotropic and therefore inducing four-fold magnetic anisotropy. For a cubic crystal, the magneto-elastic energy is equal in all the directions. Hence, it does not contribute to anisotropy and the resultant biaxial magnetic anisotropy is due to strong influence from magneto-crystalline effects. Lecoeur, P. *et al* [1] observed biaxial anisotropy in LSMO film with buckled remanent spin states along $<110>_{pc}$ crystallographic axis. However, other studies also reported that LSMO films can exhibit a uniaxial anisotropy and it is because of the steps transferred into film from the mis-cut

Content from this work may be used under the terms of the Creative Commons Attribution 3.0 licence. Any further distribution of this work must maintain attribution to the author(s) and the title of the work, journal citation and DOI.
Published under licence by IOP Publishing Ltd

angle of STO (001) substrate[2]. In addition to above studies, several authors artificially tuned uniaxial anisotropy of LSMO along the step edge direction by growing thin films on vicinal STO (001) substrates[3].

Here, we report thickness and angular dependent magnetic anisotropy studies of strained epitaxial LSMO thin films of various thicknesses (50, 25 and 12 nm) epitaxially grown onto STO (001) substrates by pulsed laser deposition technique.

2. Experimental details

The LSMO thin films were epitaxially grown on single crystal STO (001) substrates by pulsed laser deposition (PLD) technique. The films of three different thicknesses (50, 25 and 12 nm) were grown at 720°C and 0.35 mbar oxygen pressure. After deposition, films were immediately cooled down to room temperature at the rate of 10°C per minute at $7x10^2$ mbar oxygen background pressure. Room temperature magnetic anisotropy measurements in thin films were carried out by using high resolution angular dependent vectorial Magneto-Optical Kerr (v-MOKE)[4] magnetometry.

3. Results and discussion

The structural, electrical and magnetic transport measurements were studied and the results are tabulated in table 1.Thickness and angular dependent *in-plane* magnetic anisotropy studies of LSMO thin films grown on STO (001) substrates were performed at room temperature by v-MOKE magnetometry in longitudinal geometry. For all the measurements, the samples were mounted on eucentric goniometer that allows keeping constant the reflection plane during the angular measurements. For each film thickness,

Table 1: Structural, electrical, magnetic and magnetic anisotropy properties of LSMO thin films

Substrate	LSMO film Thickness	Out-of-plane lattice parameter (nm) and strain (%)	FWHM (°)	RMS Roughness (nm)	Tp (K)	Tc (K)	Magnetic Anisotropy
STO	50 nm	0.38511(-0.642)	0.119°	0.159	>425	348	Biaxial + off-axis Uniaxial
STO	25 nm	0.38512 (-0.645)	0.117°	0.127	369	345	Uniaxial
STO	12 nm	0.38492 (-0.6914)	0.159°	0.17	349	340	Uniaxial

hysteresis loops were measured at different magnetic field directions within the film plane with an interval of 4.5° each for about 360°. Figure 1 shows the angular dependence of normalized remanence magnetization of LSMO/STO (001) films of varying thickness with easy axis (e.a) and hard axis (h.a) indicated by arrows respectively. Figure 1(a) shows the 2D plot of angular dependent remanence fields of LSMO film of 50 nm thick has biaxial anisotropy with easy axes aligned along <110> axis. Figure 1(b) shows the polar plot of remanence magnetization that has butterfly structure with four lobes showing a biaxial anisotropy. One should observe that the strength of easy axes is not same in all the directions and also the periodicity between easy axis and immediate hard axis is not exactly 45°, suggesting that there is an additional anisotropy present in the film (unless very weak) that is induced by the surface symmetry breaking due to the step formation. Therefore, there exists a competition between a biaxial (strong) due to magnetocrystalline anisotropy and uniaxial (weak) due to steps formation with a small offset angle. Moreover, as the thickness of the thin film decreases from 50 to 12 nm, there is a change in anisotropy from biaxial to uniaxial. The results are clearly visualized in the angular dependent 2D plots of the remanence magnetization (figure 1(d)) of the LSMO film of 25 nm thick. There are periodic oscillations observed with 180° periodicity and the easy axis is aligned along [100], whereas the hard axis is aligned along [010] axis. Figure 1(g) shows the remanence field at different angles for LSMO film of 12 nm thickness also shows uniaxial anisotropy. Also, there is a change in easy axis direction as compared to 25 nm film and this can due to the different mis-cut angle and direction of STO (001) substrate. Figure 1(e and h) shows the remanence polar plots of 25 and 12 nm respectively, with symmetrical lobes clearly indicating a well-defined uniaxial anisotropy.

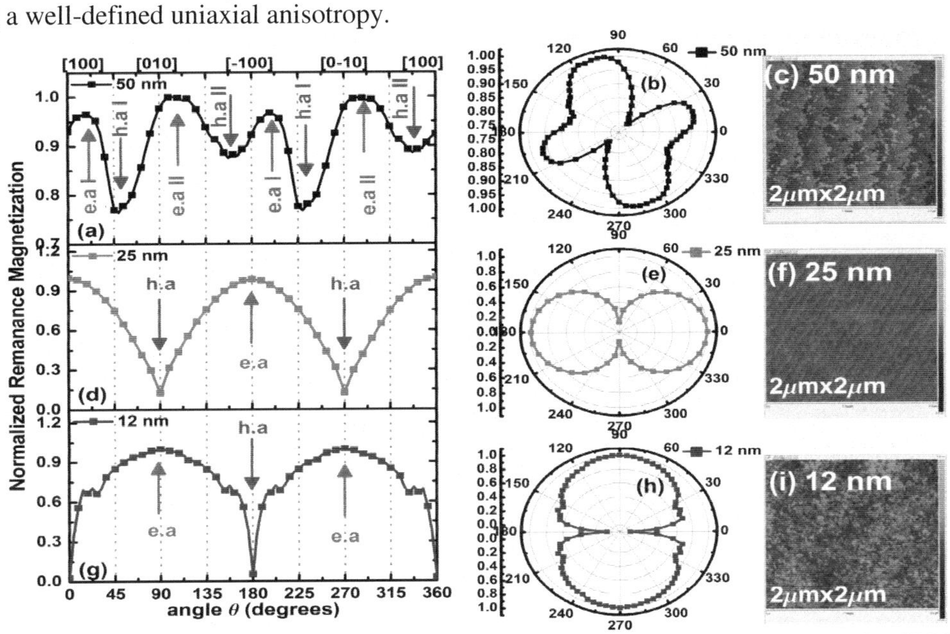

Figure 1: (a, d and g) v-MOKE angular dependent remanence magnetization of LSMO films of varying thickness grown on STO (001) substrate; (b, e, h and c, f, i) are the corresponding polar plots and AFM topography respectively.

As STO (001) is cubic, the LSMO film undergoes an *in-plane* biaxial tensile strain. The magneto-elastic energy (due to strain) is constant in the film plane. Therefore, the cause for anisotropy is not due to magneto-elastic effects but it is owing to defects in crystal structure. Due to step formation, there are broken bonds along the step edges and direction (as observed in AFM topography in figure 1(c, f and i). As the thickness of LSMO film decreases, magnetic anisotropy exhibits a strong uniaxial anisotropy. This suggests that the film at lower thickness has strong influence of steps and mis-cut angle from STO (001) substrates in determining the magnetic anisotropy. In order to obtain clear insight about the anisotropy, low temperature v-MOKE and *in-plane* XRD studies are underway.

Conclusion

In summary, we have presented the angular dependent *in-plane* magnetic anisotropy properties of LSMO thin films of different thicknesses grown on STO (001). The magnetic anisotropy behavior of LSMO thin is thickness dependent. At lower thickness, uniaxial anisotropy is more predominant in LSMO films and is due to the step edges and mis-cut angle of STO (001) substrate. In the intermediate thickness range, there exists a competition between uniaxial and biaxial anisotropy. Depending upon their relative strength, either a uniaxial or biaxial or mixture of both is present. At higher thicknesses, the magneto-crystalline anisotropy dominates all other effects and leads to a bi-axial anisotropy.

References

[1] P. Lecoeur, P. L. Trouilloud, G. Xiao, A. Gupta, G. Q. Gong, and X. W. Li, *J. Appl. Phys.*, vol. 82, no. 8, 3934, 1997.
[2] D. S. Chuang, C. A. Ballentine, and R. C. O'Handley, *Phys. Rev. B*, vol. 49, no. 21, 15084, 1994.
[3] P. Perna, L. Mechin, M. Saib, J. Camarero, and S. Flament, *New J. Phys.*, vol. 12, 2010.
[4] E. Jiménez, N. Mikuszeit, J. L. F. Cuñado, P. Perna, J. Pedrosa, D. Maccariello, C. Rodrigo, M. A. Niño, J. Camarero, and R. Miranda, *Rev. Sci. Instrum.*, vol. 85, 053904, 2014.

IOP Conf. Series: Journal of Physics: Conf. Series **903** (2017) 012022 doi:10.1088/1742-6596/903/1/012022

Preparation of high-quality planar FeRh thin films for *in situ* TEM investigations

Trevor P. Almeida[1]*, Damien McGrouther[1], Yevheniy Pivak[2], Hector Hugo Perez Garza[2], Rowan Temple[3], Jamie Massey[3], Christopher H. Marrows[3] and Stephen McVitie[1]

[1]SUPA, School of Physics and Astronomy, University of Glasgow, Glasgow, G128QQ, UK.
[2]DENSsolutions, Delft, 2628ZD, The Netherlands.
[3]School of Physics and Astronomy, University of Leeds, Leeds, LS29JT, UK.

Abstract. The preparation of a planar FeRh thin film using a focused ion beam (FIB) secondary electron microscope (SEM) for the purpose of *in situ* transmission electron microscopy (TEM) is presented. A custom SEM stub with 45° faces allows for the transfer and milling of the sample on a TEM heating chip, whilst Fresnel imaging within the TEM revealed the presence of the magnetic domain walls, confirming the quality of the FIB-prepared sample.

1. Introduction

Equiatomic iron-rhodium ($Fe_{48}Rh_{52}$ to $Fe_{56}Rh_{44}$) has attracted much interest due to its magnetostructural transition from its antiferromagnetic (AF) to ferromagnetic (FM) phase. This ordered α' alloy adopts a CsCl structure and is AF at room temperature, but makes a first-order phase transition to a FM state at ~ 350 to 380 K, and hence can present phase co-existence and hysteresis [1]. The co-existing phases are separated by a phase-boundary domain wall (DW) and effective control over the creation and motion of these phase boundary DWs are considered desirable for potential application in a new generation of novel nanomagnetic or spintronic devices [2]. Previous studies have shown that the DWs can be nucleated in FeRh films and driven through combining heating with differential gradients of chemical doping [3]. Current understanding of the behaviour of DWs in FeRh is based upon bulk magnetic measurements and imaging, *e.g.* Kerr imaging or X-ray circular dichroism. Significant advances in knowledge are potentially enabled using the transmission electron microscopy (TEM) techniques of Fresnel imaging, off-axis electron holography and differential phase contrast imaging, which allow for the visualising of magnetic induction at nanometre spatial resolution. Further, the development of TEM holders which include micro-electro-mechanical systems (MEMS) has led to the ability to apply a range of external stimuli to samples *in situ* within the TEM, *i.e.* temperature, electrical bias, etc., whilst maintaining stability and minimising drift. TEM lamellae of bulk samples have previously been prepared on these MEMS-based chips [4] and planar magnetic thin films have been made for magnetic imaging within the TEM [5]. However, limited work has been reported on the preparation of planar magnetic films on MEMS chips for *in situ* TEM studies.

In this context, the preparation of a planar FeRh thin film on a MEMS-based heating chip using a focused ion beam (FIB) secondary electron microscope (SEM) is presented. The use of a SEM stub with 45° faces, custom-designed by DENSsolutions®, provides a new and innovative method for the transfer and preparation of planar magnetic thin films for the purpose of *in situ* TEM investigations.

*Author to whom correspondence should be addressed: trevor.almeida@glasgow.ac.uk.

Content from this work may be used under the terms of the Creative Commons Attribution 3.0 licence. Any further distribution of this work must maintain attribution to the author(s) and the title of the work, journal citation and DOI.
Published under licence by IOP Publishing Ltd

2. Experimental

A thin film of ordered α'-FeRh alloy (~ 55 nm thick) was grown epitaxially on a clean (001) MgO substrate (~ 500 µm thick) by conventional DC magnetron sputter co-deposition, as described in [6]. The sample was fixed onto a glass slide (FeRh film face-down) using Kemdent® wax and cut into ~ 1.5 mm square sections using a circular diamond saw. The glass slide was then placed on a hot plate to melt the wax and release the square sections, which were cleaned with acetone to remove any waxy residue. A dried square section was then fixed, again with FeRh film face-down, onto a Pyrex® specimen stub using ethyl cyanoacrylate (Loctite® 'super glue') and cured at 80°C for 2 hours. It was then mounted onto a tripod polisher and thinned to ~ 5 µm with a Struers® Labopol wheel polisher and diamond paper, using progressively finer grades (30 – 1µm grit). An Omniprobe® Cu grid was then overlaid onto the sample and fixed using an epoxy adhesive (Araldite®) and cured at 80◦C for 12 h. The stub / sample / Cu-grid combination was then immersed fully in acetone for 2 hours to dissolve the 'super glue' and release the remaining thinned sample / Cu-grid combination. The sample was then inserted in a FEI Dual Beam FIB Nova 200 for transfer to a DENSsolutions® MEMS-based heating chip (mounted on the custom SEM stub with 45° faces), as well as final thinning of the FeRh planar sample for the purpose of *in situ* TEM investigation. The FIB instrument comprises a 30 kV electron (e^-) column for SEM and a 30 kV sidewinder ion column (Ga^+), mounted at 52° to the e^- column. The system is equipped with an *in situ* micro-manipulator and a gas injector for the *in situ* deposition of Pt. Magnetic imaging was performed using a JEOL ARM-200cF TEM in Lorentz mode and Fresnel fringes in over- or under-focus provides contrast from in-plane DWs.

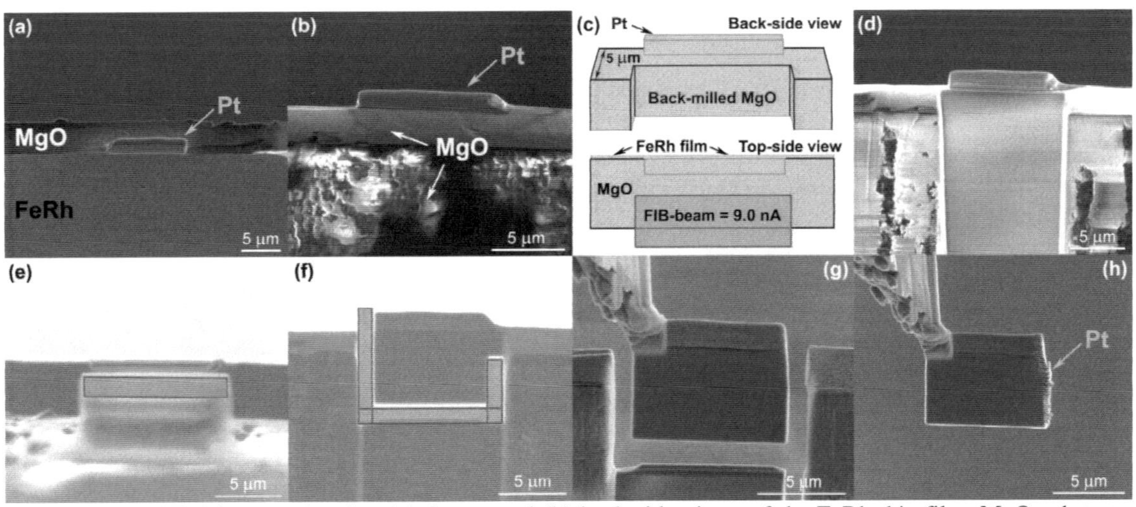

Figure 1 (a,b) SEM images showing (a) front-; and (b) back-side views of the FeRh thin film, MgO substrate and protective Pt layer. (c) Schematics of the back- and top-side views of the FIB-milling of the substrate using a current of 9.0 nA (red boxed region). (d) SEM image of the FIB back-milling shown in (c). (e,f) FIB images showing the (e) top-side view of back-milling; and (f) undercutting of the lamella (boxed regions), both using a current of 0.9 nA. (g,h) SEM images of (g) the welding of the micro-manipulator to the lamella and its release from the bulk sample; and (h) deposition of the Pt layer to allow for its attachment to the MEMS heating chip.

3. Results

This section illustrates our hydrid mechanical and FIB-based polishing approach for the preparation of an electron-transparent planar thin film FeRh sample on a MEMS heater chip, for the purpose of *in situ* TEM investigations.

Fig. 1 presents the initial milling and release of a thick TEM lamella, with dimensions of ~ 9 µm × 5 µm × 1 µm, from the polished sample. The sample was tilted until the e^- beam was parallel to the face of the FeRh film and the angle was noted as 6°. The SEM image of Fig. 1a shows the FeRh thin film grown on the MgO substrate and a protective Pt layer. The Pt was deposited using the FIB (30.0 kV) at a current of 0.93 pA and at an angle of 10° away from the face of the FeRh film to prevent unwanted Pt deposition on its surface. The sample was then rotated 180° to ensure the FeRh film undergoes limited exposure to both the e^- beam and FIB (Fig. 1b), as well as a favourable orientation for transfer to the heating chip. The sample was then tilted so that the face of the FeRh film (was positioned 1° away from the ion-beam path and back-milling of the MgO substrate was performed at a current of 9.0 nA (Fig. 1c). The FIB milling was monitored using the e^- beam until the milled section reached a sufficient depth (Fig. 1d). Back-milling continued using progressively lower currents of 2.7 nA and 0.9 nA until a lamella thickness of 1μm was achieved (Fig 1e). The sample was then tilted to 0°, rotated by 180° and three undercuts were made simultaneously (Fig. 1f), also using a FIB current of 0.9 nA, to free most of the lamella from the polished sample (Fig. 1g). The sample was rotated by 180°, tilted back to 1° away from the FIB path and both the micro-manipulator and gas injector were inserted. The micro-manipulator was brought into contact with the top left corner of the lamella and welded together with Pt using the e-beam (5.0 kV, 1.6 nA). The FIB was then used to release the lamella from sample using a current of 0.9 nA, as well as deposit a layer of Pt on the right-hand side of the lamella (I = 90 pA) to allow for its attachment to the MEMS heating chip (Fig. 1h).

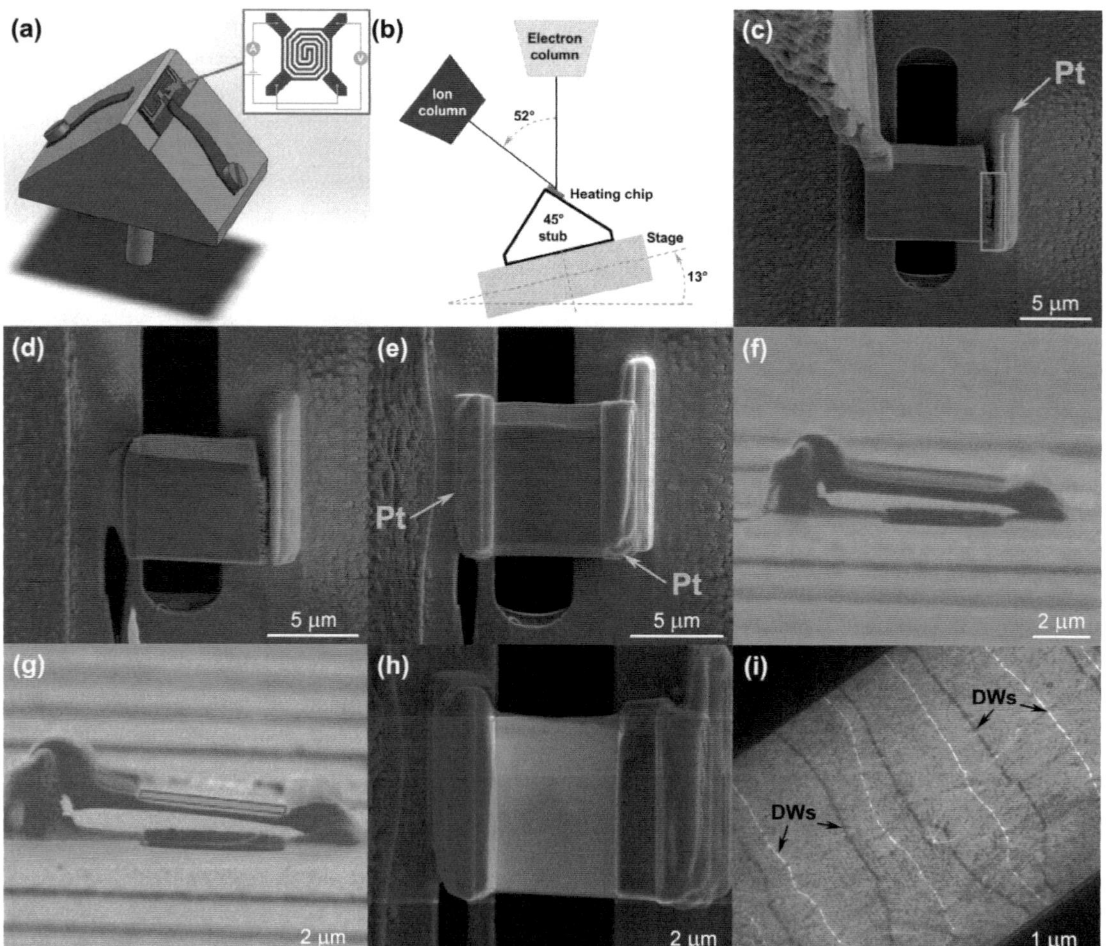

Figure 2 (a) 3D model showing the positioning of the MEMS chip and heating coil (inset), on the custom-designed SEM stub with 45° faces. (b) Schematic diagram illustrating the configuration of the 45° stub / heating chip in relation to the e^- and FIB columns within the FEI Helios workstation. (c-e) SEM images of the lamella (c) brought into contact with the pre-deposited Pt step (boxed region); (d) welded to the Pt step with the e^- beam and released from the micro-manipulator; and (e) reinforced with additional Pt deposition. (f,g) FIB images showing the side-view and thinning of the lamella using a current of (f) 0.26 nA; and (g) 90 pA, as denoted by the red boxed region in (g). (h) SEM image of the electron-transparent planar FeRh thin film. (i) Fresnel TEM image showing the presence of DWs in the FeRh thin film at T = 200°C, in the ferromagnetic state.

Fig. 2 presents the transfer of the TEM lamella onto the MEMS heating chip and final FIB milling into an electron-transparent FeRh thin film. The 3D model of Fig. 2a shows how the MEMS-based heating chip, incorporating a heating coil (Fig. 2a, inset), is positioned on the 45°-faced SEM stub. Fig. 2b illustrates the configuration of the 45° stub / heating chip in relation to the e^- and FIB columns, to allow for placement of lamella on the heating chip. The tilting of the stage to 13° provides a 7° difference in angle between the ion column and heating chip surface, which allows an effective milling path for the ion beam without damaging the heating element beyond the lamella. The micro-manipulator brings the Pt layer on right-hand side of the lamella in contact with a pre-deposited Pt step (~10 μm × 1 μm × 1 μm) (Fig. 2c). SEM imaging is then magnified to 500kx and rastered along the Pt-Pt interface for 5 min using the e^- beam shifts to weakly weld the surfaces together (boxed region). The FIB is then used to release the lamella from the micro-manipulator (Fig. 2d) (I = 0.9 nA) and, after tilting to 45°, deposit additional Pt to firmly weld both sides of the lamella to the heating chip (Fig. 2e) (I = 90 pA). The stage is tilted back to 13° and 12° for final milling using progressively lower currents of 0.26 nA (Fig. 2f) and 90 pA (Fig. 2g), respectively, until the lamella is ~ 100 nm thick. A final low energy FIB polish was carried out at 5.0 keV and 47 pA at an incidence angle of +7° (stage tilt of 20°), until the SEM image revealed a clean planar TEM lamella with an electron-transparent area of ~ 6 μm × 4 μm (Fig. 2h). *In situ* heating to 200°C within the TEM in Lorentz mode (Fig. 2i) revealed the presence of DWs in the planar FeRh thin film, visible as black and white lines (arrowed).

4. Discussion

This study has provided a simple methodology for the preparation of large planar magnetic thin films on MEMS-based chips for the purpose of *in situ* TEM investigations. The SEM images of Fig.1a,b confirm the preliminary polishing as an effective approach to reduce the MgO substrate thickness from ~ 500 μm to ~ 5 μm, before insertion into the FIB SEM for the main preparation. The subsequent FIB back-milling using gradually lower currents of 9.0 nA, 2.7 nA and 0.9 nA, whilst monitoring the progress using the e^- column, offered excellent control over the milling-rate and thickness of the lamella section as it approached 1 μm, before being undercut and transferred (Fig. 1f,g).

It is evident the 45°-faced SEM stub provides a favourable geometrical arrangement for the near-parallel placement of the lamella on the heating chip, with the magnetic thin film face-down away from exposure of the e^- beam and FIB. One unconventional caveat of this setup is the need to use the e^- beam for both Pt deposition and welding Pt interfaces together. This is due to the FIB not having the necessary access to the point of contact between the lamella and micro-manipulator, where the e^- beam supplied an adequate Pt weld instead (Fig. 1g). In addition, the gas injector collides with the heating chip when on the 45° face and so the e^- beam was rastered along the interface of the Pt on the right-hand side of the lamella and the Pt step to provide a temporary weld whilst the micro-manipulator was removed. Nevertheless, the lamella was transferred and welded onto the heating chip successfully to enable FIB back-milling of the MgO substrate into an electron-transparent planar FeRh thin film. This was confirmed through TEM investigation in Lorentz mode, where the sample was heated to 200°C *in situ* using the MEMS chip and Fresnel fringes were clearly visible, indicating the presence of DWs.

In summary, a planar FeRh thin film sample was prepared on a MEMS-based TEM heating chip using a novel 45°-faced SEM stub. TEM investigation revealed the presence of DWs and confirms the applicability of this method for preparing planar samples for *in situ* TEM studies.

References

[1] Lewis L H, Marrows C H and Langridge S 2016 *J. Phys. D: Appl. Phys.* **49** 323002.
[2] Marti X, Fina I *et al* 2014 *Nature Mat.* **13** 367-374.
[3] Le Graët C, Charlton T R *et al* 2015 *APL Mater.* **3** 041802.
[4] Duchamp M, Xu Q and Dunin-Borkowski R E 2014 *Microsc. Microanal.* **20** 1638–1645.
[5] O'Shea K J, McGrouther D *et al* 2014 *Micron* **66** 9-15.
[6] Le Graët C, de Vries M A *et al* 2013 *J. Vis. Exp.* **80** e50603.

Acknowledgements

The authors thank the EPSRC for funding (grants EP/M019020/1 & EP/M018504/1).

IOP Conf. Series: Journal of Physics: Conf. Series **903** (2017) 012023 doi:10.1088/1742-6596/903/1/012023

Magnetic and structural properties of RTA-processed $L1_0$ FePt-Mn thin films

N. Y. Safonova, M. Riepp, O. Klein, and M. Albrecht

Institute of Physics, University of Augsburg, 86159 Augsburg, Germany

nataliia.safonova@physik.uni-augsburg.de

Abstract. In this study, we have investigated the effect of Mn addition on texture, phase formation, and magnetic properties of $L1_0$ FePt thin films. FePt/Mn/FePt trilayers were processed by rapid thermal annealing in the temperature range of 650 – 800°C for 30 s in N_2 atmosphere. It is shown that Mn addition up to 10 at.% supports $L1_0$(001)-texture formation resulting in large perpendicular magnetic anisotropy. However, it is observed that Mn is only partially incorporated into the FePt lattice. It was found that Mn diffuses through the FePt top layer to the sample surface, promoting the ordering process and forming a Mn oxide layer.

1. Introduction

Chemically ordered $L1_0$ FePt(001) thin films exhibit strong perpendicular magnetic anisotropy (PMA) and make it the most promising material for heat assisted magnetic recording (HAMR) [1, 2]. It was reported that addition of third elements to FePt allows controlling its magnetic properties [1-3]. For instance, a reduction of the Curie temperature and saturation magnetization M_S with Mn addition is accompanied by a moderate decrease in PMA [4] required for HAMR application. In that case, stoichiometric $Fe_{1-x}Mn_xPt$ alloy thin films were prepared by epitaxial growth on MgO(001) at elevated temperatures [5].

In this work, we are focusing on the influence of Mn addition on the phase and texture formation in FePt thin films processed by rapid thermal annealing (RTA) on thermally oxidized silicon substrates.

2. Experimental part

$Fe_{52}Pt_{48}$/Mn/$Fe_{52}Pt_{48}$ trilayers with a total thickness of about 10 nm were deposited by dc magnetron sputtering at room temperature on SiO_2(100 nm)/Si(001) substrates. A $Fe_{52}Pt_{48}$ alloy target was used for deposition. Variation of the Mn content is achieved by altering the thickness of the Mn and FePt layers. Detailed information about the sample compositions is listed in Table 1. Post-annealing was performed by

Table 1. Layer thicknesses and compositions of the $Fe_{52}Pt_{48}$/Mn/$Fe_{52}Pt_{48}$ sample series

	Layer 1	Thickness	Layer 2	Thickness	Layer 3	Thickness	Expected composition
1	$Fe_{52}Pt_{48}$	4.9 nm	Mn	0.23 nm	$Fe_{52}Pt_{48}$	4.9 nm	$(Fe_{52}Pt_{48})_{95}Mn_5$
2	$Fe_{52}Pt_{48}$	4.8 nm	Mn	0.46 nm	$Fe_{52}Pt_{48}$	4.8 nm	$(Fe_{52}Pt_{48})_{90}Mn_{10}$
3	$Fe_{52}Pt_{48}$	4.6 nm	Mn	0.87 nm	$Fe_{52}Pt_{48}$	4.6 nm	$(Fe_{52}Pt_{48})_{80}Mn_{20}$

Content from this work may be used under the terms of the Creative Commons Attribution 3.0 licence. Any further distribution of this work must maintain attribution to the author(s) and the title of the work, journal citation and DOI.
Published under licence by IOP Publishing Ltd

RTA in the temperature range of 650 – 800°C for 30 s in N_2 atmosphere [6]. Structural properties were investigated with x-ray diffraction (XRD) and for selected samples cross-section transmission electron microscopy (TEM) studies were performed. For the characterization of the magnetic properties, superconducting quantum interference device - vibrating sample magnetometry (SQUID-VSM) was carried out at room temperature.

3. Results

XRD θ/2θ-scans of FePt films RTA processed at 750°C with different Mn content are presented in Fig. 1a. All film samples exhibit a pronounced $L1_0$ (001)-texture. In addition, a significant (111)-reflection is observed for the pure FePt film. The intensity of the (111) reflection decreases with addition of Mn up to 10 at.%. Further increase of the Mn content up to 20 at.% alters the film orientation as indicated by a reduced (001)-reflection intensity as well as the reappearance of the (111)- and (110)-reflection (Fig. 1a). Thus, a nominal Mn content of 10 at.% seems to be the optimum in terms of (001)-texture formation.

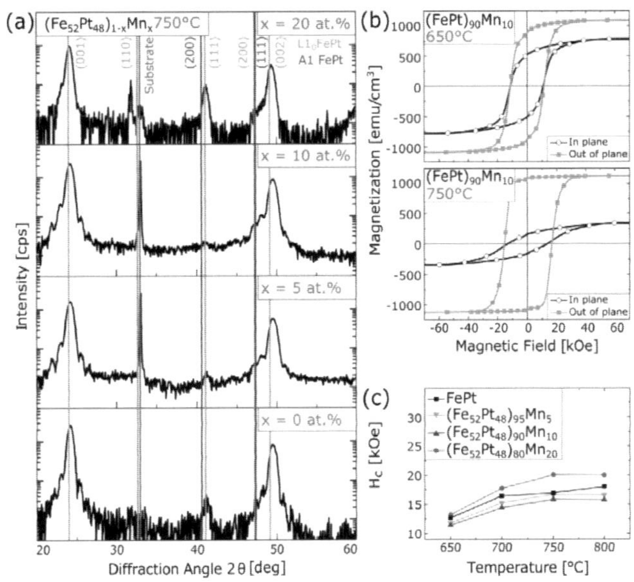

Figure 1. (a) XRD (θ–2θ) scans of FePt films with different nominal Mn content RTA processed at 750°C. (b) Out-of-plane and in-plane hysteresis loops for the FePt film with 10 at.% Mn annealed at 650°C and 750°C. (c) Annealing temperature dependence of the coercivity (out-of-plane).

Exemplarily, magnetic hysteresis loops taken in out-of-plane and in-plane geometry for FePt films with 10 at.% Mn after RTA processing at 650°C and 750°C are shown in Fig. 1b. Increasing the annealing temperature leads to an increase in remanent magnetization due to the formation of a high fraction of $L1_0$-ordered (001) grains. In the in-plane direction, a reduced remanence is observed, which confirms its hard-axis character. Furthermore, the occurrence of a significant coercivity in the in-plane direction indicates the presence of (111)-oriented grains. Please note that the magnetic moments cannot be fully saturated even with 70 kOe field in the in-plane direction, which makes it difficult to extract the magnetic anisotropy values. Nevertheless, the PMA shows a strong increase with annealing temperature, achieving values in the range up to 3.5 MJ/m^3.

Furthermore, only a small reduction in M_s value from 1125 to 900 emu/cm^3 by adding 20 at.% of Mn is observed for RTA processed samples at 750°C (not shown). This behavior is rather unexpected as an antiparallel coupling between Mn and Fe magnetic moments was observed experimentally in epitaxially grown $Fe_{1-x}Mn_xPt$ thin films [5]. There, a linear reduction in M_s with Mn content was found. For instance, for a $Fe_{0.92}Mn_{0.08}Pt$ sample, which corresponds in our case to a Mn content of 4 at.%, a saturation magnetization of 930 emu/cm^3 was observed. This leads to the conclusion that for our RTA processed FePtMn sample with nominal Mn content of 20 at.% only a small amount of about 4 at.% Mn is in fact incorporated into the FePt lattice.

Coercivity values in out-of-plane direction are slightly increasing with annealing temperature and remain almost unchanged up to 800°C (Fig. 1c). Films with a nominal Mn content of 10 at.% annealed at 750°C show a continuous film morphology with a route-mean-square roughness of about 1.8 nm as shown in the atomic force microscope (AFM) image of Fig. 2a. The corresponding magnetic force microscope

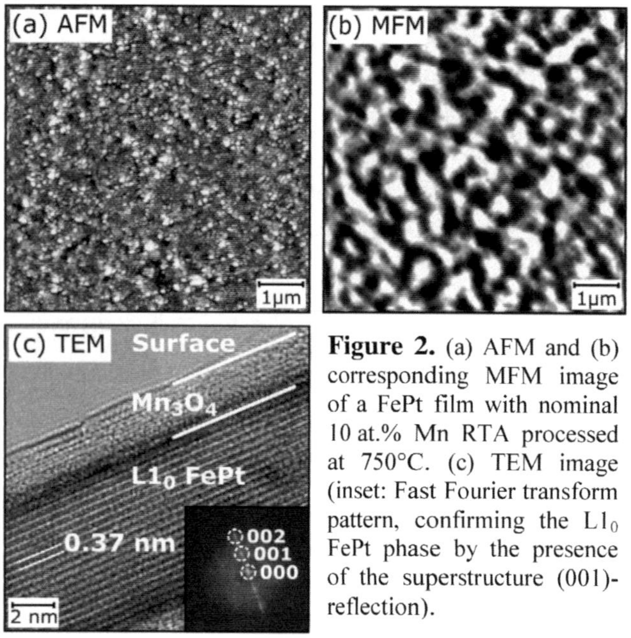

Figure 2. (a) AFM and (b) corresponding MFM image of a FePt film with nominal 10 at.% Mn RTA processed at 750°C. (c) TEM image (inset: Fast Fourier transform pattern, confirming the $L1_0$ FePt phase by the presence of the superstructure (001)-reflection).

(MFM) image (Fig. 2b) shows bright and dark contrast, revealing magnetic domains in the size range of about 500 nm pointing up and down, respectively. This is consistent with the strong PMA. The addition of 20 at.% Mn leads to an enhanced coercivity (Fig. 1c) due to increase in roughness and formation of small holes in the film, which act as defect sites for domain wall pinning (not shown).

In Fig. 2c, a cross-section TEM image for the film with 10 at.% Mn annealed at 750°C is shown, revealing large and well-ordered $L1_0$ FePt(001) grains. Interestingly, we observe the presence of a non-continuous Mn oxide layer (most likely Mn_3O_4) at the free surface. This result suggests that most of the Mn diffuses through the top FePt layer towards the free surface and is not incorporated into the FePt lattice during RTA processing. This behavior is consistent with the observation of a rather unaffected saturation magnetization with Mn content, as discussed before. Please note that an additional study on RTA processed stoichiometric $Fe_{1-x}Mn_xPt$ thin films did not improve the texture formation and revealed rather soft magnetic properties.

4. Conclusions

FePt/Mn/FePt thin films with a total thickness of about 10 nm were deposited at room temperature on SiO_2(100 nm)/Si(001) substrates. In order to transform them into the $L1_0$ FePt phase, RTA post-processing at different temperatures was carried out. For films with a nominal Mn content up to 10 at.%, the (001)-texture formation is enhanced, resulting in strong PMA. Larger Mn contents up to 20 at.% lead to a degradation of the (001)-texture in particular at high annealing temperatures.

However, only a small amount of Mn will be actually incorporated into the FePt lattice. Most of the Mn diffuses through the FePt layer, promoting the ordering process, and forms a Mn oxide layer at the free surface. Additional studies on RTA processed stoichiometric $Fe_{1-x}Mn_xPt$ thin films did not lead to strong (001) texture or PMA. These findings are in contrast to $Fe_{1-x}Mn_xPt$ films prepared by epitaxial growth on MgO(001) at elevated temperatures, where up to 68 at.% of Mn can be incorporated into the FePt lattice [5].

5. References

[1] Weller D, Parker G, Mosendz O, Champion E, Stipe B, Wang X, Klemmer T, Ju G, and Ajan A 2014 *IEEE Trans. Magn.* **50** 3100108
[2] Weller D, Parker G, Mosendz O, Lyberatos A, Mitin D, Safonova N Y and Albrecht M 2016 *J. Vac. Sci. Technol. B* **34** 060801
[3] Brombacher C, Schletter H, Daniel M, Matthes P, Jöhrmann N, Maret M, Makarov D, Hietschold M, and Albrecht M 2012 *J. Appl. Phys.* **112** 073912
[4] Park J, Hong Y-K, Kim S-G, Gao L, and Thiele J-U 2015 *J. Appl. Phys.* **117** 053911
[5] Meyer G and Thiele J-U 2006 *Phys. Rev. B* **73** 214438
[6] Albrecht M and Brombacher C 2013 *Phys. Status Solidi A* **210** 1272

IOP Conf. Series: Journal of Physics: Conf. Series **903** (2017) 012024 doi:10.1088/1742-6596/903/1/012024

Evolution of the structural and electronic properties of thin Bi films on Ge(111)

C Zucchetti,* F Bottegoni, A Calloni, G Bussetti, L Duò, M Finazzi and F Ciccacci

LNESS-Dipartimento di Fisica, Politecnico di Milano, Piazza Leonardo da Vinci 32, 20133 Milano, Italy

E-mail: *carlo.zucchetti@polimi.it

Abstract. We have incrementally grown bismuth thin films onto a n-doped Ge(111) substrate. Low energy electron diffraction reveals that the first Bi atomic layer is characterized by the $(\sqrt{3} \times \sqrt{3})R30^\circ$ reconstruction. By angle-resolved photoemission spectroscopy we observe Rashba-split bands that do not cross the Fermi level. At higher coverages, where a Rashba type of splitting should still be present, the density of occupied states close to the Fermi energy gradually increases, while extra diffraction spots, related to Bi(110) islands, appear.

1. Introduction

Electrons in bismuth, thanks to its high atomic number ($Z_{\mathrm{Bi}} = 83$), are subject to a huge spin-orbit interaction. This makes Bi one of the most appealing materials for spintronic applications. Despite bulk Bi is a non-magnetic material and a semimetal, in reduced-symmetry environments its electronic structure can be dramatically changed and its spin-related properties boosted. For instance, in Bi thin films it is possible to induce a transition into a semiconductor as a function of thickness [1, 2] or into a semiconductor or a topological semimetal as a function of strain [3]. Furthermore, confinement promotes the formation of quantum well-related states [4, 5], while the removal of inversion symmetry at surfaces or interfaces yields a non null Rashba term in the hamiltonian, resulting in spin-polarized surface bands [6–9].

In this context, particularly important for both fundamental and technological reasons is the deposition of Bi films onto semiconductors. Noteworthy, the Rashba effect is found to be greatly enhanced in one monolayer (ML) of Bi deposited on Ge(111) [7, 8] with respect to the (111) surface of bulk Bi [10]. This highlights the active role played by the substrate and calls for a more detailed study of the interplay between the two media. Moreover, Ge is one of the key semiconductor for spintronic applications [11–13], due to inversion symmetry, absent in III-V semiconductors [14, 15], which results in long electron spin lifetime. In this work we present an analysis of the structural and electronic properties of ultrathin Bi layers grown on n-doped Ge(111), based on X ray photoelectron spectroscopy (XPS), low energy electron diffraction (LEED), and angle-resolved photoemission (ARPES).

2. Materials and methods

A Sb-doped Ge(111) substrate ($N_{\mathrm{D}} = 1.1 \times 10^{16}$ cm^{-3}) was cleaned by cycles of Ar^{+} sputtering (1 kV, normal incidence) and thermal annealing (1175 K for few minutes) until a sharp $c\,(2 \times 8)$

Content from this work may be used under the terms of the Creative Commons Attribution 3.0 licence. Any further distribution of this work must maintain attribution to the author(s) and the title of the work, journal citation and DOI.
Published under licence by IOP Publishing Ltd

reconstruction was visible at LEED [7, 17, 18]. Bismuth was then evaporated from a boron nitride crucible at a rate of about 1 Å/min, calibrated by a quartz microbalance. The base pressure was in the low 10^{-8} Pa. As a standard procedure [8], the first Bi single layer was obtained by room temperature deposition of an amount of Bi slightly above 1 ML.[1] The excess material was then removed by heating the sample for 5 min at 725 K [8]. Further Bi layers were grown by depositing the exact quantity at room temperature without annealing.

XPS and ARPES measurements were performed by illuminating the sample with unmonochromatized Al Kα radiation ($h\nu$ = 1486.6 eV) and unpolarized He Iα radiation ($h\nu = 21.2$ eV), respectively. Photoelectrons were collected by a 150 mm hemispherical analyzer (from SPECS GmbH) equipped with a standard multichannel detector, operating with a pass energy of 0.5 eV (20 eV) and providing a full width at half maximum resolution of 40 meV (0.9 eV) for ARPES (XPS) [19]. Wave-vector selection was obtained by tilting the sample around the [11$\bar{2}$] direction [see Fig. 1(d)], whereas Fermi level (ε_F) was determined from the metallic Fermi edge of a Ta foil. All measurements were performed with the sample at 100 K.

3. Structural evolution

The LEED measured on 1 ML Bi/Ge(111) [Fig. 1(a)] shows both the (1×1) pattern related to the germanium substrate (red dashed line) and the $(\sqrt{3} \times \sqrt{3})\,R30°$ [from now on: $(\sqrt{3} \times \sqrt{3})$] reconstruction (green dashed line) typical of the bismuth first layer, which is known to form

[1] Bismuth monolayer thickness (2.55 Å) is the deposed bismuth quantity which matches the atomic density of Ge(111) surface, *i.e.* $7.21 \cdot 10^{14}$ atoms/cm^2.

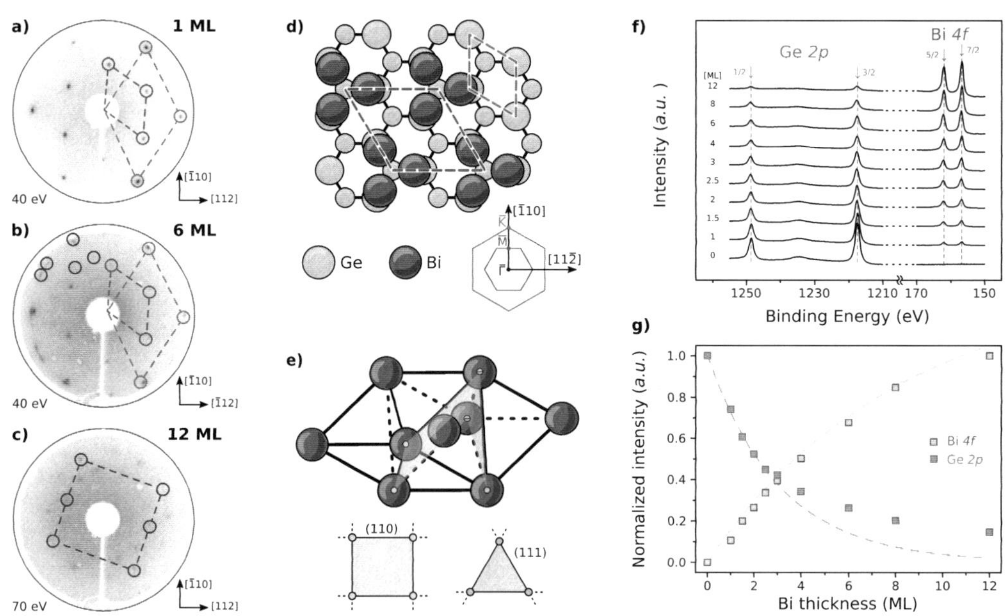

Figure 1. (a-c) LEED spectra of 1, 6 and 12 ML of Bi/Ge(111). Green [red] dashed lines show the unit cell of the Ge(111) surface [$(\sqrt{3} \times \sqrt{3})$ reconstruction], while blue ones are related to the (1×1) pattern of Bi(110). (d) Real lattice configuration of the first Bi layer, with a sketch of the (1×1) and $(\sqrt{3} \times \sqrt{3})$ surface first Brillouin zone. (e) Bulk Bi unit cell in which the pseudosquare (110) and hexagonal (111) planes are highlighted, reported underneath as top views. (f) Bi 4f and Ge 2p XPS spectra and (g) peak intensities vs. Bi coverage (lines: exponential fits performed with inelastic mean free paths consistent with Ref. [16]).

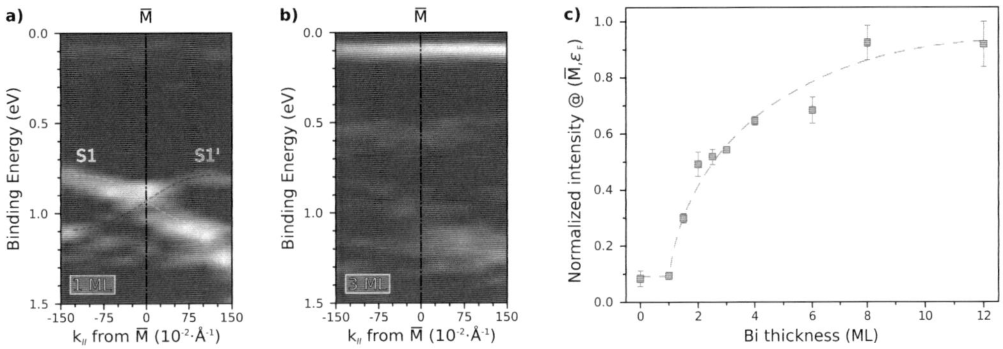

Figure 2. (a,b) Valence band map – obtained as in Ref. [8] – along the $\overline{\Gamma K}$ direction around the $\overline{M}$ point of the $(\sqrt{3}\times\sqrt{3})$ SBZ of 1 and 3 ML Bi/Ge(111). (c) Intensity of ARPES spectra measured at ε_F as a function of the Bi thickness, probed at the $\overline{M}$ point of the $(\sqrt{3}\times\sqrt{3})$ SBZ.

trimers with both Ge(111) and Si(111) substrates [20, 21]. The corresponding direct-space lattice is shown in Fig. 1(d) with the same color convention. For a thickness $\geq$ 2 ML, weak LEED spots appear [blue circles in Fig. 1(b) for 6 ML], with increasing intensity at higher coverages, forming a fully-developed (1×1) reconstruction around 12 ML [Fig. 1(c)]. According to the STM analysis by Hatta *et al.* [20], such spots are indicative of the presence of pseudosquare Bi(110) islands [see Fig. 1(e)]. It is worth noticing that the stability of these islands, as recently pointed out in other ultrathin films [22, 23], is driven by the epitaxy with the substrate, since, for higher coverages Bi thin films on Ge shows the (111) surface [20]. Notably, the 6 ML pattern partially retains the hexagonal Bi reconstruction, whereas at 12 ML it is barely visible.

XPS spectra [Fig. 1(f)] show the decrease (increase) of Ge $2p$ (Bi $4f$) peak with increasing Bi coverage, as reported in Fig. 1(g). It is worth mentioning that the Ge $2p$ intensity deviates from the exponential attenuation curve above $\approx$ 4 ML, as expected in the case of island nucleation [24, 25]. In this case, in fact, the signal from the substrate is less quenched than in a layer-by-layer growth, since regions not covered by islands provide a signal that reaches a non vanishing saturation value [26]. Notably, Bi $4f$ peak intensity does not show the same deviation from the exponential curve because the inelastic mean free path (IMFP) of Bi $4f$ electrons ($\approx$ 11 ML) is three times larger than the one of Ge $2p$ electrons [16], making the former less sensitive to the overlayer inhomogeneities. Similarly, Ge $2p$ IMFP limits XPS sensitivity, so that 4 ML have to be considered as an upper bound of island nucleation starting coverage, in agreement with the 2 ML revealed at LEED. Hence, XPS and LEED analysis suggest a Stranski-Krastanov growth mode for this system, with the formation of a single $(\sqrt{3}\times\sqrt{3})$ wetting layer and Bi(110) islands. Very recent STM data seems to support this interpretation.

4. Electronic structure

As reported in our previous work [7], for the single layer we observe a strong Rashba splitting of Bi surface resonance band, resulting in two spin-polarized branches [S1 and S1' in Fig. 2(a)] symmetric with respect to the $\overline{M}$ point of the $(\sqrt{3}\times\sqrt{3})$ surface first Brillouin zone (SBZ), and located at around 1 eV below ε_F. In this case, no band is observed to cross ε_F, in agreement with Sandomirskii's prediction [27], according to which bismuth should maintain a semiconductor-like character up to a critical thickness, later estimated to be about 100 ML [28]. Nevertheless, as reported in Fig. 2(b), the second derivative of the experimental spectra measured for 3 ML clearly indicates the presence of states close to ε_F. This is in qualitative agreement with both experiments [in Bi/Si(111) films thicker than 10 ML] [4] and *ab initio* calculations [thin Bi(110) films] [1], which contradict the prediction of Ref. [28]. To study the density of states (DOS) at ε_F as a function of the coverage we have normalized the acquired spectra to the same background value at a binding energy of 5 eV, where no spectral feature is observed. It can be noticed that the DOS, evaluated at $\overline{M}$, begins to increase for a Bi thickness just larger than 1 ML [Fig. 2(c)]. Notably, according to Refs. [1] and [9], the Rashba effect should be significant also for these coverages, *i.e.* for bands crossing ε_F, paving the way to potential applications in spintronics.

5. Conclusions

We have studied ultrathin Bi/Ge(111) films as a function of the Bi coverage. Bi atoms in the 1 ML-thick film are arranged in trimers while the film itself has a semiconducting character. At higher thickness we measured a nonvanishing DOS at the Fermi level, together with the observation of a Bi(110) island nucleation. Notably, the Rashba splitting should be present also for high coverages, making the states crossing the Fermi level spin-polarized.

Acknowledgments

The authors thanks A Picone for having made available recent and unpublised STM data. Partial funding is acknowledged to SEARCH-IV project (Cariplo foundation, grant 2013-0623).

References

[1] Koroteev Y M, Bihlmayer G, Chulkov E V and Blügel S 2008 *Phys. Rev. B* **77** 045428–7
[2] Xiao S, Wei D and Jin X 2012 *Phys. Rev. Lett.* **109** 166805–5
[3] Aguilera I, Friedrich C and Blügel S 2015 *Phys. Rev. B* **91** 125129–7
[4] Hirahara T, Nagao T, Matsuda I, Bihlmayer G, Chulkov E V, Koroteev Y M and Hasegawa S 2007 *Phys. Rev. B* **75** 035422–9
[5] Hirahara T 2015 *J. Electron. Spectrosc. Relat. Phenom.* **201** 98–7
[6] Bychkov Y A and Rashba E I 1984 *JETP Letters* **39** 78–4
[7] Bottegoni F, Calloni A, Bussetti G, Camera A, Zucchetti C, Finazzi M, Duò L and Ciccacci F 2016 *J. Phys.: Condens. Matter* **28** 195001-6
[8] Hatta S, Aruga T, Ohtsubo Y and Okuyama H 2009 *Phys. Rev. B* **80** 113309–4
[9] Hsu C H, Chang H R, Chuang F C, Liu Y T, Huang Z Q, Lin H, Ozolinš V and Bansil A 2014 *Surf. Sci.* **626** 68–8
[10] Ast C R, Henk J, Ernst A, Moreschini L, Falub M C, Pacilé D, Bruno P, Kern K and Grioni M 2007 *Phys. Rev. Lett.* **98** 186807–4
[11] Bottegoni F, Ferrari A, Isella G, Finazzi M and Ciccacci F 2012 *Phys. Rev. B* **85** 245312–4
[12] Ferrari A, Bottegoni F, Isella G, Cecchi S and Ciccacci F 2013 *Phys. Rev. B* **88** 115209–6
[13] Bottegoni F, Celebrano M, Bollani M, Biagioni P, Isella G, Ciccacci F and Finazzi M 2014 *Nat. Mater.* **13** 790–5
[14] Bottegoni F, Drouhin H J, Wegrowe J E and Fishman G 2012 *J. Appl. Phys.* **111** 07C305–3
[15] Bottegoni F, Drouhin H J, Fishman G and Wegrowe J E 2012 *Phys. Rev. B* **85** 235313–7
[16] Tanuma S, Powell C J and Penn D R 1994 *Surf. Interface Anal.* **21** 165–12
[17] Bringans R D and Höchst H 1982 *Phys. Rev. B* **25** 1081–9
[18] Knapp B J and Tobin J G 1988 *Phys. Rev. B* **37** 8656–60
[19] Berti G, Calloni A, Brambilla A, Bussetti G, Duò L and Ciccacci F 2014 *Rev. Sci. Instrum.* **85** 073901–4
[20] Hatta S, Ohtsubo Y, Miyamoto S, Okuyama H and Aruga T 2009 *Appl. Surf. Sci.* **256** 1252–6
[21] Gierz I, Suzuki T, Frantzeskakis E, Pons S, Ostanin S, Ernst A, Henk J, Grioni M, Kern K and Ast C R 2009 *Phys. Rev. Lett.* **103** 046803–4
[22] Riva M, Picone A, Giannotti D, Brambilla A, Fratesi G, Bussetti G, Duò L, Ciccacci F and Finazzi M 2015 *Phys. Rev. B* **92** 115434–9
[23] Lu Y, Wang C, Shi R, Cui Y, Shi Z, Yang S, Cui Y and Liu X 2016 *APL Mater.* **4** 076102–7
[24] Argile C and Rhead G E 1989 *Surf. Sci. Rep* **10** 277–99
[25] Huang W, Jiang Z, Dong F and Bao X 2002 *Surf. Sci.* **514** 420–5
[26] Ossicini S, Memeo R and Ciccacci F 1985 *J. Vac. Sci. Technol. A* **3** 387–5
[27] Sandomirskii V B 1967 *Sov. Phys. — JETP* **25** 101–6
[28] Hoffman C A, Meyer J R, Bartoli F J, Di Venere A, Yi X J, Hou C L, Wang H C, Ketterson J B and Wong G K 1993 *Phys. Rev. B* **48** 11431–4

IOP Conf. Series: Journal of Physics: Conf. Series **903** (2017) 012025 doi:10.1088/1742-6596/903/1/012025

Magneto-optical reflection spectroscopy on graphene/Co in the soft x-ray range

H.-Ch. Mertins[1], C. Jansing[1], M. Gilbert[1], M. Krivenkov[2], J.Sanchez-Barriga[2], A. Varykhalov[2], O. Rader[2], H. Wahab[3], H. Timmers[3], A. Gaupp[2], M. Tesch[2], A. Sokolov[2], D. Legut[4] and P M. Oppeneer[5]

[1]University of Applied Sciences Münster, Stegerwaldstr. 39, D-48565 Steinfurt
[2]HZB, Albert-Einstein-Str. 15, D-12489 Berlin, [3]University of New South Wales, Canberra, ACT 2600, [4]IT4-Innovations Centre, VSB-TU Ostrava, CZ 70833 Ostrava
[5]Department of Physics and Astronomy, Uppsala University, Box 516, S-751 20 Uppsala

mertins@fh-muenster.de

Abstract. The existence of ferromagnetic ordering in graphene on cobalt is demonstrated by means of resonant magnetic reflection spectroscopy exploiting the transversal magneto-optical Kerr-effect (T-MOKE). Using linearly polarized synchrotron radiation in the soft x-ray range with energies spanning the carbon 1s edge, the π- and σ- bonds of graphene were excited individually, showing that magnetism in graphene is carried by the π – orbitals. Magnetic signals were detected over a wide energy range from 257 – 340 eV with a T-MOKE peak value of 1.1 % at the π – resonance energy near 285 eV. By comparison with corresponding spectra measured at the 2*p* edges of the Co substrate, a large induced magnetic moment of 0.14 μ_B was derived for graphene. Individual hysteresis curves monitored at the Co 2*p* and C 1*s* edges show that the carbon magnetism is induced by the Co substrate.

1. Introduction

Graphene is a promising candidate material for novel electronic devices [1]. Its properties result from the layered structure of carbon atoms which gives rise to a large anisotropy in its structural and electronic properties due to the anisotropic bonding characteristics [1]. The carbon atoms in the basal plane are bound together by strong covalent σ-bonds while the π-bonds are oriented perpendicular to the substrate. This anisotropy can be probed best by polarization- and reflection spectroscopy using linearly polarized light. Setting the electric field vector parallel to the respective orbitals allows to probe individually the π-states and σ-states, respectively [2]. Recently magnetism was detected in carbon based materials, graphite and graphene which could offer access to new technological applications, e.g. in the field of spintronics or data storage [3, 4]. However, the origin of this magnetism has still to be identified. In graphite (HOPG) it is believed to result from the defect structures, *i.e.* from localized electron states at grain boundaries which form arrays of point defects [4]. In single layer graphene grown on SiC ferromagnetic behavior at room temperature could be detected after hydrogenization [5]. For technical applications having graphene grown on magnetic,

Content from this work may be used under the terms of the Creative Commons Attribution 3.0 licence. Any further distribution of this work must maintain attribution to the author(s) and the title of the work, journal citation and DOI.
Published under licence by IOP Publishing Ltd

metallic substrates seems to be a fruitful approach [6, 7]. However, a clear understanding of the interaction of carbon atoms with the metallic substrate is still missing.

Up to now, the applied magnetic detection techniques were not able to 1) unambiguously identify the localization of magnetic moments at the carbon atoms and 2) to identify the influence of π- or σ-bonds. A promising method for solving point 1), i.e. separating the magnetic response of possible metallic impurities from that of carbon, is the element-selective excitation of carbon at the C 1*s* state with polarized synchrotron radiation (SR) [8]. Up to now only x-ray magnetic circular dichroism (XMCD) measurements of graphene on Ni exist which suggest a magnetic moment of carbon in the order of 0.05 – 0.1 μ_B [7]. However, these XMCD experiments suffer from the fact that the polarization state of circularly polarized SR varies dramatically when tuning the photon energy across the C 1*s* edge caused by the unavoidable carbon contamination of the optical elements in the beamline [9]. To solve point 2) a spectroscopy is necessary which allows to probe the orientation of bonds. Both tasks can be tackled by our magneto-optical polarization- and reflection spectroscopy.

2. Experiment

The experimental setup for our magneto-optical reflection spectroscopy is shown in figure 1. The incident intensity I_0 is reflected partially at the graphene surface and partially, after transmission of the graphene layer, at the metallic substrate. Upon reflection a phase shift appears at the graphene surface as well as at the metallic substrate with difference α between these phase shifts. An additional phase shift due to the optical path length is negligible for the extremely thin graphene layer. The total reflected intensity I is described by $I/I_0 = R_G + (1-R_G)T_G^2 R_M + 2|r_G r_M|(1-R_G)^{0.5} T_G \cos\alpha$ with reflectance $|r_G|^2 = R_G$ and transmittance T_G of graphene and $|r_M|^2 = R_M$ the reflectance of the substrate acting as a mirror and α the phase shift of light reflected at the graphene surface and the substrate, respectively. The transmission through the graphene dominates over the reflectance at the graphene top surface since $R_G \approx R_M/10$. This result, obtained from independent reflection experiments of graphene, has been confirmed by model calculations using the computer code REFLEC [10] and optical constants of carbon and cobalt from Henke data table [11].

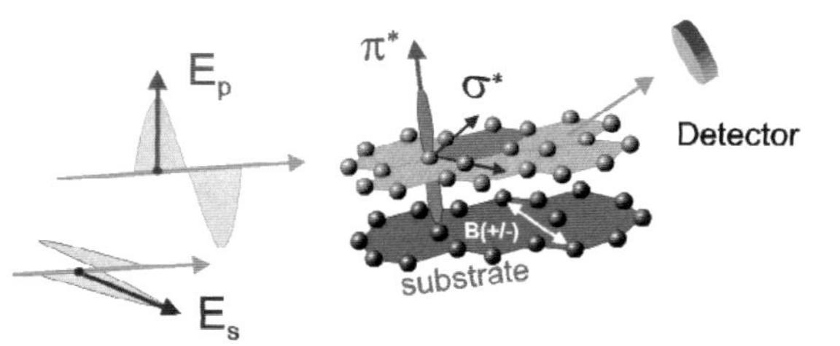

Figure 1. Experimental setup for detecting reflection and T-MOKE, applying transversal magnet fields *B*(+/-) to graphene on the Co substrate.

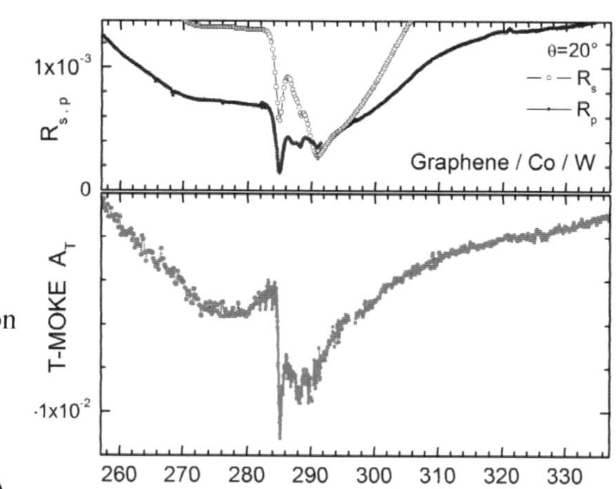

Figure 2. Top: Reflectance for *p* and *s*-geometry, bottom: T-MOKE spectrum deduced from R_p (B+/-).

The experiments were performed at the undulator beamline UE56-2-PGM2 at BESSY [12] using the BESSY ultrahigh-vacuum polarimeter chamber [13]. The spectral resolution near 290 eV was set to $E/\Delta E$=2000 and the accuracy of the absolute energy calibration was better than 0.1 eV. The electric field vector E of the linearly polarized incoming light could be set alternatively parallel to the π^*–orbitals (E_p in figure 1, p-geometry) or perpendicular to the π^* – orbitals (E_s, s-geometry) with fixed angle of incidence θ = 20°, measured to the surface. The samples were monolayer graphene deposited on 6.0 nm Co (0001) grown on W(110) substrate characterized prior to measurement by STM and LEED [6]. All recorded spectra were corrected for contributions from high order light applying a procedure described in Ref. 14. Any in-plane magnetization geometry could be set by *in situ* rotation of permanent magnets with maximum field strength of 250 mT [15]. Hysteresis curves were achieved by applying a solenoid creating in-plane fields of up to +/-50 mT.

3. Results

The absolute reflectance R_p is plotted in figure 2, top, for p-polarization geometry (E_p in figure 1). At the π*-resonance (285.0 eV) the measured reflectance decreases due to the decrease of transmission of the graphene layer and accompanied decrease of light reflected from the metallic Co substrate. A second minimum in the reflectance appears at the σ*-resonance (291.2 eV). Since the electric field E_p is oriented parallel to the π*-orbitals excitations from C 1*s* to π*-states dominate. For electric field vector E_s oriented parallel to the σ*-orbitals (s-geometry, figure 1) the minimum at 291.2 eV is more pronounced (R_s in figure 2, top). The magnetization of graphene was probed by T-MOKE [8] scanning across the C 1*s* edge. The normalized difference signal $A_T = (R_p(B+) - R_p(B-))/(R_p(B+) + R_p(B-))$ of the reflectance $R_p(B(+/-)$ is monitored for two anti-parallel in-plane magnetization directions $B(+/-)$ perpendicular to the light's scattering plane (figure 1). The obtained T-MOKE spectrum A_T (figure 2, bottom) clearly shows magnetic signals over a wide range across the C 1*s* edge. Strongest effects of 1.1% occur at the π*-resonance while at the σ*-resonance only small effects are detected. This clearly shows that the magnetism in graphene is carried by the π-orbitals. The appearance of magnetic signals even far below the π*-resonance is explained by the typical enhancement of magnetic signals in reflection spectroscopy at low reflectance values due to normalization [16]. Exactly this is the case for carbon showing small R_G below the π*-resonance [17].

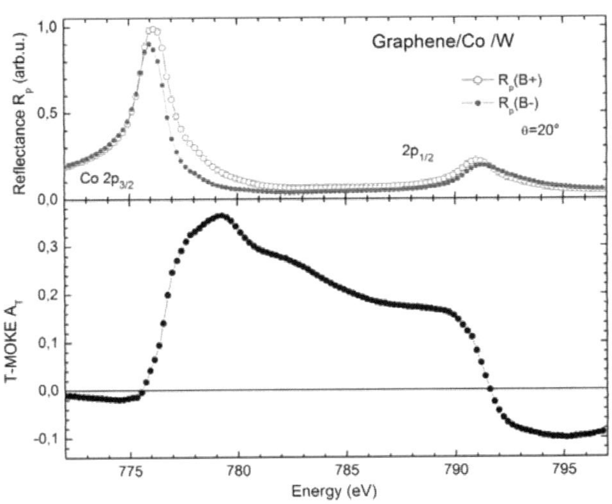

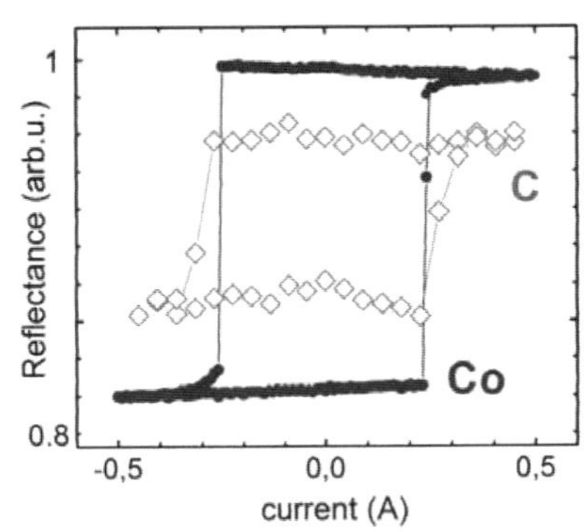

Figure 4. Hysteresis curves of graphene and Co recorded with R_p as function of the coil current at the C 1*s* and at the Co $2p_{3/2}$ edge.

Figure 3. Top: Reflected Intensity across the Co 2*p* edge of graphene on Co/W of *p*-polarized light for two transversal magnetizations $B(+/-)$ at incidence angle θ – 20°. Bottom: Deduced T-MOKE A_T.

For comparison of magnetism in graphene and the underlying Co substrate T-MOKE spectra where recorded across the Co 2*p* edges (figure 3). The reflectance R_p is resonantly enhanced at the Co $2p_{3/2}$ and $2p_{1/2}$ edges. Strong differences at and in between these edges appear for inverted transversal magnetic fields (figure 3, top). These lead to large T-MOKE values of up to 37% in between the edges (figure 3, bottom). Note that the magnetic T-MOKE value A_T is increased in between the edges due to normalization on small reflectance values as typical for Co films [18]. However, resonant excitation of

Co atoms occurs at the Co $2p_{3/2}$ edge (776.35 eV). At this energy the relevant T-MOKE signal A_T = 0.1 has to be taken for comparing magnetic moments as discussed below.

Exploiting the T-MOKE, two element specific hysteresis loops where recorded by setting the photon energy first to the Co $2p_{3/2}$ resonance at 775.75 eV and then to the C 1*s* resonance energy (286.3 eV) and detecting the reflected intensity as function of the applied magnetic field strength (figure 4). These hysteresis loops show the individual magnetic behaviour of carbon atoms in graphene demonstrating that Co induces the ferromagnetism in graphene which is harder than in Co.

The magnetic moment of carbon atoms in graphene was deduced by comparing the strength of T-MOKE data at the C 1*s* edge of graphene (A_T(C) = 0.011) with T-MOKE data exactly at the Co $2p_{3/2}$ edge A_T (Co) = 0.1. From the ratio of the asymmetry parameter A_T (C)/A_T (Co) and the magnetic moment of Co μ_{Co}= 1.76μ_B we obtain the magnetic moment of carbon μ_C= 0.14±0.03μ_B. In this calculation a correction factor of 0.87 enters which takes into account the small difference in the polarization power between carbon at the C 1s edge and cobalt at the 2p edge. This polarization power enters the T-MOKE signal and must be considered according to the procedure described in Ref. 16. The obtained magnetic moment of graphene is a factor of three larger than the value obtained for amorphous C-layer in a Fe/C multilayer [19]. This is surprising in view of the larger magnetic moment of Fe compared to that of Co. We explain this larger magnetic moment due to the collectively aligned π*-orbitals in graphene, confirmed by polarization analysis [17], in contrast to the situation in the polycrystalline C-layers of Fe/C [19].

In summary we observed ferromagnetic order in graphene on Co by T-MOKE spectroscopy across the C 1s edge and determined experimentally an induced magnetic moment of 0.14 μ_B on carbon which is based on the hybridization of Co 3*d* with carbon π*-orbitals.

We acknowledge financial and technical support by the Helmholtz-Zentrum Berlin, by the DAAD exchange programs No. 56270051 and No. 57215199 and by the Swedish Research Council (VR). D.L. was supported by The Ministry of Education, Youth and Sports from the Large Infrastructures for Research, Experimental Development and Innovations project „IT4Innovations National Supercomputing Center – LM2015070“ and acknowledges also the COST Action MP1306 "Modern Tools for Spectroscopy on Advanced Materials: a European Modelling Platform"

References

[1] K.S. Novoselov, V.I. Fal'ko, L. Colombo, P.R. Gellert, et al. 2012, Nature **490** 192.
[2] H.-Ch. Mertins, P.M. Oppeneer et al., 2004, Phys. Rev. B **70**, 235106.
[3] P. Esquinazi et al., 2002, Phys. Rev. B **66**, 024429 and references there in.
[4] J. Cervenka et al.,2009, Nature Physics **5**, 840.
[5] A.J.M. Giesbers et al., 2013, Phys. Rev. Lett. 111, 166101.
[6] A. Varykhalov et al., 2009, Phys. Rev. B **80**, 035437.
[7] M. Weser et al., 2010, Appl. Phys. Lett. **96**, 012504.
[8] M. Tesch, M. Gilbert, H.-Ch. Mertins, D.E. Bürgler et al., 2013, Appl. Opt. **52**, 4294.
[9] A. Gaupp et al., J. Phys. Conf. Ser. **425**, 122013 (2013).
[10] F. Schäfers, M. Krumrey1996, Reflec, Tech. Rep. BESSY, TB 201.
[11] B.L. Henke et al., http://www-cxro.lbl.gov/optical_constants/ asf.html
[12] S. Sasaki, K. Miyata, and T. Takada, 1992, Jpn. J. Appl. Phys. **31,** L1794.
[13] F. Schäfers, H.-Ch. Mertins, A. Gaupp, W. Gudat et al., 1999, Appl. Opt. **38**, 4074.
[14] C. Jansing, H.-Ch. Mertins, A. Gaupp et al., 2016, J. Physics Conf Ser. **712**, 012031.
[15] M. F. Tesch, D. Legut, H.-Ch. Mertins et al., 2014, Phys. Rev. B **89**, 140404(R).
[16] H.-Ch. Mertins, D. Abramsohn, A. Gaupp, F. Schäfers et al. 2002, Phys. Rev. B **66**, 184404.
[17] C. Jansing, H.-Ch. Mertins et al. 2016, Phys. Rev. B **94**, 045422
[18] H.-Ch. Mertins, S. Valencia, D. Abramsohn, A. Gaupp et al., 2004, Phys. Rev. B **69**, 064407.
[19] H.-Ch. Mertins, S. Valencia, W. Gudat et al. 2004, Euro. Phys. Lett. **66**, 743.

Magnetization measurement of a solid CO_2 grain based on its field-induced translation under microgravity conditions

Wakana Yamaguchi, Keiji Hisayoshi, Chiaki Uyeda

Institute of Earth and Space Science, Graduate School of Science, Osaka University, Toyonaka, Osaka, Japan

uyeda@ess.sci.osaka-u.ac.jp

Abstract. Diamagnetic susceptibility χ_{DIA} per unit mass was obtained in a dry-ice (CO_2) grain using a simple method in which translation of the grain induced by a field-gradient force was observed in an area of T =196 K using a short-microgravity (μg) condition. It was not necessary to know the value of the sample mass m in the measurement because the field-gradient force induced in the sample was proportional to m. The proposed method is hence effective in measuring χ of a small volatile grain that would require additional operations to obtain its m value. By comparing the obtained χ_{DIA} value with a list of published values, the material of an unidentified volatile grain can be determined in a simple manner without consuming the sample.

1. Introduction

The spatial variation of the field-induced potential that is produced by a static electromagnet has been conventionally used in a magnetic balance to measure the magnetization of a material. Diamagnetic susceptibility χ_{DIA} was recently detected in small samples based on the translation in a fluid medium due to the effect of a large field gradient at a level of ~10^4 Oe/cm [1][2]. This translation proceeded under the balance between the field-gradient force and the viscous drag of the medium. Acceleration of diamagnetic solids caused by a field-gradient force [3][4] was recently observed in a diffuse area using a short μ g condition (< 0.6 s); here, the field gradient was produced by a centimetre-size magnetic circuit and [emu/g] was obtained from this translation without knowing the value of the mass m. This was possible because in this translation, the variance of the field-induced potential between two arbitrary sample positions was completely converted to kinetic energy. In general, magnetization measurement of a weak magnetic material becomes more difficult with the reduction of sample size because of the difficulty in measuring m. Additionally, the interference of background signals emitted from the sample holder is not negligible when the sample size is at the sub-millimetre level. Hence, the above-mentioned method based on field-induced translation could be a breakthrough in solving these two problems [4], although currently the field-induced translation is realized only at room temperature.

In this report, the efficiency of the detection of χ_{DIA} by the field-induced acceleration is examined at a reduced temperature of T=196 K. Based on the results obtained on the CO_2 grains, the remaining problems in measuring the magnetic properties of single volatile materials of reduced sizes are considered.

2. Experimental Methods

The field gradient used in the present study was produced by a compact magnetic circuit composed of a pair of NeFeB plates, as shown in Fig. 1. The [+z]-axis of a coordinate was defined as the direction of the magnetic line of force produced between the N pole and S pole of the circuit, and the [+x]-axis was defined as the direction in which the gradient of the monotone-decreasing field is largest.

Content from this work may be used under the terms of the Creative Commons Attribution 3.0 licence. Any further distribution of this work must maintain attribution to the author(s) and the title of the work, journal citation and DOI.
Published under licence by IOP Publishing Ltd

A solid CO_2 block was separated into small sample pieces with diameters of less than 1 mm, and the sample pieces were placed at a point where the field-gradient force was expected attain its maximum value, according to field-distribution measurements. At this position, the sample pieces were expected to be spontaneously released in the area of monotone-decreasing field due to the field-gradient force. The sample, together with the magnetic circuit, was contained in a two-fold glass tube, which was used to maintain the cryogenic conditions; in the present experiment, the interior of the tube was filled with dry ice blocks to keep the experimental area at T= 196±1 K.

Translational motions of the samples were recorded by a high-speed camera (CASIO EX-F1), which could observe the sample motions with a spatial resolution of 0.004 cm and a time resolution of 0.033 fps. The camera was set outside the transparent glass tube.

As the capsule descends, the residual gravity caused by air resistance is expected to increase inside the capsule. Therefore, a two-fold box system was introduced in the present experiment. Specifically, the (inner) drop box is enclosed by an exterior box composed of cardboard. Immediately after the release of this two-fold box, the samples were released into the μg environment and translated in the [+x] direction by the field-gradient force.

3. Results and Discussion

For a diamagnetic particle that is translating in an area of monotone-decreasing field, the sum of the kinetic energy and the field-induced potential is conserved, and χ_{DIA} of the of the translating particle is described as

$$\chi_{DIA} = \frac{v_T^2}{B_0^2},$$

where B_0 denotes the field intensity at the initial position, and v_T denotes the terminal velocity in the area of B=0. The conservation rule was adopted in the analyses of the field-induced translation in the previous studies [3][4].

An example of a field-induced translation observed in a CO_2 grain is shown in Fig. 2 as a sequence of photographs. It is confirmed by the experiments that by using the method adopted for the proposed apparatus, the acceleration of a volatile solid released in an area of monotone-decreasing field is observable under reduced-temperature conditions. Using the two-fold capsule, the samples reached the area of B=0 with the terminal velocity that was expected from the above-mentioned energy conservation rule. When the single capsule was used in the experiment, the residual gravitation in the setup gradually increased, and the maximum velocity was only 70~80% of the expected v_T value.

As shown in Table 1, the experimental χ_{DIA} values of the three materials agreed fairly well with the published values, which indicates that the proposed method for detecting χ_{DIA} is successful at T = 196 K. The deviations of the measured χ_{DI} values from the published values are attributed to the attachment of some micro-sized magnetic contamination on the sample. The χ_{DIA} value of the translating particle is simply obtained by inserting the numerical v_T and B_0 values into the above equation without measuring m. As mentioned before, this is possible because the field-gradient force that is induced in a particle is proportional to m. Accordingly, the method has an advantage in the measurement of volatile solids in general; if the volatile grains sublimate at room temperature, the measurement of the grain mass m will require some additional treatment compared to the refractory grains. It is expected that the principle of the present system for detecting χ will be operational at lower temperatures, and it can be applied in observing the field-induced translation of other volatile grains, for example, CO, CH_4, and N_2.

To realize the non-destructive material identification using the χ_{DIA} measurement as previously proposed [4], the precision of the χ_{DIA} values should be improved to the standard level; i.e., the values should be obtained by three orders of magnitude. This can be achieved by reducing the erratic reading of the scale recorded in the images of the translation. It is also important to extend the μ g duration so that a translating sample with a small χ_{DIA} value can reach its terminal velocity. These technical improvements are now in progress for measuring the CO_2 grains as well as the solid H_2O grains.

References

[1] Suwa M and Watarai H, 2002, *Anal. Chem.* **74** 5027Another reference
[2] Takahashi K, Mogi I, Awaji S and Watanabe K, 2009, *J. Phys. Conf. Ser.* **156** 0120203
[3] Hisayoshi K, Kanou S and Uyeda C, 2009, *J.Phys.:Conf.Ser.* **156** 012021
[4] Uyeda C, Hisayoshi K and Kanou S, 2010, *J.Phys.Soc.Jpn.* **79** 064709
[5] Gputa R, 1983, *Diamagnetism, Landort Bornstein, New Seires II* (Springer Verlg Berlin) 445.

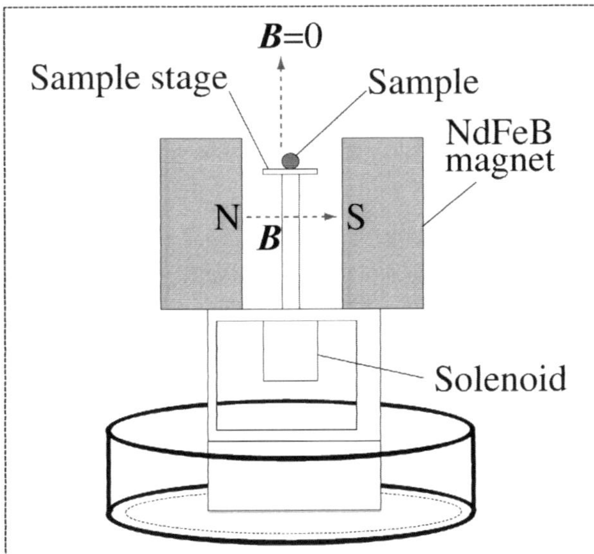

Figure 1. Schematic view of an apparatus for observing field-induced translation of a volatile particle of sub-millimetre size.

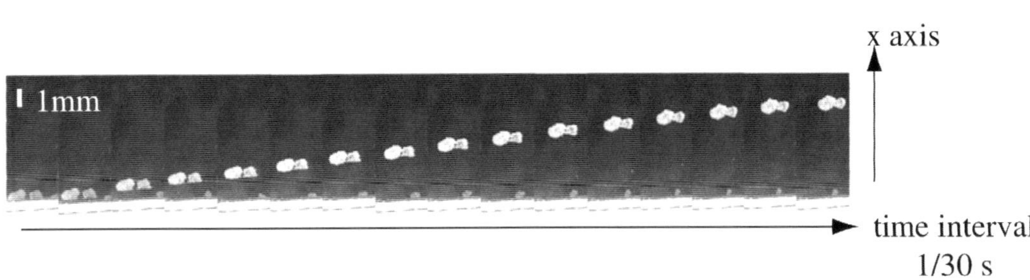

Figure 2. Visual images of translational motion of a CO_2 particle. The field decreases monotonously in the direction of the +x-axis. The images are arranged in sequence of time from left to right; the time interval between the images is 1/30 s.

Table 1. Numerical data of graphite, bismuth and CO_2 measured in the present study.

Material	mass [g]	measured χ_{DIA} [x10^{-7}emu/g]	published χ_{DIA} [x10^{-7}emu/g] [5]
Graphite	0.8x10^{-3}	-54±5	-52
Bismuth	2.7x10^{-3}	-17±1	-16
CO_2 a	4.8x10^{-3}	-4.8±1.2	-4.7
CO_2 b	2.4x10^{-3}	-5.8±1.4	-4.7

IOP Conf. Series: Journal of Physics: Conf. Series **903** (2017) 012027 doi:10.1088/1742-6596/903/1/012027

Ab-Initio Calculation of the Magnetic Properties of Metal-Doped Boron-Nitrogen Nanoribbon

J. Rufinus

Science Division, Widener University, Chester, PA 19013, USA

rufinus@cs.widener.edu

Abstract. The field of spintronics has been continuously attracting researchers. Tremendous efforts have been made in the quest to find good candidates for future spintronic devices. One particular type of material called graphene is under extensive theoretical study as a feasible component for practical applications. However, pristine graphene is diamagnetic. Thus, a lot of research has been performed to modify the graphene-based structure to achieve meaningful magnetic properties. Recently, a new type of graphene-based one-dimensional material called Boron Nitrogen nanoribbon (BNNR) has been of interest, due to the theoretical predictions that this type of material shows half-metallic property. Here we present the results of the theoretical and computational study of M-doped (M = Cr, Mn) Zigzag BNNR (ZBNNR), the objective of which is to determine whether the presence of these dopants will give rise to ferromagnetism. We have found that the concentration and the atomic distance among the dopants affect the magnetic ordering of this type of material. These results provide a meaningful theoretical prediction of M-doped ZBNNR as a basic candidate of future spintronic devices.

1. Introduction

The exploration of new materials for spintronic [1] – electronics based on electron spin rather than charge – continues to attract great interests. From the practical point of view, any candidates of spintronic materials need to have room-temperature ferromagnetism. In addition, it would also be a major advantage if these new materials already have an existing technology-base in other applications (e.g. optoelectronic). Among the many materials fitted as candidates for spintronic devices, graphene-based and other light p-block elements nanostructures have received considerable attention. This all began in 2004 with the first isolation of a truly two-dimensional structure of sp^2-bonded densely packed Carbon graphene [2-3] which has a unique electronic structure (linear band dispersion at the Fermi level). The discovery of graphene has since attracted attention due to its many important technological applications

Although the ideal graphene is non-magnetic, its nanostructures and derivative materials show promising magnetic properties. One of these materials, the quasi one-dimensional graphene nano-ribbons (GNRs), have been experimentally synthesized and characterized successfully. Due to its honeycomb structure, the edges of the GNRs may exhibit either the armchair edge or the zigzag edge. Naturally, depending on the type of edge and the combination of elements connected to them, the electronic structure and magnetic properties of these two types of GNRs are different. Thus, the rich variations of these edges will affect the physical and chemical properties of the nanostructures. It also has been shown that applied external stimuli could affect the spin-transport properties of these

Content from this work may be used under the terms of the Creative Commons Attribution 3.0 licence. Any further distribution of this work must maintain attribution to the author(s) and the title of the work, journal citation and DOI.
Published under licence by IOP Publishing Ltd

materials. Applied electric fields might generate half-metallicity in zig-zag graphene nanoribbons according to theoretical investigation [4]. Additionally, it also has been predicted that a high transition temperature might be naturally generated in materials containing sp-elements due to the high spin-wave stiffness [5-6].

In recent years, however, experimental as well as theoretical investigations of nanoribbons have shifted to other types of materials. Examples of materials include nanoribbons of Boron Nitride (BN) [7], mixed of graphene/BN [8], BNC [9], etc. Recent theoretical and experimental studies showed, e.g., edge modified Boron Nitrogen Nanoribbon (BNNR) has a wide range of magnetic, electronic, as well as optical characteristics [10-12]. While modifying the edges could change their physical properties, the present investigations of possible ferromagnetism in graphene and nanoribbon structures are also focused in defect vacancies [13] and doping of foreign atom(s) [14-15]. The doping of transition metal atoms in this type of structure is especially interesting because of the possible generation of ferromagnetism due to the unpaired d (or f) electrons of the atoms. Recently, the realization of well isolated, high spin state, transition metal atoms at the edge or divacancy of the graphene layer was demonstrated [15]. This has opened up an avenue to control the spin state in metal-doped graphene and nanoribbon structures for future spintronic devices.

In a recent paper Luo et. al. [16] presented a study of M-chain-embedded (M = Fe, Cr, Mn, Co) Zigzag BNNR (ZBNNR) with different dimers in pentagon-octagon-pentagon line defects and found some of their ground states to be ferromagnetic. They have also predicted that the hydrogen-passivated systems with C_2 dimer result in half-metallic behavior. Even though the present paper and Ref. [16] have a similar spirit, i.e. to investigate the occurrence of ferromagnetism in ZBBNR systems, they differ in several important points: Firstly, we have used different structures to substitute the atoms, i.e. an M-doped ZBNNR (this paper) vs an M-chain-embedded ZBNNR with line defects (Ref. [16]); secondly, the locations of metal dopants in our study are arbitrary vs the more centered M-chain-embedded dopants in Ref. [16]; thirdly we also studied the effect of dopant concentrations and distance on the ferromagnetism of the M-doped ZBNNR system which had not been done in Ref. [16] However, since the system we considered in this paper is small in term of the number of atoms, we are not able to compare the results of more dopant concentration levels.

2. Computational Method

First-principles total energy calculations of transition metal (Mn, Cr)-doped ZBNNR were performed using the ABINIT code [17-18]. ABINIT is a plane-wave-based pseudopotential density functional theory (DFT) code. Specifically, the generalized gradient approximation (GGA) of DFT is used in our calculations. The GGA is a widely used approximation for the exchange correlation functional in DFT and has proven to be very successful in predicting many material properties.

For our calculations, the GGA-PBE from Fritz-Haber-Institute (Troullier-Martins scheme) [19] has been used for all elements. The automatic generation of k points in the Irreducible Brillouin Zone of 5 X 1 X 1 Monkhorst-Pack grids has been performed. The cutoff energy of the plane-wave basis set was fixed at 70 Ha.

For our calculations, the supercell of ZBNNR is flat in the x-y plane. The two edges are passivated by Hydrogen atoms. In our notation, the ZBNNR consists of N atoms of Boron Nitrogen and 2 atoms of Hydrogen. For example, a 8-ZBNNR contains 8 atoms of Boron, 8 atoms of Nitrogen and 2 atoms of Hydrogen. Figure 1 shows an example of the structure of ZBNNR with $N = 5$. The dashed line in the figure represents the supercell of the nanoribbon being considered

3. Results and Discussion

We substituted (Mn, Cr) atoms to replace only the Boron atoms: $(Mn_x, Cr_x)\ B_{(1-x)}N$ with the consideration that in the limit of x = 100% we should have MnN or CrN compounds. For each case, the total energy for the ferromagnetic (FM), antiferromagnetic (AFM) and non-magnetic spin alignment were computed. We have performed calculations for $N = 8$ (i.e. a system with 8 atoms of Boron, 8 atoms of Nitrogen and 2 atoms of Hydrogen). To study the effect of metal concentration on

ZBNNR, we considered the following two cases: (i) x = 12.5% (equal to 2 dopants per 16 atoms in the supercell) and (ii) x = 25% (equal to 2 dopants per 8 atoms in the supercell). Note that since the number of atoms used in the present study is small it is not possible to study more dopant concentration levels (e.g. x = 3.125% or x = 6.25%).

Table 1 shows the results of our calculations. The table lists the relative total energy per atom for Mn and Cr dopants. The relative total energy was calculated as the difference of total energy (for each magnetic structure): $\Delta E = E_{total}$ (FM or AFM) $- E_{total}$ (non-magnetic) For both dopants the FM state is the preferred state. However, as the dopant concentration is increased from 12.5% to 25% the FM energy difference also increases.

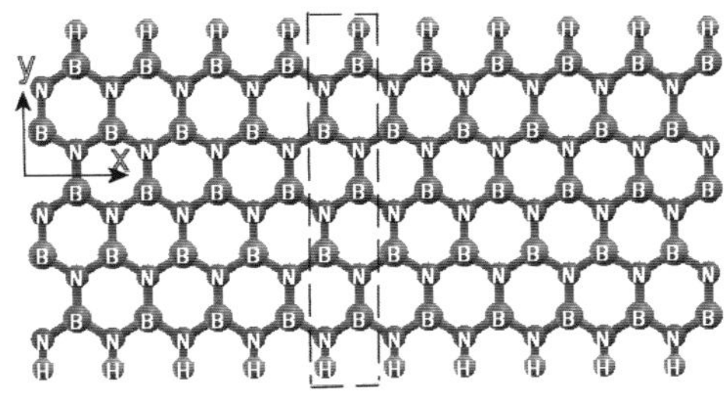

Figure 1. An example of ZBNNR structure with $N = 5$. Dashed line is the supercell.

The occurrence of ferromagnetism can be explained as follows. Assuming that metal atoms are M^{3+} when they substitute for Boron atom, Mn^{3+} and Cr^{3+} have 4 and 3 d electrons, respectively. In this case, FM state is preferred because there are plenty of donors available.

To investigate the interaction between two dopants, we vary the locations of two dopants in the supercell with 25% concentration. Increasing the distance by ~1.84 Angstrom, we found, significantly reduces the relative total energy of Mn-doped ZBNNR from 0.213 eV/atom to 10^{-7} eV/atom, while for Cr-doped BNNR the relative energy decreases from 0.242 eV/atom to 0.017 eV/atom. This result shows the source of ferromagnetism originates from the interaction among dopants located closely.

Table 1. The relative total energy per atom (in eV/atom) for (Mn or Cr)-doped ZBNNR. FM means ferromagnetic configuration and AFM means antiferromagnetic configuration. The preferred state is given in bold.

Dopant	Concentration = 12.5%	Concentration = 25%
Mn	**0.002 (FM)** ; < 10^{-5} (AFM)	**0.213 (FM)** ; < 10^{-4} (AFM)
Cr	**0.010 (FM)** ; < 10^{-5} (AFM)	**0.242 (FM)** ; < 10^{-4} (AFM)

4. Conclusion

Using the first-principles DFT within the GGA, we have calculated the magnetic properties of (Mn, Cr)-doped ZBNNR. Our calculations show that increasing the concentration of the dopants will increase the chance of obtaining ferromagnetism in this type of nanoribbon. This result provides a meaningful theoretical prediction of M-doped ZBNNR as a candidate for future spintronic devices.

5. Acknowledgement

The author is very grateful to the referees for all their recommendations and suggestions to improve the content of this paper.

References

[1] Dietl T, Ohno H, Cibert J and Ferrand D 2000 *Science* **287** 1019
[2] Novoselov K S, Geim A K, Morozov S V, Jiang D, Dubonos S V, Girgorieva I V and Firsov AA 2004 *Science* **306** 666
[3] Novoselov K S, Geim A K, Morozov S V, Jiang D, Katsnelson M I, Girgorieva I V, Dubonos S V and Firsov AA 2005 *Nature* **438** 197
[4] Son Y W, Cohen M L and Louie S G 2006 *Nature* **444** 347
[5] Edwards D M and Katsnelson 2008 *J. Phys.: Condens. Matter* **18** 7209
[6] Yazyev O V and Katsnelson 2008 *Phys. Rev. Lett.* **100** 047209
[7] Lou P and Lee J Y 2009 *J. Phys. Chem* C **113** 12637
[8] Barone V and Peralta J E 2008 *Nano Lett.* **8** 2210
[9] Liu Y, Wu X, Zhao Y, Zeng X C and Yang J 2011 *J. Phys. Chem.* C **115** 9442
[10] Topsakal M, Akturk E and Ciraci S 2009 *Phys. Rev.* B **79** 115442
[11] Basheer E A, Parida P and Pati S K 2011 *New J. Phys.* **13** 053008
[12] Erickson K J, Gibb A L, Sinitskii A, Rousseas M, Alem N, Tour J M and Zettl A 2011 *Nano Lett.* **11** 3221
[13] Yazyev O V and Helm L 2007 *Phys. Rev.* B **75** 125408
[14] Xu J, Zhang W, Wei C, Huang J, Mao Z and Yu G 2016 *RSC Adv.* **6** 10017
[15] Lin Y C, Teng P Y, Chiu P W and Suenaga K 2015 *Phys. Rev. Lett.* **115** 206803
[16] Luo K-W, Xu L, Wang L-L, Li Q, Huang W-Q, Huang G-F and Li X-F 2015 *Physica E* **74** 431
[17] Gonze X, Beuken J M, Caracas R, Deutraux F, Fuchs M, Rignanese G M, Sindic L, Verstraete M, Zerah G, Jollet F, Torrent M, Roy A, Mikami M, Ghosez Ph., Raty J Y and Allan D C 2002 *Comput. Mater. Sci.* **25** 478
[18] Gonze X, Rignanese G M, Verstraete M, Beuken J M, Pouillon Y, Caracas R, Jollet F, Torrent M, Zerah G, Mikami M, Ghosez Ph, Veithen M, Raty J Y, Olevano V, Bruneval F, Reining L, Godby R, Onida G, Hamann D R and Allan C 2005 *Z. Kristallogr.* **220** 558
[19] Troullier N and Martins J L, *Phys. Rev. B* **43**, 1993 (1991).

IOP Conf. Series: Journal of Physics: Conf. Series **903** (2017) 012028 doi:10.1088/1742-6596/903/1/012028

Spin density wave behaviour in the $(Cr_{98.4}Al_{1.6})_{100-y}Mo_y$ and $(Cr_{100-x}Al_x)_{95}Mo_5$ alloy series

B Muchono[1,2], ARE Prinsloo[1], CJ Sheppard[1], AM Venter[3] and KD Liss[4]
[1]Physics Department, UJ, P.O. Box 524, Auckland Park 2006, Johannesburg, RSA
[2]Applied Physics Department, NUST, Box AC939, Ascot, Bulawayo, Zimbabwe
[3]Research and Development Division, Necsa SOC Ltd., P.O. Box 582, Pretoria 0001, RSA
[4]Bragg Institute, Ansto, Locked Bag 2001, Kirrawee DC NSW 2232, Australia

E-mail address: alettap@uj.ac.za

Abstract. Neutron diffraction studies of the spin-density-wave (SDW) behaviour in antiferromagnetic $(Cr_{98.4}Al_{1.6})_{100-y}Mo_y$ with y = 0, 3.7 and 4.2, as well as $(Cr_{100-x}Al_x)_{95}Mo_5$ with x = 0 and 4.6 alloy systems are reported. The results confirm an incommensurate (I) SDW to paramagnetic (P) phase transition at the quantum critical point of the $(Cr_{98.4}Al_{1.6})_{100-y}Mo_y$ series. In addition two further quantum critical magnetic phase transitions are proposed for the $(Cr_{100-x}Al_x)_{95}Mo_5$ series, one from ISDW to P and the other from P to commensurate (C) SDW.

1. Introduction

The combination of properties of Cr and its dilute alloys associated with the Fermi surface, magnetic structure and spin-density-wave (SDW) state provides for an ideal case-study of itinerant electron antiferromagnetism (AFM) in metals [1]. Alloying Cr with Al, a group 3 non-transition metal element, decreases the Néel temperature (T_N) reaching a minimum near 2 at.% Al, where after T_N increases on further Al addition [1, 2, 3]. This minimum occurs at a critical concentration where the incommensurate (I) SDW, commensurate (C) SDW and paramagnetic (P) phases co-exist [2, 3] on the magnetic phase diagram (MPD). Recent research suggests that this triple point might indeed be quantum critical (QC) in nature [3].

Addition of 5 at.% Mo to the $Cr_{100-x}Al_x$ alloy system completely suppresses AFM in the concentration range $2 \leq x \leq 5$, after which it reappears on further Al addition [1, 4]. This results in two possible quantum critical points (QCP) on the MPD of the $(Cr_{100-x}Al_x)_{95}Mo_5$ system, one expected to be an ISDW-P and the other a P-CSDW phase transition [5]. On the other hand, the $(Cr_{98.4}Al_{1.6})_{100-x}Mo_x$ series indicate only one QCP, expected to be an ISDW-P transition [6, 7].

In the present study, neutron diffraction measurements have been performed to confirm the nature of the magnetic phases around the QCPs predicted from previous studies of $(Cr_{100-x}Al_x)_{95}Mo_5$ [1, 4, 5] and $(Cr_{98.4}Al_{1.6})_{100-y}Mo_y$ [6, 7] alloys.

2. Experimental

Two sets of polycrystalline ternary $(Cr_{98.4}Al_{1.6})_{100-y}Mo_y$ with y = 0, 3.7 and 4.2, and $(Cr_{100-x}Al_x)_{95}Mo_5$ with x = 0 and 4.6 alloys were prepared as was previously described [5, 6, 7]. Analyses confirmed that the samples were homogeneous, with actual concentrations of (1.6 ± 0.1) at.% Al in the $(Cr_{98.4}Al_{1.6})_{100-y}Mo_y$, and (5.0 ± 0.2) at.% Mo in the $(Cr_{100-x}Al_x)_{95.0}Mo_{5.0}$ series, respectively. Temperature dependence

Content from this work may be used under the terms of the Creative Commons Attribution 3.0 licence. Any further distribution of this work must maintain attribution to the author(s) and the title of the work, journal citation and DOI.
Published under licence by IOP Publishing Ltd

of the electrical resistivity (ρ) and magnetic susceptibility (χ) measurements gave T_N values corresponding with that previously obtained [5, 6, 7]. Investigations were extended to neutron diffraction measurements at the Australian Nuclear Science and Technology Organisation (ANSTO) using the Wombat powder diffractometer [8].

3. Results and discussion

Examples of neutron diffraction patterns for the $(Cr_{98.4}Al_{1.6})_{100-y}Mo_y$ and $(Cr_{100-x}Al_x)_{95}Mo_5$ alloys taken at the (100) reciprocal lattice positions (rlp) are shown in figure 1. The presence of the diffraction peaks are purely of magnetic origin [1]. Diffraction patterns were obtained at various set temperatures on cooling from the P phase of each alloy to 4 K. A monochromatic wavelength of 2.41 Å was used for the sample with $y = 0$, with 1.54 Å for the samples with $x = 0$ and 4.6. Magnetic satellites at the $(1\pm\delta,0,0)$ rlp for the alloys with $y = 0$ and $x = 0$ indicate that these alloys order in the ISDW phase [1, 2]. In this phase the magnetic moments are perpendicular to the wave vector Q directed along the (100) cubic axis, given by [1, 2]: $Q = \frac{2\pi}{a}(1 \pm \delta, 0,0)$, where a is the lattice parameter and δ is a parameter that measures the deviation from a CSDW state. Figure 1(c) shows a weak single diffraction peak at the (100) rlp for the $x = 4.6$ alloy. The absence of magnetic satellites indicates that this alloy orders in the CSDW phase. i.e. $\delta = 0$. Alloys with $y = 3.7$ and 4.2 did not show magnetic reflections, despite the presence of a low temperature minimum in the $\rho(T)$ curves. These concentrations are close to the expected QCP [6], which leads to weak intensities that were possibly indistinguishable above the background.

Figure 2 shows representative graphs of the temperature dependences of the integrated neutron intensities. Results of the solid rod and powdered samples for the alloy with $y = 0$ are shown in figure 2(a), with figure 2(b) showing that of the solid rod sample with $x = 0$. Typically, the integrated neutron intensities decrease with temperature, levelling-off in the P region. T_N was subsequently taken at the temperature where a quadratic fit to the data points in the AFM phase intersects with a linear fit in the P phase. For the $x = 0$ sample the T_N-value corresponds well with that obtained from $\rho(T)$ and $\chi(T)$ measurements, despite the limited number of points in the P phase. T_N-values obtained for the powdered and solid rod samples for the $y = 0$ alloy are similar. Surprisingly it indicates that crushing the sample didn't affect T_N significantly, contrary to results previously reported for pure Cr [9, 10].

Investigations of the alloy with $x = 4.6$ at several set temperatures revealed the presence of a central peak of magnetic origin at 4 K and 25 K, that was absent for $T \geq 50$ K. With the limited number of data points in the AFM phase, the temperature dependence of the integrated intensity curve could not be plotted. It should be noted that the neutron intensities in the CSDW are weaker than in the ISDW phase, as was also observed in other experimental results on these alloys [3, 5].

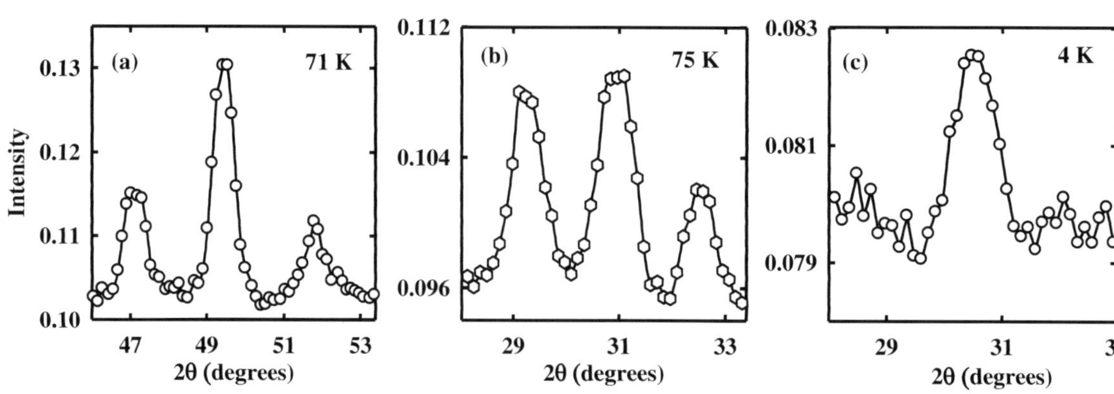

Figure 1. Neutron diffraction patterns for the antiferromagnetic $(Cr_{98.4}Al_{1.6})_{100-y}Mo_y$ and $(Cr_{100-x}Al_x)_{95}Mo_5$ alloys with (a) $y = 0$ (rod), (b) $x = 0$ (powder) and (c) $x = 4.6$ (powder). The temperatures at which the patterns were taken are indicated in each panel.

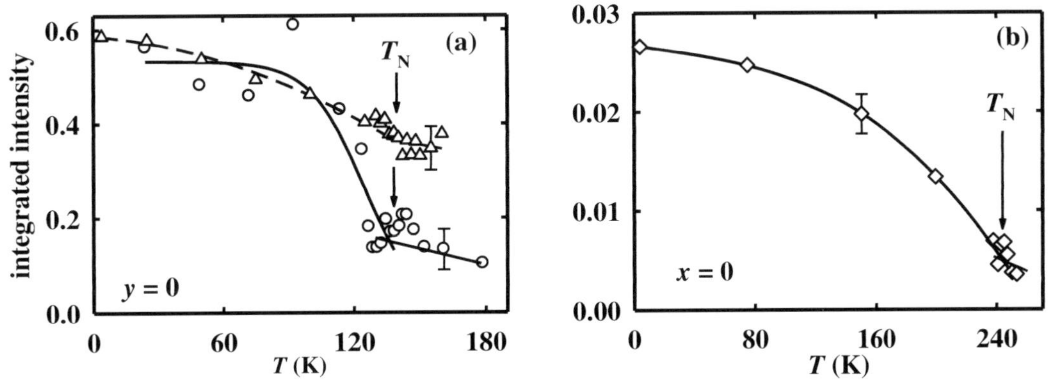

Figure 2. Temperature dependence of the integrated neutron intensities of (a) powdered (△) and solid rod (○) samples of the $(Cr_{98.4}Al_{1.6})_{100-y}Mo_y$ alloys with $y = 0$ and (b) solid rod sample of the $(Cr_{100-x}Al_x)_{95}Mo_5$ alloys with $x = 0$. The positions of Néel temperatures, T_N, are shown by the arrows.

4. Conclusions

T_N-values obtained from $\rho(T)$, $\chi(T)$ and neutron diffraction measurements for each of the samples with $y = 0$ and $x = 0$ are comparable. No additional intensity anomalies that can be attributed to magnetic changes were observed in the neutron diffraction results of the samples with $y = 3.7$ and 4.2 despite low temperature minima observed in $\rho(T)$ curves. T_N for the alloy with $x = 4.6$ could not be determined from the neutron diffraction measurements since measurements were only taken at two temperature set points in the AFM phase. Previous [5] electrical transport and specific heat measurement indicated that the AFM behaviour on the MPD of the $(Cr_{100-x}Al_x)_{95}Mo_5$ alloy system is fully suppressed down to 2 K in the concentration range $2 < x < 4.4$, after which it then reappears at larger x. The present neutron diffraction data indicates that an alloy with $x = 4.6$ orders in the CSDW phase for $T \leq 25$ K. The concentration of this alloy lies close to the second critical concentration $x = 4.4$ [5] where AFM reappears on the MPD of this system. This is a first clear indication that AFM in this alloy system reappears in the CSDW phase. The results furthermore confirm an ISDW-P phase transition, previously inferred from $\rho(T)$ and $\chi(T)$, at the QCP of the $(Cr_{98.4}Al_{1.6})_{100-y}Mo_y$ series [6]. Integrated intensities as a function of temperature were measured for the $y = 0$ alloy, both in powder and solid state form, to obtain T_N for this alloy. Contrary to expectation, crushing of the $y = 0$ sample was not found to influence T_N significantly.

Acknowledgements

Financial support from UJ and the SA NRF (Grant numbers: 80928, 93551 and 88080). Use of UJ Spectrum, as well as access to the Wombat instrument under proposal P1816 is acknowledged.

References

[1] Fawcett E, Alberts H L, Galkin V Y, Noakes D R and Yakhmi J V 1994 *Rev. Mod. Phys.* **66** 25
[2] Baran A, Alberts H L, Strydom A M and Du Plessis P de V 1992 *Phys. Rev.* B **45** 10473
[3] Sheppard C J, Prinsloo A R E, Alberts H L, Muchono B and Strydom A M 2014 *J. Alloys and Compounds* **595** 164
[4] Smit P and Alberts H L 1986 *J. Phys. F* **16** L191
[5] Muchono B, Prinsloo A R E, Sheppard C J, Alberts H L and Strydom A M *Proc. SAIP2011* 220
[6] Muchono B, Sheppard C J, Prinsloo A R E, Alberts H L and Strydom A M 2014 *J. Magn. Magn. Mater.* **354** 222
[7] Muchono B, Sheppard C J, Prinsloo A R E and Alberts H L *Proc. SAIP2014* 85
[8] Studer A J, Hagen M E, Noakes T J 2006 *Physica B* **385-386** 1013
[9] Shull C G and Wilkinson M K 1953 *Rev. Mod. Phys.* **25** 100
[10] Bacon G E and Cowland N 1969 *J. Phys. C. Ser 2.* **2** 238

IOP Conf. Series: Journal of Physics: Conf. Series **903** (2017) 012029 doi:10.1088/1742-6596/903/1/012029

Bulk and local properties of intermetallic GdAgSn compound

K Łątka[1], **A W Pacyna**[2]

[1] M. Smoluchowski Institute of Physics, Jagiellonian University, Prof. Łojasiewicza 11, 30-348 Kraków, Poland
[2] Henryk Niewodniczański Institute of Nuclear Physics, Polish Academy of Sciences, Radzikowskiego 152, 31-342 Kraków, Poland

uflatka@cyf-kr.edu.pl

Abstract. The results of magnetic susceptibility as well as electric resistivity measurements complemented with ^{155}Gd Mössbauer spectroscopy investigations carried out within wide temperature range for GdAgSn compound are discussed.

1. Introduction
The ternary equiatomic stannides (R, A)TSn (R or A = rare earth or actinide element, respectively; T = $3d$, $4d$ or $5d$ transition element) have intensively been studied with respect to their largely varying magnetic and electrical properties [1-3]. Among them the title compound reveals interesting magnetic behaviour as it was shown by the AC susceptibility and Mössbauer measurements at low temperatures [4, 5]. It has been found that on cooling down this compound orders first anitiferromagnetically at T_N=34.3(2) K, while at T_f = 15.3 K undergoes spin glass-like transition (SG).

2. Experimental
Our polycrystalline GdAgSn sample was synthesized by arc melting the constituent elements with purity better than 99.9% in argon atmosphere and consecutive annealing procedure for better homogeneity as previously described [4]. The X-ray powder diffraction pattern was recorded at room temperature with a Siemens D-501 diffractometer using the Ni-filtered CuK$_\alpha$ and the results has already been published elsewhere [4, 5]. Magnetic susceptibility and magnetization measurements were carried out with a 7225 Lake Shore AC Susceptometer/DC Magnetometer operating in AC mode on powdered samples in the temperature range 4.2 K – 200 K and in the external magnetic fields up to 60 kOe. The electrical resistivity (ρ) measurements were performed by means of Physical Property Measurement System (Quantum Design, PPMS-9T) on a bulk probe using a steady-current, standard four point technique and applying an excitation current of 80 mA in zero external magnetic field. Our specimen had of approximate dimensions 1x1x5 mm^3. (86.5 keV (I_g = 3/2, $E1$, I_e = 5/2 transition) ^{155}Gd Mössbauer spectra were collected in a transmission geometry cryostat using a conventional constant-acceleration spectrometer of the Kankeleit type. The 20 mCi ^{155}Eu:SmPd$_3$ source was kept at 4.2 K to increase the efficiency of the resonance emission. The temperature of the absorber was varied between 4.2 and 50 K with a stabilisation better than 0.05 K. In contrast to the fitting method applied in our previous investigations (where Lorentzian approximation was used [5]), for the proper description of the recorded ^{155}Gd Mössbauer spectra a least-squares fitting procedure which includes the diagonalization of the full hyperfine Hamiltonian within the transmission integral approximation was used.

Content from this work may be used under the terms of the Creative Commons Attribution 3.0 licence. Any further distribution of this work must maintain attribution to the author(s) and the title of the work, journal citation and DOI.
Published under licence by IOP Publishing Ltd

IOP Conf. Series: Journal of Physics: Conf. Series **903** (2017) 012029 doi:10.1088/1742-6596/903/1/012029

3. Results and discussion

X-ray diffraction investigations made on the polycrystalline GdAgSn sample revealed the existence of only single phase and confirmed that this compound crystallizes in the hexagonal $CaIn_2$-type of structure with space group $P6_3/mmc$ and the derived lattice parameters are following: a = 470.9(1) pm and c = 742.5(3) pm [2]. The temperature dependence for the absolute value of the magnetic AC susceptibility $\chi_{AC} = [(\chi')^2 + (\chi'')^2]^{1/2}$ and its reciprocal are displayed in figure 1.

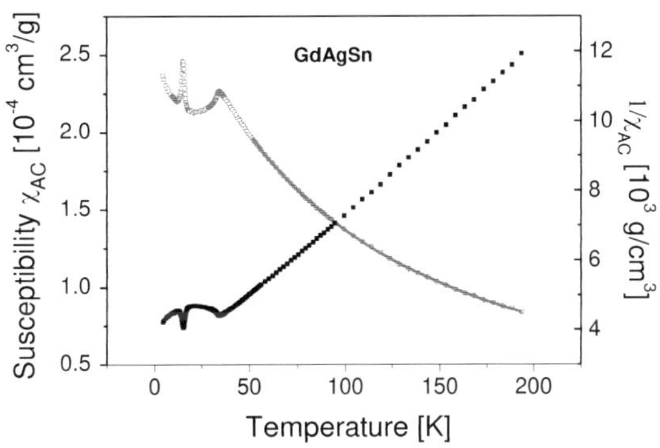

Figure 1. Temperature dependence for the absolute value of the magnetic ac susceptibility $\chi_{AC} = [(\chi')^2 + (\chi'')^2]^{1/2}$ (left-hand scale) and its reciprocal (right-hand scale) recorded for GdAgSn in zero external magnetic field at the internal frequency f = 375 Hz and the applied oscillating field strength H_{AC} = 5 Oe. The solid line is a result of the fitting procedure in the paramagnetic range, for details see the text.

The temperatures observed here for both maxima are in perfect agreement with those previously published [4, 5] i.e. a phase transition from a paramagnetic to an antiferromagnetic state (PM→AFM) takes place at the Néel temperature T_N = 34.3(2) K while the second one at T_f = 15.3(2) is that which have been ascribed to the spin-glass transition. Above 52 K (see Fig. 1), the recorded susceptibility obeys fairly well a modified Curie-Weiss law in the form: $\chi_{AC} = \chi_0 + C/(T - \theta_p)$ with the temperature independent factor $\chi_0 = -1.21\cdot10^{-6}$ cm^3/g, the Curie constant C = $2.092\cdot10^{-2}$ Kcm^3/g, and the paramagnetic Curie temperature θ_p = −53.7 K. The negative value of θ_p is indicative for antiferromagnetic correlations. The effective magnetic moment was derived from the formula $\mu_{eff} = 2.83(MC)^{1/2}$ where M is the molar mass. The experimental value μ_{eff} = 8.01 μ_B is somewhat higher but agrees well with the theoretical Russel-Saunders value $\mu_{eff} = g\mu_B[J(J+1)]^{1/2}$ = 7.94 μ_B for the free Gd^{3+} ion. Figure 2 shows the temperature variation of electrical resistivity recorded at zero external magnetic field.

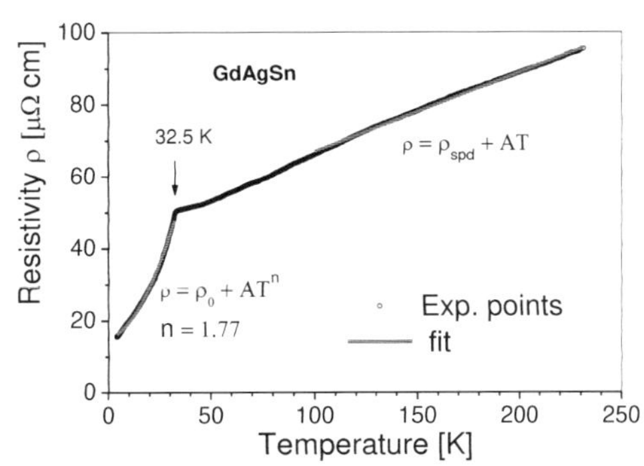

Figure 2. Temperature dependence of electrical resistivity recorded for GdAgSn. The continuous lines present the least-squares fits to the low and high temperature parts of $\rho(T)$ according to the given formulas below each part, respectively.

The linear behavior at high temperatures and the positive temperature coefficient of resistivity ($d\rho/dT>0$) indicate the metallic nature of the compound. The striking feature is a lack of any resistivity anomaly around T_f =15.3 K while a drastic change of resistivity slope is registered at T = 32.5 K, i.e. close to T_N = 34.3. It has to be underlined here, that such an observation is in line with the proposed spin glass-like character of GdAgSn below T_f and its ordered antiferromagnetic character below T_N [4,5]. As a matter of fact, the transport properties do not reveal any transition at spin glass temperature but only a gradual transformation which is associated with the lack of ordered state (see [6]). Below T_N we have found that the best fit to the experimental data in an expanded range (see Fig. 2.) may be approximated by the simple power low in the form: $\rho(T) = \rho_0 + AT^n$ with the ρ_0 = 15.44 μΩcm, A = 0.069 μΩcm/K$^{1.77}$ and n = 1.77. At high temperatures ($T>T_N$) the resistivity $\rho_{mag}(T) \equiv \rho_{spd}(T)$ = constant, where ρ_{spd} is the spin-disorder contribution and the phonon contribution $\rho_{ph}(T)$ is fairly well linear function of T [7]. Thus, in this temperature region the formula: $\rho(T) = \rho_{spd} + AT$ can be fitted. The estimated in such a way values of ρ_{spd} and A are 45.27 μΩcm, and 0.218 μΩcm/K, respectively.
Low temperature ^{155}Gd Mössbauer spectra obtained around and below T_N for GdAgSn can be fitted with the single magnetic hyperfine component rather well, and no drastic change can be registered in the shape of these spectra below T_f in spite of expected field distribution characteristic for spin glass state. The obtained quadrupole interaction constant $\Delta E_Q = eQ_gV_{zz}$ = –1.592(2) mm/s can be easily converted into value of the electric field gradient (EFG) V_{zz} at the gadolinium nuclei by the formula: $V_{zz} = 2.2206\cdot10^{21}\,\Delta E_Q$ [mm/s] V/m^2 = –3.535(4)x10^{21} V/m^2. Furthermore, knowledge of the EFG at the Gd site allows an estimation of the quadrupolar term B_2^0 in the Stevens expansion of the crystal field Hamiltonian: $\sum B_n^m\, O_n^m$. This term is directly related to $\Delta E_Q(^{155}\text{Gd})$ by the expression:

$$B_2^0\,[\text{K}] = -\,\alpha_J * \langle r^2\rangle_{4f} * 90.2 * \Delta E_Q\,(^{155}\text{Gd})$$

Here, α_J is the appropriate Stevens factor, the mean squared radius $\langle r^2\rangle_{4f}$ is expressed in atomic units and ΔE_Q in mm/s. Under assumption that B_2^0 is the leading term in the crystal field Hamiltonian, then when translated for other rare earth ions in the family of isostructural RAgSn compounds, where the rare earths metals have non-zero angular momentums its sign gives information about the magnetocrystalline anisotropy, since then H_2^0 (CEF) = B_2^0 $[3J_z^2 - J(J+1)]$. For a negative B_2^0 the ground state doublet is built up of states with a maximum absolute value of J_z which means that the magnetic moment is pointed in the direction of the z-quantization axis. In contrary, for positive values of B_2^0 only states with the possible smallest $|J_z|$ contribute to the electronic ground state and the magnetic moment is aligned in the basal plane ($\perp z$).

Conclusions

All carried measurements confirm very well the transition observed at T_N, however, the transition at T_f has no noticeably influence on the shape of ^{155}Gd spectra what could be explained by an assumption that the local nearest-neghbour Gd^{3+} spin configuration below T_f is not significantly modified.

Acknowledgments

The research was carried out with the PPMS-9T (Quantum Design) and the Mössbauer equipments purchased thanks to the financial support of the European Regional Development Fund in the framework of the Polish Innovation Economy Operational Program (contract no. POIG.02.01.00-12-023/08).

References

[1] Pöttgen R and Łątka K 2010 *Z. Anorg. Allg. Chem.* **636** 2244
[2] Kruk R, Kmieć R, Łątka K, Tomala K, Troć R and Tran VH 1997 *Phys. Rev.* B **55** 5851
[3] Łątka K, Kmieć R, Gurgul J, Pacyna A.W., Rams M., Schmidt T and Pöttgen R 2006 *J. Magn. Magn. Mater.* ***301*** 359

[4] Łątka K, Görlich E A, Chajec W, Kmieć R and Pacyna AW 1997 *J. Alloys Comp.* **262-263** 108
[5] Görlich E A, Łątka K, Kmieć R and Warkocki W 1998 *Molecular Physics Reports* **22** 35
[6] Mydosh JA 1993 *Spin glasses: an experimental introduction* (London,Washington, DC: Taylor&Francis)
[7] Gratz E and Zuckerman M J 1982 *Handbook on the Physics and Chemistry of Rare Earth* vol 5, ed K A Gschneidner Jr and LeRoy Eyring (Amsterdam, New York, Oxford: North-Holland Publishing Company) p. 117

IOP Conf. Series: Journal of Physics: Conf. Series **903** (2017) 012030 doi:10.1088/1742-6596/903/1/012030

An *ab initio* study of the effects of vacancies on the static and dynamic magnetic properties of Co_2MnSi

B. Pradines, R. Arras, L. Calmels

CEMES, Université de Toulouse, CNRS, UPS, 29 rue Jeane Marvig, F-31055, Toulouse, France

E-mail: barthelemy.pradines@cemes.fr

Abstract.

The full-Heusler alloy Co_2MnSi is a promising highly spin-polarized magnetic metal for spintronic applications. However, significant differences have been reported between the computed properties of the ideal material and the properties of real samples measured in experiments. In this paper, we study the influence of atom vacancies on the electronic structure and on the magnetic properties of Co_2MnSi, as these defects could explain the disagreement between the expected and measured behavior of this alloy. The effects of atom vacancies have been calculated from first principles, using the fully relativistic Korringa-Kohn-Rostoker (KKR) method in conjunction with the coherent potential approximation (CPA) and the linear response formalism.

1. Introduction

The quest of magnetic materials presenting a half metallic behaviour at high temperatures has intensified during the last 30 years. Defined by the presence of an energy gap in the band structure of only one of the two spin states, this interesting property is sought by the spintronics community, since it would maximize the efficiency of spintronic devices for many applications. With their 100% spin-polarization at the Fermi level, half- metals would for instance provide efficient magnetic electrodes for magnetic tunnel junctions or spin-torque oscillators.

Among all the compounds which have been theoretically predicted as being half-metallic, the full-Heusler alloy Co_2MnSi [1, 2] has managed to get noticed because of its high Curie temperature of 985 K [3] and its Gilbert damping parameter α, lower than 1×10^{-3} [4]. However, the theoretically predicted 100% spin-polarization at E_F of Co_2MnSi has not been yet experimentally confirmed. A value of 93% has been recently measured [5], but most of the previous works announced spin polarization values around 50-60%, at low or room temperature [6, 7, 8, 9]. Furthermore, the values of the Gilbert damping are generally more than 10 times higher than the computed results [10].

Several explanations have been suggested to clarify these qualitative differences between theory and experiment. After confirmation of the presence of atomic disorder in some experimental samples by neutron diffraction experiments [11, 12], chemical disorder became a popular hypothesis to explain the loss of half-metallicity and the high value of α. This hypothesis has further been widely studied in the literature [13, 14, 15]. Another hypothesis advanced by the experimentalist is the presence of vacancies in this material. This latter thesis is well less considered in literature. Özdogan *et. al* studied numerically the effect of vacancies on each of

Content from this work may be used under the terms of the Creative Commons Attribution 3.0 licence. Any further distribution of this work must maintain attribution to the author(s) and the title of the work, journal citation and DOI.
Published under licence by IOP Publishing Ltd

the atomic sites of Co_2MnSi [16], and concludes that the half-metallic character of the alloy loses its robustness (large reduction of the minority spin band-gap) when there are vacancies on the Co atomic sites. But numerical studies on the effects of atom vacancies on α are still missing.

For a better understanding of the modification of the physical properties of Co_2MnSi induced by atom vacancies, we have calculated the effect of an increasing density of vacancies, and we only consider here the case of stoichiometric Co_2MnSi alloys, in which vacancies are equitably distributed over all the atomic sites of the alloy. We focus on the effects of vacancies on the electronic structure, the magnetic properties and the Gilbert damping parameter.

2. Calculation details

Our results were obtained using the Korringa-Kohn-Rostoker (KKR) formalism implemented in the spin-polarized relativistic KKR (SPRKKR) code [17]. This program uses the multiple scattering theory to calculate the one-electron Green function of crystals including possible defects via the coherent potential approximation (CPA) [18]. This latter approximation plays a major role in our calculations because it allows to handle the presence of defects like chemical disorder or, in our case, the presence of vacancies with adjustable concentrations on selected atomic sites. Here, we chose to only consider cases where the vacancy concentration is the same for all the atomic sites and where chemical disorder between atomic sites is neglected. A parameter-free expression for α was recently derived by Ebert *et. al.* from the linear response formalism [19]. This technique, contrary to previous methods for calculating α [20, 21], is fully *ab initio* and does not involve any phenomenological parameter accounting for the finite lifetime of the electron states.

The crystal structure of Co_2MnSi is described by the space group $Fm\bar{3}m$ for the perfectly ordered $L2_1$ phase. In this phase, the Co (X site), Mn (Y site) and Si (Z site) atoms occupy respectively the 8c, 4a and 4b positions of Wyckoff. To account for the effects of vacancies on each atomic site, the following chemical formula was used: $([Co_{1-x}]_X)_2[Mn_{1-x}]_Y[Si_{1-x}]_Z$ where $x \leq 5\%$ is the percentage of vacancies by atomic site. The results presented in the next section were calculated with the local spin-density approximation (LSDA), with the exchange-correlation potential of Vosko, Wilk and Nusair [22] in fully relativistic mode at 0 K. We sampled, after convergence tests, the first Brillouin zone with 3000 k vectors and set up the selected l-cut-off l_{max} of the KKR formalism to 4. A volume optimization was then performed for the perfect $L2_1$ phase and for the phase with 5% of vacancies per atomic site, the minimum value of the ground state energy was found for $a_0^{DFT} = 5.48$ Åin both case and will be used throughout this paper.

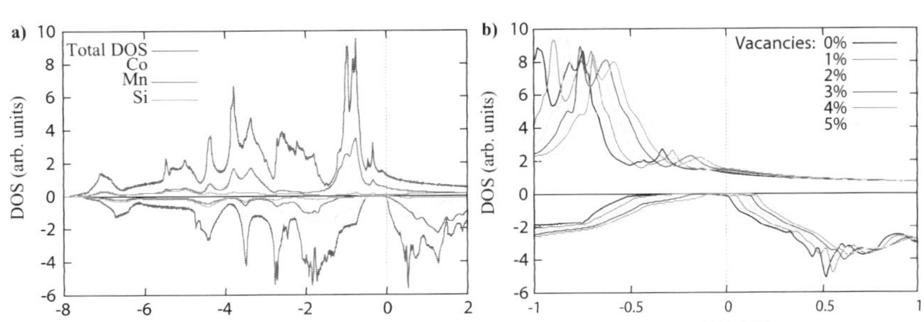

Figure 1. Spin-resolved DOS of Co_2MnSi, a) for the perfect $L2_1$ phase (with partial contributions of Co, Mn and Si atoms), b) near the Fermi level E_F, as a function of the percentage of vacancies x. Upper and lower panels respectively correspond to majority and minority spin electrons.

3. Results

3.1. Influence of vacancies on the static magnetic properties

The density of states (DOS) of the perfectly ordered $L2_1$ phase of Co_2MnSi is plotted on Fig. 1a. We found a ferromagnetic half-metallic behaviour, in agreement with previous results from the literature, with a total (spin+orbital) magnetic moment M_{tot} of 4.94 μ_B per unit cell. In Fig. 1b, the DOS is plotted at energies near the Fermi level, as a function of the atom vacancy concentration in the atomic sites. Vacancies may induce impurity states at the low energy side of the minority spin band-gap. The half-metallic character of Co_2MnSi is however preserved, whatever the percentage of atom vacancies we considered, but the minority spin band gap is considerably narrowed by vacancy states or by the shift of electronic states, in particular for the highest contents of these defects: this band gap decreases from 0.43 eV for the perfect crystal to 0.17 eV with 5% of vacancies on each atomic site. In Fig. 2a we can see that the total magnetic moment of the unit cell decreases linearly from about 5 μ_B without vacancies to 4.1 μ_B with 5% of vacancies per atomic site. It should be noted that the value of M_{tot} decreases because magnetic atoms are remove from the crystal, but also because the magnetic atoms which are still present in the crystal have a smaller magnetic moment than in perfect Co_2MnSi, due to the vicinity of vacancies: the magnetic moment decreases from 2.89 μ_B to 2.76 μ_B for Mn atoms and from 1.05 μ_B to 0.80 μ_B for Co atoms, when the vacancy concentration on each atomic site goes from 0 to 5%. This reduction of the magnetic moments is due to a reorganisation of the electron density near vacancies, which is different for majority and minority spins.

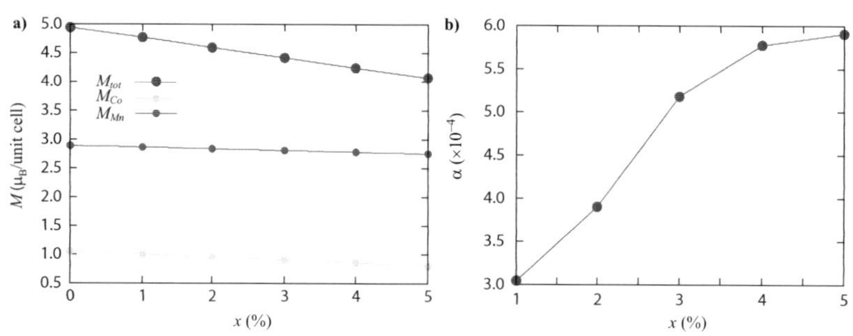

Figure 2. a) Total magnetic moment M_{tot}/unit cell of Co_2MnSi as a function of the percentage of vacancies x, b): α of Co_2MnSi as a function of the percentage of vacancies x.

3.2. Influence of vacancies on the Gilbert damping parameter

The Gilbert damping parameter α is plotted on Fig. 2b as a function of the percentage of atom vacancies x. It is worth noting that since the linear response method cannot calculate α for a totally ordered alloy at 0 K, the Gilbert damping parameter of the perfect $L2_1$ phase ($x = 0$) is not reported here. The Gilbert damping parameter α increases almost linearly from 3×10^{-4} to 6×10^{-4} when x increases from 1 to 5%. The variation of α can often be connected to changes in the electronic structure of the crystal. The total DOS at the Fermi level Z_F is usually considered as a measurement of the number of channels available for magnetic relaxation. However, below 5% of atom vacancies in Co_2MnSi, there are no major changes in the DOS at the Fermi level, and we must assign the modifications of alpha induced by atom vacancies to another cause. Kambersky proposed, in order to quantitatively rationalize α, the two following contributions [20]: $\alpha_{sf} = \frac{\pi\gamma\hbar}{\mu_0 M} Z_F \frac{(g-2)^2}{\tau}$ and $\alpha_o = \frac{\pi\gamma}{\mu_0 M} Z_F \lambda_{SO}(g-2)^2\tau$, where α_{sf} is the spin-flip scattering contributions and α_o describes the ordinary scattering. In these equations, γ is the gyromagnetic ratio, M the saturation magnetization, λ_{SO} the spin-orbit coupling parameter, g

the Landé factor and τ the electron scattering time. Relying on this model we can first conclude that the decrease of the total magnetic moment M_{tot} with vacancies (which, in this case is totally proportional to the saturation magnetization since we did not find changes of the unit cell volume induced by vacancies), contribute to an increase of the total magnetic damping α. We can also postulate that atom vacancies are responsible for an increase of the total damping via ordinary and spin-flip scattering, that can be attributed to a lowering of the electron scattering time and/or to an increase of the spin-orbit coupling parameter.

4. Conclusions

We have shown that the presence of vacancies in the $L2_1$ phase of Co_2MnSi causes a reduction of the total magnetic moment M_{tot} and a reduction of the minority spin band-gap that can be directly related to the increase of the Gilbert damping parameter. This first-principles study of the modifications of the properties of this alloy induced by atoms vacancies could help experimentalists for the interpretation of the physical properties measured on real (imperfect) samples.

Acknowledgments

This work was granted access to the HPC resources of CALMIP supercomputing center under the allocation p1252 (2014-2016).

References

[1] Heusler F 1903 *Verhandlungen der Deutschen Physikalischen Gesellschaft* **5** 219
[2] Fujii S, Sugimura S, Ishida and Asano S 1990 *Journal of Physics: Condensed Matter* **2** 8583
[3] Webster P J 1971 *Journal of Physics and Chemistry of Solids* **32** 1221–1231
[4] Shi-Zhu Q, Jie Z, Yu-Feng Q, Run-Run H, Hai Z, Da-Peng Z, Yun K, Shi-Shou K, Yu Shu-Yun, Guang-Bing H, Shi-Shen Y and Liang-Mo M 2015 *Chinese Physics Letters* **32** 057601
[5] Jourdan M, Minr J, Braun J, Kronenberg A, Chadov S, Balke B, Gloskovskii A, Kolbe M, Elmers H J, Schnhense G, Ebert H, Felser C and Klui M 2014 *Nature Communications* **5** 3974
[6] Sakuraba Y, Hattori M, Oogane M, Ando Y, Kato H, Sakuma A, Miyazaki T and Kubota H 2006 *Applied Physics Letters* **88** 192508
[7] Raphael M P, Ravel B, Willard M A, Cheng S F, Das B N, Stroud R M, Bussmann K M, Claassen J H and Harris V G 2001 *Applied Physics Letters* **79** 4396–4398
[8] Schmalhorst J, Thomas A, Kmmerer S, Schebaum O, Ebke D, Sacher M D, Reiss G, Htten A, Turchanin A, Glzhuser A and Arenholz E 2007 *Physical Review B* **75** 014403
[9] Fetzer R, Wstenberg J P, Taira T, Uemura T, Yamamoto M, Aeschlimann M and Cinchetti M 2013 *Physical Review B* **87** 184418
[10] Oogane M, Kubota T, Naganuma H and Ando Y 2015 *Journal of Physics D: Applied Physics* **48** 164012
[11] Raphael M P, Ravel B, Huang Q, Willard M A, Cheng S F, Das B N, Stroud R M, Bussmann K M, Claassen J H and Harris V G 2002 *Physical Review B* **66** 104429
[12] Ravel B, Raphael M P, Harris V G and Huang Q 2002 *Physical Review B* **65** 184431
[13] Picozzi S, Continenza A and Freeman A J 2004 *Physical Review B* **69** 094423
[14] Pandey H, Prasad R and Budhani R C 2014 Correlation between Site Disorder and Magnetic Moment in Full-Heusler Co2mnsi (Journal of the Physical Society of Japan)
[15] Sakuma A 2015 *Journal of Physics D: Applied Physics* **48** 164011
[16] Özdoğan K, Şaşoğlu E and Galanakis I 2007 *physica status solidi (RRL) Rapid Research Letters* **1** 184–186
[17] Ebert H The Munich SPR-KKR package,version 5.4 URL `http://olymp.cup.uni-muenchen.de/ak/ebert/SPRKKR`
[18] Soven P 1967 *Physical Review* **156** 809–813
[19] Ebert H, Mankovsky S, Kdderitzsch D and Kelly P J 2011 *Physical Review Letters* **107** 066603
[20] Kamberský V 1970 *Canadian Journal of Physics* **48** 2906–2911
[21] Kamberský V 1976 *Czechoslovak Journal of Physics B* **26** 1366–1383
[22] Vosko S H, Wilk L and Nusair M 1980 *Canadian Journal of Physics* **58** 1200–1211

Uniaxial pressure effect on magnetization process of ferromagnetic domains in a tensile deformed Pt_3Fe antiferromagnet

Satoru Kobayashi[1], **Yasunori Konno**[1]

[1] Faculty of Engineering, Iwate University, Morioka, Japan

E-mail: `koba@iwate-u.ac.jp`

Abstract. Pt_3Fe antiferromagnet with $L1_2$-type structure exhibits ferromagnetism even at room temperature after plastic deformation. We have studied effects of elastic uniaxial stress along the [100] axis on magnetization hysteresis loops for a Pt_3Fe single crystal with 11.6% plastic strain, in order to investigate magnetic nature of ferromagnetic domains around the {111} slip plane under uniaxial pressure. We observed an exchange bias effect below $T_{N1} \sim 170$ K up to the measured maximum pressure of 150 MPa and found that the exchange bias field is almost constant at low uniaxial pressures, but it steeply increases above 100 MPa and shows $\sim$ 6% increase at 150 MPa. Correspondingly, a slight increase of coercivity above 100 MPa was observed. These observations indicate the strengthening of interfacial coupling between ferromagnetic domains and an antiferromagnetic matrix, and/or modification of deformation-induced magnetic anisotropy under uniaxial pressure.

$L1_2$-type Pt_3Fe ordered alloy exhibits successive antiferromagnetic (AFM) transitions at $T_{N1} \sim 170$ K and $T_{N2} \sim 100$ K and becomes ferromagnetic even at room temperature if the alloy is plastically deformed [1]. The plastic deformation induces the formation of ferromagnetic (FM) domains around superlattice dislocations near the {111} glide plane, where local nearest-neighbor Fe-atom pairs with ferromagnetic coupling are formed. Our recent magnetization measurements under magnetic fields for a crystal deformed in tension along the [100] axis revealed an exchange bias effect below T_{N1}, where a coupling between FM domains and AFM matrix was inferred [2]. Further, measurements varying magnetization direction with respect to [100] tensile axis revealed an uniaxial-type magnetic anisotropy along the [100] axis, from which homogeneous distribution of FM domains around {111} glide plane was suggested [3]. Nevertheless, the morphology of FM domains such as size, shape, as well as the detailed magnetization mechanism were not fully understood.

In this study, we have investigated effects of uniaxial pressure on magnetization process of FM domains for a Pt_3Fe single crystal with 11.6% plastic strain, in order to get the detailed information about magnetic interactions of Fe atoms near {111} glide plane.

We used a single crystal of 25.6 at.% of iron for Fe-Pt alloy, which was pre-strained up to 11.6% in tension [2, 3]. Magnetization measurements were performed using a commercial SQUID magnetometer (Quantum Design MPMS-XL) for a rectangular single crystal with dimensions of $3.0 \times 2 \times 2$ mm^3. The sample has a flat surfaces normal to [100], [035], and [0$\bar{5}$3] crystal axes and the long axis corresponds to the [100] axis. Uniaxial pressure was applied using a stick-type piston-cylinder pressure cell, which can be inserted in the magnetometer. We tested several

Content from this work may be used under the terms of the Creative Commons Attribution 3.0 licence. Any further distribution of this work must maintain attribution to the author(s) and the title of the work, journal citation and DOI.
Published under licence by IOP Publishing Ltd

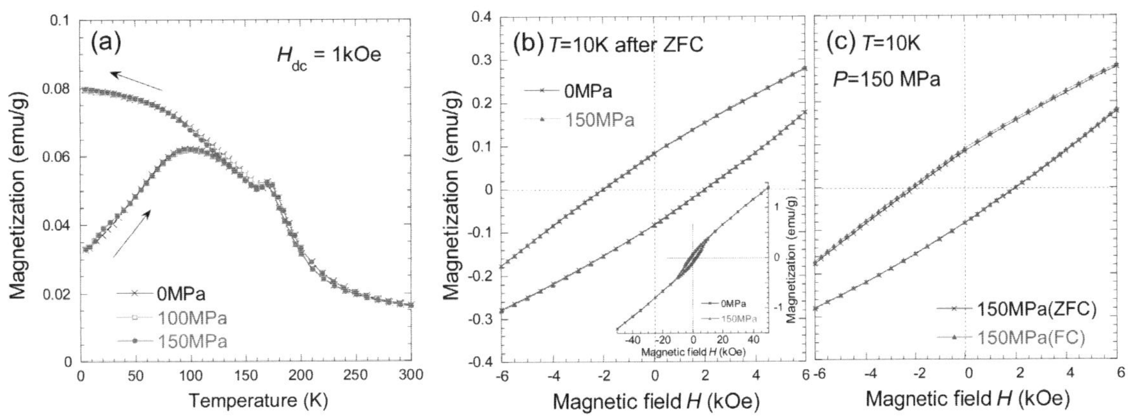

Figure 1. (a) Magnetization as a function of temperature after ZFC and FC before and after the application of uniaxial pressure, taken at H=1 kOe. (b) M-H curves at T=10K after ZFC with and without uniaxial pressure of 150 MPa. The inset shows an overview of the M-H curves. (c) M-H curves at T=10K under uniaxial pressure of 150 MPa, taken after ZFC and FC at H=50 kOe.

Pt_3Fe single crystals under uniaxial pressure along the [100] axis and found that the application of uniaxial pressure above 200 MPa yields further plastic deformation. Therefore, the applied maximum uniaxial pressure was limited to 150 MPa in the elastic region in this study. Magnetic field of up to H = 50 kOe and uniaxial pressure of up to P = 150 MPa were applied along the [100] axis. M-H loops were measured at two different temperatures of 10 K ($< T_{N2}$) and 120 K ($< T_{N1}$), after zero-field cooling (ZFC) or field cooling (FC) at H = 50 kOe from T = 200 K.

Figure 1(a) shows magnetization as a function of temperature after ZFC and FC, taken at H = 1 kOe under uniaxial pressure of up to 150 MPa. While there exists no significant change in the shape of M-T curves and AFM transition temperatures of T_{N1} and T_{N2} after the application of uniaxial pressure, a slight decrease of magnetization was observed at temperatures between T_{N1} and T_{N2}.

Figures 1(b) shows M-H curves at T=10 K after ZFC, with and without uniaxial pressure of 150 MPa. In Fig. 1(c), M-H curves at T=10 K and P = 150 MPa, taken after ZFC and FC, are shown. As in the case of no uniaxial pressure, we confirmed an exchange bias effect under uniaxial pressure of up to 150 MPa as shown in Fig. 1(c). A change of M-H curves due to uniaxial pressure, however, was found to be very small.

Figures 2(a) and 2(b) show the uniaxial pressure dependence of exchange bias field, H_{eb}, and coercivity, H_c, respectively, taken at T = 10 and 120 K. Here, both H_{eb} and H_c were determined from magnetization curves, where contribution of an AFM matrix to observed M-H curves was subtracted [3]. H_{eb} is almost constant at low uniaxial pressures and then steeply increases above 100 MPa. The increase of H_{eb} at 150 MPa attains $\sim$ 6% at T = 10 K and $\sim$ 4% at T = 120 K. On the other hand, H_c is very weakly dependent on uniaxial pressure at T = 10 K, whereas an increase at 150 MPa is at most $\sim$ 4% at T = 120 K. Generally, both H_{eb} and H_c are in proportion to $J_{AFM/FM}/d_{FM}$, where $J_{AFM/FM}$ and d_{FM} are interfacial exchange coupling and FM domain size, respectively [4]. Therefore, these observations indicate the strengthening of exchange coupling between FM domains and AFM matrix, shrinking of FM domains, and/or modification of deformation-induced [100] uniaxial anisotropy, by uniaxial pressure. Since no significant change in saturation magnetization of FM domains with uniaxial pressure was observed, the second possibility can be excluded.

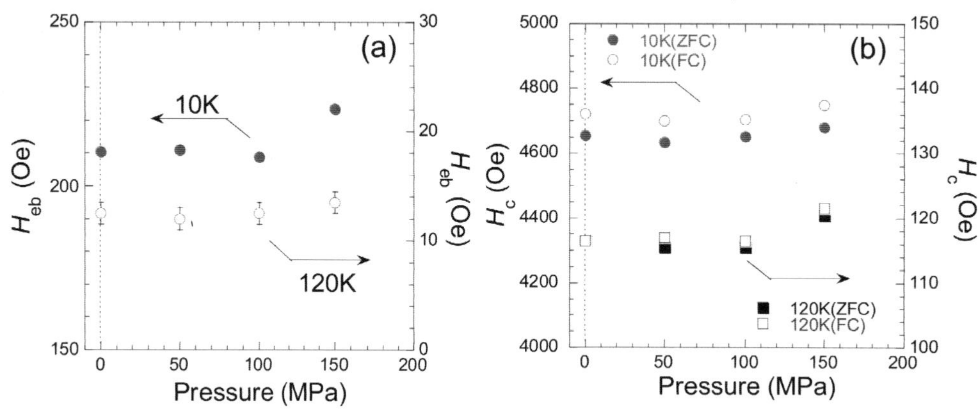

Figure 2. (a) Exchange bias field as a function of uniaxial pressure at T = 10 and 120 K. (b) Coercivity as a function of uniaxial pressure at T = 10 and 120 K. The data taken after ZFC and FC at H = 50 kOe are given.

In conclusion, we performed magnetic measurements for a tensile-deformed Pt_3Fe single crystal under uniaxial pressure of up to 150 MPa and demonstrated the uniaxial pressure control of an exchange bias effect for FM domains formed in an AFM matrix. Further magnetic measurements under uniaxial pressure, varying pressure direction, are planned.

Acknowledgments
We thank Prof. S. Mitsuda of Tokyo University of Science for technical support for uniaxial pressure measurements.

References

[1] G.E. Bacon and J. Crangle, Proc. Roy. Soc. (London) **A272** (1963) 387.
[2] S. Kobayashi, S. Takahashi, Y. Kamada, H. Kikuchi, IEEE Trans. Mag. **44** (2008) 4225.
[3] S. Kobayashi, R. Morita, J. Appl. Phys. **117** (2015) 17D120.
[4] C. Leighton, J. Nogues, B.J. Jonsson-Akerman, Ivan K. Schuller, Phys. Rev. Lett. **84** (2000) 3466.

Synthesis and magnetic properties of the thin film exchange spring system of MnBi/FeCo

S. Sabet, E. Hildebrandt and L. Alff

Institute of Materials Science, Technische Universität Darmstadt, 64287 Darmstadt, Germany

E-mail: sabet@oxide.tu-darmstadt.de

Abstract. Manganese bismuth thin films with a nominal thickness of ~40 nm were grown at room temperature onto quartz glass substrate in a DC magnetron sputtering unit. In contrast to the usual multilayer approach, the MnBi films were deposited using a single sputtering target with a stoichiometry of $Mn_{55}Bi_{45}$ (at. %). A subsequent in-situ annealing step was performed in vacuum in order to form the ferromagnetic LTP of MnBi. X-ray diffraction confirmed the formation of a textured LTP MnBi hard phase after annealing at 330 °C. This film shows a maximum saturation magnetization of 530 emu/cm^3 and high out-of-plane coercivity of 15 kOe induced by unreacted bismuth. The exchange coupling effect was investigated by deposition of a second layer of FeCo with 1 nm and 2 nm thickness onto the LTP MnBi films. The MnBi/FeCo double layer showed as expected higher saturation magnetization with increasing thickness of the FeCo layer while the coercive field remained constant. The fabrication of the MnBi/FeCo double layer for an exchange spring magnet was facilitated by deposition from a single stoichiometric target.

1. Introduction

Synthesis of rare earth free permanent magnet materials has drawn much attention recently since the unreliable global market for rare earth elements made their application critical [1, 2]. Rare earth free magnets must have high magnetic anisotropy, high temperature stability, and high energy products to be considered as promising substitutes for rare-earth containing magnets [3]. Despite its relatively low theoretical saturation magnetization of 710 emu/cm^3 compared to the rare earth magnets, MnBi potentially has the required magnetic properties to be qualified as suitable candidate [4, 5, 6]. Particularly, it shows a positive temperature coefficient for coercivity and magnetic anisotropy which makes it even more interesting for high temperature applications [7, 8, 9]. It is possible to improve the magnetic moment and also the energy density for MnBi using it as hard magnetic component in combination with a soft magnetic phase in an exchange spring magnet configuration [10].

Although synthesis of MnBi thin films has been studied for long, the exchange coupling with a soft magnetic phase as a mean to improve its performance has been barely investigated so far. In contrast to the previous approach with multiple depositions of separate Bi and Mn layers used for preparation of MnBi thin films followed by a post annealing step [11, 12, 13, 14, 15, 16, 17], in this work we have established a novel more straightforward synthesis route. We have deposited MnBi thin films via a single step deposition from a $Mn_{55}Bi_{45}$ (at. %) stoichiometric sputtering target which provides us an easier way to deposit additional layers for spring magnet hetero-structure configurations.

Content from this work may be used under the terms of the Creative Commons Attribution 3.0 licence. Any further distribution of this work must maintain attribution to the author(s) and the title of the work, journal citation and DOI.
Published under licence by IOP Publishing Ltd

IOP Conf. Series: Journal of Physics: Conf. Series **903** (2017) 012032 doi:10.1088/1742-6596/903/1/012032

2. Material and methods

Manganese bismuth thin films with a typical thickness of 40 nm were deposited on quartz glass substrates in a DC magnetron sputtering unit with a base pressure of $\sim 4.0\times10^{-6}$ Pa. We have used single stoichiometric sputtering targets with composition of $Mn_{55}Bi_{45}$ (at.%) and $Fe_{65}Co_{35}$ (at.%) for our depositions. The optimized growth rate was 0.04 nm/s for MnBi layer, and 0.008 nm/s for FeCo layer. The MnBi thin films were deposited at room temperature and subsequently annealed for 1 hr *in situ* under vacuum ($\sim 1.0\times10^{-5}$ Pa) at different annealing temperature. For the double layer system, the FeCo layer was deposited at room temperature on top of LTP MnBi annealed at 330 °C. The films were capped with a 10 nm thick tantalum layer to prevent oxidation. We have determined the phase composition by X-ray diffraction, the film thickness by a Bruker Dektak-XT surface profilometer, and the magnetic properties were measured using a SQUID magnetometer.

3. Results and discussion

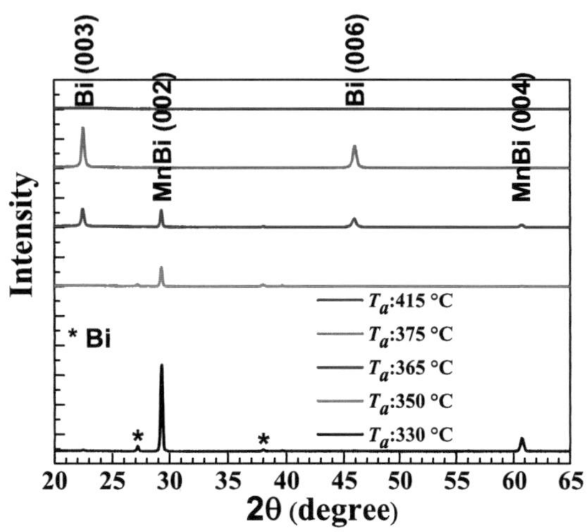

Figure 1. X-ray Diffraction patterns for MnBi thin films annealed at different temperatures, T_a, between 330 °C and 415 °C. For the purpose of clarity the spectra are vertically offset.

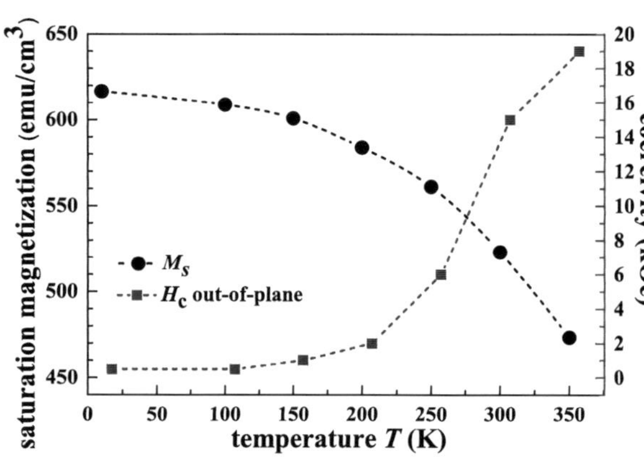

Figure 2. Saturation magnetization and out-of-plane coercivity as a function of temperature for a MnBi single layer (40 nm) annealed at T_a = 330 °C measured at different temperatures between 10 K to 350 K.

Fig. 1 contains room-temperature XRD patterns of MnBi thin films annealed at different temperatures in the range of 330 °C to 415 °C. The indexed peaks belong to hexagonal MnBi(with space group of $P63/mmc$) and also show traces of residual bismuth in the films. The MnBi (002) and (004) peaks appear with the highest intensity when T_a reaches 330 °C which indicates the formation of ferromagnetic LTP MnBi with c-axis texture. The Bi (003) and Bi (006) peaks were also observed but with much lower intensities. By increasing the annealing temperature, the intensity of MnBi ($00l$) peaks drops which is accompanied by an increase in the intensity of Bi ($00l$) reflections. This shows that the LTP MnBi starts to decompose into crystalline Bi and amorphous Mn. This trend continues until T_a above 415 °C at which the LTP MnBi phase is completely decomposed and no more reflections are visible. In comparison to a previous similar study, we have observed that the optimized temperature depends on the vacuum level where the annealing takes place: a better vacuum is associated with a lower optimal annealing temperature.

Fig. 2 shows the changes in volume saturation magnetization and out-of-plane coercivity at different temperatures between 10 K and 350 K for a MnBi thin film annealed at 330 °C. As it can bee seen, the saturation magnetization (M_s) decreases with increasing temperature from 610 emu/cm^3 at 10 K down to 470 emu/cm^3 at 350 K. On the other hand, the out-of-plane coercivity (H_c) is nearly zero below 150 K and grows drastically with further increase in temperature, as it is expected for LTP MnBi, and reaches 19 kOe at 350 K. The room-temperature M_s and H_c for the MnBi thin film annealed at 330 °C were 530 emu/cm^3 and 15 kOe, respectively. According to the XRD patterns in Fig. 1, some unreacted Bi is still present in this sample acting as an hindrance for domain wall motion leading to a high coercivity and a high energy density $(BH)_{max}$= 8.7 MGOe for this sample.

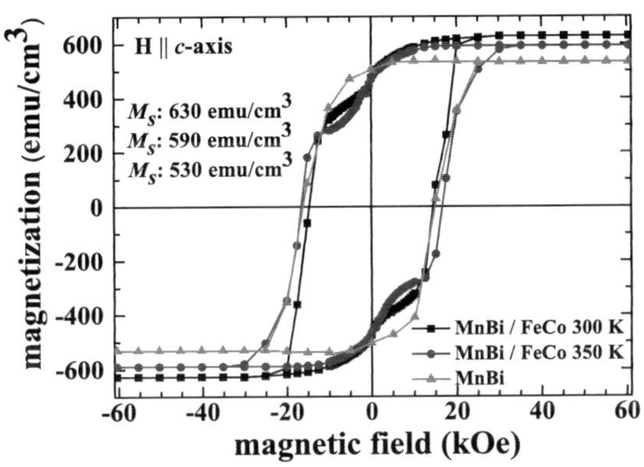

Figure 3. Out-of-plane magnetization data for a MnBi single layer (40 nm) measured at 300 K (red) and a MnBi(40 nm)/FeCo(1 nm) double layer measured at 300 K (black) and at 350 K (blue).

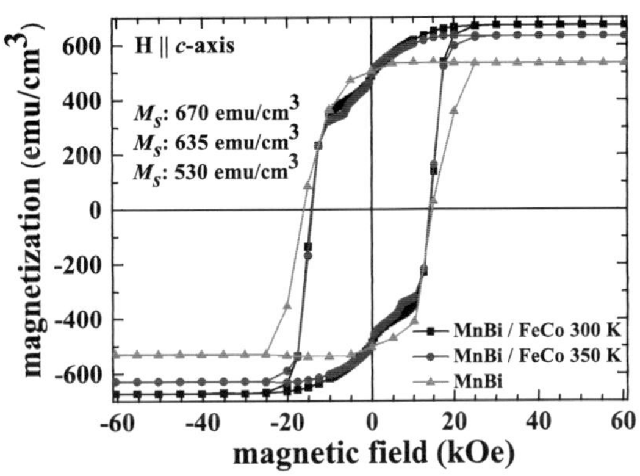

Figure 4. Out-of-plane magnetization data for a MnBi single layer (40 nm) measured at 300 K (red) and a MnBi(40 nm)/FeCo(2 nm) double layer measured at 300 K (black) and at 350 K (blue).

Fig. 3 and Fig. 4 show the hysteresis loops for the double layer of MnBi plus FeCo measured at 300 K and 350 K with 1 nm and 2 nm of FeCo as soft layer, respectively. In order to be able to compare these results, the magnetization data for the single MnBi film annealed at 330 °C is also included. We have grown the second layer of FeCo on top of MnBi at room-temperature with two different thicknesses. By addition of 1 nm and 2 nm of FeCo the saturation magnetization has increased as expected from 530 emu/cm^3 to 630 emu/cm^3 and 670 emu/cm^3, respectively. The coercivity, on the other hand, stays nearly constant. Comparing the graphs in Fig. 3 and Fig. 4 with the summarized graph in Fig. 2 for single MnBi layer at 300 K shows that the thermal stability has improved for the coupled layers. The kink in the hysteresis curve of the double layer indicates that the exchange coupling is incomplete, which is most likely due to the roughness of the interface. The interface roughness can be reduced by growing much thinner MnBi layers ($<$ 10 nm) at decreased deposition rates. The thickness of the hard magnetic layer in an exchange spring system anyhow should be as thin as possible, since this results in the largest increase of total magnetization.

4. Conclusions

LTP MnBi thin films with c-axis texture, high coercivity of 15 kOe and a maximum room-temperature energy product of $\sim$ 8.7 MGOe were deposited directly from a single stoichiometric $Mn_{55}Bi_{45}$ (at. %) sputtering target onto quartz glass substrates and subsequently annealed at $\sim$ 330 °C. In order to investigate the exchange coupling effect we have deposited a soft magnetic layer of FeCo on top of the MnBi thin film. The magnetization of the total system increased as expected while the coercivity remains unchanged. Since smooth interfaces are required to obtain an effective exchange path, our deposition approach working with a single MnBi layer with the targeted stoichiometry seems to be a promising way to enable the desired exchange coupling.

Acknowledgement

The authors thank the LOEWE project RESPONSE funded by the Ministry of Higher Education, Research and Arts (HMWK) of the state of Hessen.

References

[1] Poudyal N and Liu J P 2013 *J. Phys. D: Appl. Phys.* **46** 043001
[2] Kamer M J, McCallum R W, Anderson I A and Constantinides S 2012 *JOM-J. Min. Met. Mat. S.* **64** 752-763
[3] Gutfleisch O, Willard M A, Brück E, Chen C H, Sankar S G and Liu J P 2011 *Adv. Mater.* **23** 821-842
[4] Park J, Hong Y K, Lee J, Lee W, Kim S G and Choi C J 2014 *Metals.* **4** 455-464
[5] McGuire M A, Cao H, Chakoumakos B C and and Sales B C 2014 *Phys. Rev. B.* **26** 1744251-8
[6] Cui J *et al.* 2014 *J. Phys. Condens. Mat.* **26** 064212
[7] Yang J B *et al.* 2011 *Appl. Phys. Lett.* **99** 082505
[8] Yang J B, Yelon W B, James W J, Cai Q, Roy S and Ali N 2002 *J. Appl. Phys.* **91** 7866-7868
[9] Curcio C, Olivetti E S, Martino L, Küpferling M and Basso V 2015 *Phys. Procedia.* **75** 1230-1237
[10] Li D, Pan D, Li S and Zhang Z 2016 *Sci. China-Phys. Mech. Astron.* **59** 617501-17
[11] Kharel P, Skomski R, Kirby R D and Sellmyer D J 2010 *J. Appl. Phys.* **107** 09E303
[12] Kharel P, Skomski R and Sellmyer D J 2011 *J. Appl. Phys.* **109** 07B709
[13] Zhang W, Kharel P, Valloppilly S, Yue L and Sellmyer D J 2015 *Phys. Status. Solidi. B.* **252** 1934-1939
[14] Kharel P, Li X Z, Shah V R, Al-Aqtash N, Tarawneh K, Sabirianov R F, Skomski R and Sellmyer D J 2012 *J. Appl. Phys.* **111** 07E326
[15] Suzuki T, Hozumi T, Barker J, Okatov S, Mryasov O and Suwa T 2015 *IEEE Trans. Magn.* **51** 2102804
[16] Hozumi T, LeClair P, Mankey G, Mewes C, Sepehri-Amin H, Hono K and Suzuki T 2014 *J. Appl. Phys.* **115** 17A737
[17] Li B, Liu W, Zhao X G, Gong W J, Zhao X T, Wang H L, Kim D, Choi C J and Zhang Z D 2014 *J. Magn. Magn. Mater.* **372** 12-15

IOP Conf. Series: Journal of Physics: Conf. Series **903** (2017) 012033 doi:10.1088/1742-6596/903/1/012033

Structure and magnetic properties of $Sm_{1-x}Zr_x Fe_{10}Si_2$ (x= 0.2-0.6) alloys

M. Gjoka [1], C. Sarafidis[2], V. Psycharis[1], E. Devlin [1], D. Niarchos[1] and G. Hadjipanayis [3]

[1] *INN, NCSR Demokritos, Athens 15310, Greece*
[2]*Department of Physics, Aristotle University of Thessaloniki, 54124 Thessaloniki, Greece*
[3]*Department of Physics, U of Delaware, Newark, USA*

e-mail:m.gjokas@inn.demokritos.gr

Abstract
Structure and magnetic properties of $Sm_{1-x}Zr_xFe_{10}Si_2$ ($0.1 \leq x \leq 0.6$) alloys have been characterized using X-ray diffraction, thermomagnetic analysis and Mössbauer spectroscopy. The formation of the tetragonal $ThMn_{12}$ -type structure was been observed in all alloys, without further annealing. The Curie temperature decreases linearly with Zr substitution from 322 ^{o}C for x=0.1 to 395 ^{o}C for x=0.6. Mössbauer spectroscopy showed the iron hyperfine field values decrease with increasing Zr content, and also confirmed changes to the magnetic anisotropy with increasing Zr content observed by XRD on oriented samples.

1. Introduction
As a transition element, Zr is of particular interest for permanent magnets because of its covalent radius (1.45 Å) which is larger than those of Fe (1.17 Å) and Si (1.11 Å), but smaller than that of Sm (1.98 Å). Structural and magnetic properties are affected differently when Zr enters into the Fe or the rare earth (R) sub-lattice. Many investigations on substitution of various R by Zr have been reported e.g. Zr has been used to replace the R in the1:5 [1], 2:17 [2, 3], 3:29 [4, 5] and 1:12 structures. The stability of the Zr substitution for the system $RFe_{10}Si_2$ was first reported by Sakurada and al. [6] and Gabay and al. [7]. Subsequently, 1:12 phase formation was reported in the series $Gd_{8-x}Zr_x(FeMo)_{100}$ [8], $Nd_{3-x}Zr_xFe_{27.8}Mo_{1.2}$ [9, 10], $CeFe_{10}Si_2$ alloys [11], and $(R_{0.3}\,Zr_{0.7})_{1.1}Fe_{10}Si$ (R= Y, Ce, Pr and Sm) [12]. We have previously reported on the structural and magnetic properties of an alloy series with nominal compositions $Nd_{1-x}Zr_xFe_{10}Si_2$, $0.1 \leq x \leq 0.\,8$ [13]. In the present work, our goal is to explore rare-earth –substitution by Zr in other families of 1:12-type structures, namely $Sm_{1-x}Zr_xFe_{10}Si_2$ (x=0.1-0.6), alloys with a wider composition range than earlier reported [7, 12].

2. Experimental
Alloys with nominal composition $Sm_{1-x}Zr_xFe_{10}Si_2$ (x=0.1-0.6) were produced in an induction-melting furnace under argon. Structural characterizations was carried out using x-ray diffraction (XRD) with the magnetic and phase properties of the alloys studied using thermomagnetic analysis (TMA), and Mössbauer spectroscopy. Powder samples oriented in epoxy resin for XRD and Mössbauer spectroscopy were obtained using a magnetic field of 2T perpendicular to the sample plane.

3. Results and discussion
3.1 Structure
XRD patterns of the $Sm_{1-x}Zr_xFe_{10}Si_2$ (x = 0.1, 0.4, 0.5, 0.6) samples are presented in Figure 1a. All samples are practically single phase and the dominant phase is the tetragonal $ThMn_{12}$-type (Space Group I4/mmm). Traces of α-Fe(Si), 4-7 wt%, are visible in the diagrams (dashed line). The

Content from this work may be used under the terms of the Creative Commons Attribution 3.0 licence. Any further distribution of this work must maintain attribution to the author(s) and the title of the work, journal citation and DOI.
Published under licence by IOP Publishing Ltd

increasing Zr content results in larger 2-theta angles (dotted line) i.e., the unit cell volume decreases with Zr content.

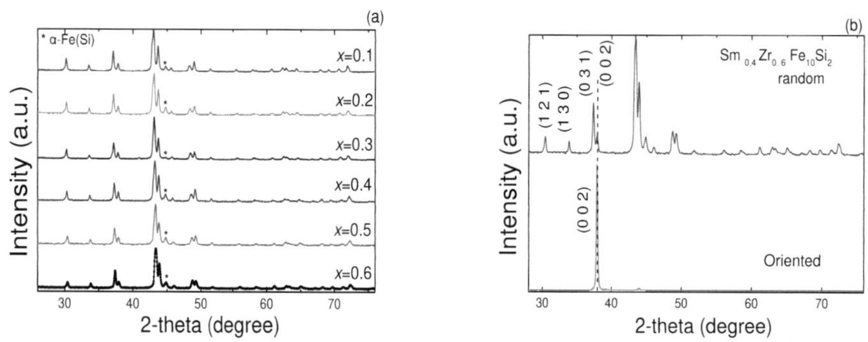

Figure 1. a) XRD raw data (Cu Kα) of $Sm_{1-x}Zr_xFe_{10}Si_2$ (x = 0.1, 0.4, 0.5, 0.6) samples. b) X-ray diffraction patterns of powder and magnetically oriented samples of the x = 0.6 compounds.

The XRD patterns were processed using Rietveld analysis, taking into account the Zr and Si atom populations in the well-established 1:12 compound [9, 13]. Following this model, Zr was placed in the tetragonal structure 1a positions, which are shared with the rare earth. Si atoms were distributed between the 8j and 8f positions. The unit cell parameters shrink with increasing Zr, due to the smaller ionic radius of Zr (Table 1). The c/a ratio is practically unchanged.

The XRD data from magnetically oriented samples can be divided into two groups. For the compounds with $0.1 \leq x \leq 0.3$ the (0 0 2) reflection is enhanced significantly, but other reflections are also visible, (0 2 2), (0 4 0), (0 3 1), indicating magnetic moments aligned close to the (0 0 2) direction i.e a cone arrangement around the c-axis. For $0.4 \leq x \leq 0.6$ a stronger uniaxial character is observed, the (0 0 2) reflection further enhanced, the (0 2 2) reflection hardly visible, and all other reflections absent, Figure 1b.

Table 1. Structural, magnetic, and Mössbauer parameters of $Sm_{1-x}Zr_xFe_{10}Si_2$ alloys.

x-Zr content	Structural					Magnetic	Mössbauer
	Phase	a(Å)	c(Å)	V(Å^3)	Weight fraction (%)	Tc (°C)	Average HF (KOe)
0.1	1:12	8.439	4.754	338.6	96	322	246
	α-Fe(Si)	2.859			3.5	711	
0.2	1:12	8.421	4.747	336.6	94,5	314	241
	α-Fe(Si)	2.858			5,5	709	
0.3	1:12	8.407	4.748	335.6	98	312	235
	α-Fe(Si)	2.859			1,5	707	
0.4	1:12	8.394	4.744	334.2	95	302	236
	α-Fe(Si)	2.860			5%	703	
0.5	1:12	8.375	4.735	332.4	93	301	233
	α-Fe(Si)	2.858			7	690	
0.6	1:12	8.360	4.735	331.0	95	295	230
	α-Fe(Si)	2.855			5	684	

3.2 Magnetic properties

Thermomagnetic data for $Zr_{1-x}Zr_xFe_{10}Si_2$ alloys are given in table 1. In accordance with the XRD data the dominant magnetic transitions are attributed to the 1:12 phase. The highest temperature transitions correspond to the α-Fe(Si) phase.

3.3 Mössbauer Spectroscopy

^{57}Fe Mössbauer Spectroscopy of isotropic samples and aligned samples were obtained at RT. The spectra were fitted in accordance with the crystallographic data with three independent subspectra for the 8i, 8j and 8f Fe-sites of the 1:12 phase, and two additional subspectra associated with the α-Fe-Si phase [14]. Site assignments for the 8i, 8j, and 8f sites follows the procedure of [15], with the hyperfine fields in the sequence 8i > 8j > 8f. The α-Fe(Si) phase contents obtained also follow the XRD data, being below 5% of the total Fe content.

The hyperfine field values decrease with the increasing Zr content, with a slight pause in the interval x=0.3-0.4, which may be due to variations in the composition arising from changes in the α–Fe(Si) content in this region.

Mössbauer spectra of the aligned samples obtained would, with perfect c-axis alignment, show a zero intensity for the Mössbauer 2,5 lines. This was not observed in any of the samples. The spectra yielded 2,5 line intensities corresponding to an easy-cone with rotation of the magnetization easy axis towards the c-axis with increasing Zr content. Deviations of the magnetization at angles of 21.5°, 21°, 22.5°, 14°, 14.5° and 14° to the c-axis for the x=0.1, 0.2, 0.3, 0.4, 0.5, and 0.6 samples, respectively, are observed i.e. there is a steady closing of the cone with increasing Zr content, with a slight pause again at the x=0.3 sample. These values indicate close to easy-axis alignment, i.e. approx 14°deviation from the c-axis for the x=0.6 sample. This is in agreement with the XRD data which indicated an anisotropy close to easy-axis (almost parallel to the c-axis) for the x=0.6 sample.

4. Conclusions

Alloys with compositions $Sm_{1-x}Zr_xFe_{10}Si_2$ ($0.1 \leq x \leq 0.6$) were synthesized. The $ThMn_{12}$- type structure was obtained for all alloys without further annealing. In all the alloys a minority phase, 2-7 wt% of α-Fe(Si) was detected. Rietveld analysis confirmed that the Zr atoms substitute only the R site and do not enter the Fe sublattice, as is expected from their larger atomic radius. Increasing Zr content results in a steady unit cell contraction, and in decreases of the Curie temperature and iron hyperfine fields. In addition, the easy-cone anisotropy narrows to a smaller cone angle with increasing Zr content.

Acknowledgement: This work is supported EU projects MSCA-RISE-*2015, INAPEM, NOVAMAG.*

References

1. Gupta H.O. and Wallace W E, 1985 J. *Magn. Magn. Mat.* **50** 339
2. Lefevre A., Cohen-Adad M Th, Mentzen B F 1997 *J. Alloys Compds.* **256** 207
3. Hao S Q, Chen N X, Shen J, 2002 *J. Alloys Compds.* **343** 53
4. Han S B, Lv J Y, Liu X F, Peng J, Li X J, Chen D F, Xue Y J, Li J H and Hu Z B 2005 *Journal of Physics: Condensed Matter*, **17** 6007
5. Han S B, Lv J Y, Liu X F, Peng J, Chen D F, Li J H, Hu Z B J 2007 *J. Alloys Compds.* **431** 68
6. Sakurada S, Tsutai A and Sahashi M 1992 *J. Alloys Compds* **187** 67
7. Gabay A M, Shchegoleva N N, Belozerov E V 2002 *Fizika Metallov i Metallovedeni* **94(3)** 41
8. Zinkevich M, Mattern Bacher N I, Puerta S 2002 *J. Alloys Compd.* **336** 320.
9. Du H L, Han J Z, Zhang W Y, Wang C S, Wang W C, Liu S Q, Chen H Y, Zhang X D, Yang Y C 2004 *J. Magn. Magn. Mater.* **283** 316.
10. Han S B, Lv J Y, Liu X F, Peng J, Chen D F, Li J H, Hu Z B 2007, *J. Alloys Compds* **431** 68
11. Gabay A and Hadjipanayis G C 2016 J. Alloys Compds. **657** 133
12. Gabay A M, Cabassi R, Fabbrici S, Albertini F, Hadjipanayis G C 2016 *J. Alloys Compds* **683** 271
13. Gjoka M, Psycharis V, Devlin E, Niarchos D, G. Hadjipanayis 2016 *J. Alloys Compds*, **687**, 240
14. Häggström L, Grånäs L, Wäppling R, Devanarayanan S 1973, *Physica Scripta* **7** 125
15. Guang'ai S., Bo C, Honglin D, Zhijian F, Tao G, Xinhua Q 2006 Science in China: Series G Physics, Mechanics & Astronomy **49** 195

IOP Conf. Series: Journal of Physics: Conf. Series **903** (2017) 012034 doi:10.1088/1742-6596/903/1/012034

Defects and magnetic structure of CuMnSb

František Máca[1], Josef Kudrnovský[1], Václav Drchal[1] and Ilja Turek[2]

[1] Institute of Physics ASCR, Praha, Czech Republic

[2] Charles University, Faculty of Mathematics and Physics, Praha, Czech Republic

E-mail: maca@fzu.cz

Abstract.

Total energy calculations show that the antiferromagnetic (111) order is not the ground state for the ideal CuMnSb in contrast to the results of neutron diffraction experiments. The magnetic phases of the CuMnSb Heusler alloy were investigated with the aim to clarify the role of defects which exist in real samples (Mn-antisites on Cu, Mn-interstitials, and Cu-Mn swaps). The full-potential supercell approach and the Heisenberg model were used to find the magnetic structure of the CuMnSb with defects. Results of both types of calculations indicate that already low defect concentrations can promote the (111) antiferromagnetic structure.

1. Introduction and formalism

The antiferromagnetic (AFM) Mn-based Heusler alloy, CuMnSb, has attracted the theoretical interest in the past, in particular as concerns its magnetic structure [1]. The situation can be quite complex as indicated in the theoretical study of (Cu,Ni)MnSb random alloys [2], where we found that the AFM100 phase has a lower total energy than the observed AFM111 phase. Real CuMnSb samples show large residual resistivities [1] indicating the presence of defects and neutron diffraction experiments confirm the existence of low defect concentrations ($\approx 2\%$) [3]. We therefore investigate the influence of defects on the magnetic structure of CuMnSb.

We consider the most probable defects, namely, Mn antisites on the Cu-sublattice, Mn_{Cu}, Mn atoms at interstitial sites for the non-stoichiometric case, and the Cu$\leftrightarrow$Mn swaps for the stoichiometric CuMnSb. Two independent approaches are used to study the magnetic structure: (i) the supercell approach employing the full-potential Vienna ab-initio simulation package (VASP) [4], and (ii) the Heisenberg model whose parameters are determined by the tight-binding linear muffin-tin orbital (TB-LMTO) method in which the effect of randomness is included using the coherent potential approximation (CPA) [5]. We include electron correlations in narrow Mn d-bands using the LDA+U approach (the effective Hubbard $U = 0.13$ Ry), use the experimental lattice constant $a = 6.088$ Å, and neglect small volume changes due to defects. It should be noted that both approaches agree reasonably for model with one single defect in the supercell. Predictions of both models can differ in more complex cases e.g. for two defects in the unit cell where the mutual defects interaction depends on their relative positions inside the supercell.

The exchange interactions among Mn magnetic moments which enter the definition of the Heisenberg model are constructed using the first-principles Lichtenstein mapping procedure adapted to random alloys in the framework of the TB-LMTO-CPA method [6]. The exchange integrals are determined from the reference disordered local moment state (DLM, the paramagnetic state) [7] so that no specific magnetic configuration is used in their construction.

Content from this work may be used under the terms of the Creative Commons Attribution 3.0 licence. Any further distribution of this work must maintain attribution to the author(s) and the title of the work, journal citation and DOI.
Published under licence by IOP Publishing Ltd

2. Results

Of many possible AFM CuMnSb structures we have considered three simplest ones, namely, the AFM111, AFM100, and AFM40 with and without defects. The first two consist of alternating [111]/[100] planes of Mn atoms with opposite spins while the AFM40 is a tetragonal structure [8] with alternating double layers of opposite spins along the [210]-direction. The supercell approach gives as the lowest total energy for ideal CuMnSb the AF40 structure. In Heisenberg model the magnetic part of total energy is estimated from the classical Heisenberg Hamiltonian [6] with exchange integrals J_{ij} between Mn moments at sites i and j. When searching for its magnetic ground state we assume that magnetic moments have fixed size, but variable directions.

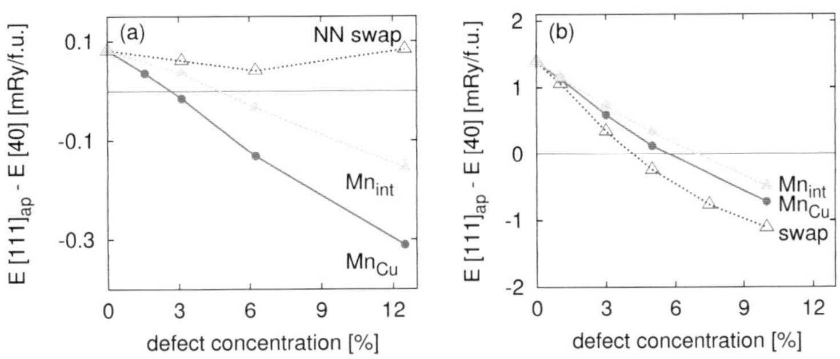

Figure 1. The total energy differences $E[111]_{\rm ap} - E[40]$ per formula unit (f. u.) between the reference AFM111$_{\rm ap}$ state and the AFM40 phase as a function of concentrations for the following defects: Mn antisites on Cu-sublattice, Mn interstitials, and Cu↔Mn swaps as obtained from the VASP (a) and Heisenberg model (b) calculations. Index ap denotes the AFM phase with the antiparallel alignment of defect Mn-moments to the moments on the native Mn-sublattice.

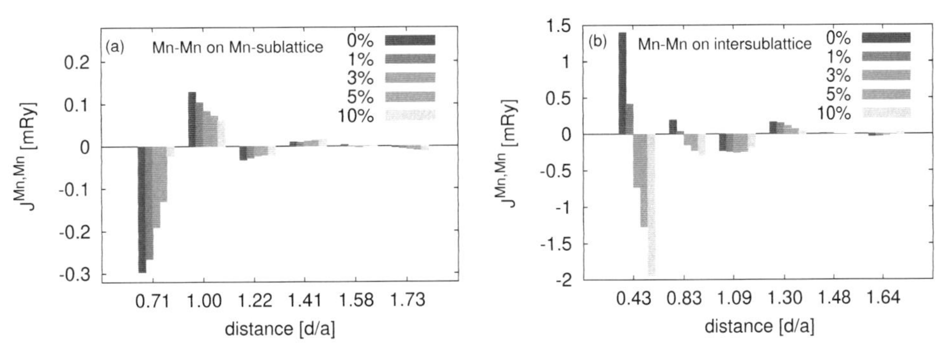

Figure 2. Exchange interactions between Mn-moments on the native Mn-sublattice (a) and between Mn moments on Mn- and Cu-sublattices (b) as a function of the distance d between them (in the units of the lattice constant) for CuMnSb with varying concentrations of Mn$_{\rm Cu}$ antisites.

Results for total energy differences based on the supercell calculations are shown in Fig. 1a with following conclusions: (i) The most favorable defect stabilizing AFM111 phase is the Mn-antisite on Cu lattice. The Mn↑ substituting Cu prefers the Mn nearest neighbors (NN) with magnetic moments ↑↓↓↓, its magnetic moment (lower than magnetic moments of atoms on the Mn-sublattice) is oriented in our naming convention *antiparallel*. The critical concentration for which the AFM111 is stabilized is less than 3%. (ii) Mn interstitials also stabilize the AFM111 phase although their effect is not so strong, the AFM111 phase is stabilized above about 6%. (iii) We have verified that lattice relaxations have a small effect on the magnetic structure and that Cu-antisites on Mn-lattice are not favorable for the stabilization of the AFM111 phase while electron correlations promote it.

Results based on the Heisenberg Hamiltonian are shown in Fig. 1b. We see a good agreement with the supercell results although the calculated critical concentrations are larger. A possible relevance of local environment effects on the magnetic stability can be illustrated on Cu↔Mn swaps in both models: while the Heisenberg model predicts the stabilization of the AFM111 by swaps the supercell approach does not. The explanation consists in very different ways of modeling of swaps in both approaches: in the supercell approach we assume NN Cu↔Mn swaps in the whole crystal, in the Heisenberg model based on the CPA we have completely uncorrelated Cu and Mn antisites (the CPA is a more realistic model in this case).

Exchange integrals allow us to understand the driving mechanism behind the AFM111 stabilization. We show corresponding integrals derived from the DLM state in Figs. 2a,b for the case of Mn_{Cu} antisites with conclusions: (i) The calculated negative (AFM-like) first NN interactions and the positive (FM-like) second NN interaction for the ideal case (x=0, Fig. 1a) contradicts the formation of the AFM111-phase [9] in the framework of the two NN model ($J_2^{\mathrm{Mn,Mn}}$ should be negative); and (ii) The leading intersublattice interactions (Fig. 2b) are much larger than those on the native lattice (the reduction of NN Mn-Mn distance) and becomes strongly AFM-like with increasing antisite concentration x. This is also true for the $J_2^{\mathrm{Mn,Mn}}$ interaction although less pronounced. The interactions, whose weight increases with increasing defect concentration x, are responsible for the crossover to AFM111 phase although weakening of the positive $J_2^{\mathrm{Mn,Mn}}$ with increasing concentration x also supports the crossover.

3. Conclusions

The controversy between the experimentally observed AFM111 ground state of the CuMnSb alloy and the total energy calculations which predict different ground state for the ideal alloy is resolved by assuming presence of defects confirmed experimentally in real samples. The main conclusions are: (i) The magnetic ground state of ideal CuMnSb crystal is not the AFM111 phase; (ii) The experimentally observed AFM111 phase is the ground state of CuMnSb samples with low concentrations of Mn defects which occupy the nearest neighbor sites of the native Mn sublattice, namely, Mn-antisites on Cu lattice, and Mn-interstitials; (iii) The crossover to the AFM111 phase is due to the very strong AFM intersublattice interactions among the NN Mn-moments with distances shorter than those on the native Mn lattice.

Acknowledgments

We acknowledge the financial support from the Czech Science Foundation (project 14-37427G) and from the National Grid Infrastructure MetaCentrum (programme CESNET LM2015042).

References

[1] Jeong T, Weht R, and Pickett W E 2005 *Phys. Rev. B* **71** 184103
[2] Kudrnovský J, Drchal V, Turek I and Weinberger P 2008 *Phys. Rev. B* **78** 054441
[3] Máca F, Kudrnovský J, Drchal V, Turek I, Stelmakhovych O, Beran P, Llobet A and Marti X 2016 *Phys. Rev. B* **94** 094407
[4] Kresse G and Furthmüller J 1996 Phys. Rev. B **54** 11169
[5] Turek I, Drchal V, Kudrnovský J, Šob M and Weinberger P 1997 *Electronic Structure of Disordered Alloys, Surfaces and Interfaces* (Boston: Kluwer)
[6] Turek I, Kudrnovský J, Drchal V and Bruno P 2006 *Philos. Mag.* **86** 1713
[7] Gyorffy B L, Pindor A J, Staunton J, Stocks G M and Winter H 1985 *J. Phys. F: Metal Phys.* **15** 1337
[8] Lu Z W, Wei S-H, Zunger A, Frota-Pessoa S and Ferreira L G 1991 *Phys. Rev. B* **44** 512
[9] Morán-López J L, Rodríguez-Alba R, and Aguilera-Granja F 1994, *J. Magn. Magn. Mater.* **131** 417

IOP Conf. Series: Journal of Physics: Conf. Series **903** (2017) 012035 doi:10.1088/1742-6596/903/1/012035

Study and characterization of soft magnetic materials for beam intensity monitors at CERN

S Aguilera[1,2], H Hofmann[2] and P Odier[1]

[1] CERN, 1211 Geneva, Switzerland

[2] École Polytechnique Fédérale de Lausanne, 1015 Lausanne, Switzerland

E-mail: silvia.aguilera@cern.ch

Abstract. At CERN, the circulating beam current measurement is provided by two types of transformers, the DC Current Transformer (DCCT) and the Fast Beam Current Transformer (FBCT). Each transformer is built based on toroidal cores made from a magnetic material. Depending on the type of measurement to be performed these cores require different magnetic characteristics for parameters such as permeability, coercivity and the shape of the magnetization curve. In order to study the effect of changes in these parameters on the current transformers, several interesting raw materials based on their as-cast properties were selected. The materials have been characterised to determine a suitable annealing processes to tailor their properties. As-cast properties such as the permeability, coercivity and Barkhausen noise have also been measured to enable the study of the effect of thermal treatment in the microstructure of the alloys, and the correlation of this with the change in the magnetic properties. The comparison of the measured properties have narrowed down the best materials for further study.

1. Magnetic materials for transformer cores at CERN

At CERN, the electric current of the charged particle beam circulating in the accelerators is measured by a family of devices that include current transformers, such as the DCCTs and FBCTs. This measurement is crucial for the correct operation of the accelerator complex and the experiments that it feeds, as well as for safety measures that rely on these readings.

Transformer cores are made out of a wound ribbon of soft magnetic material. The core couples to the magnetic field carried by the beam of charged particles as it traverses the transformer. Each type of transformer requires different magnetic characteristics in terms of permeability, coercivity and the shape of the magnetization curve. The choice of material affects the performance of the transformer, in particular in terms of the resolution in the case of the DCCT and the bandwidth in case of the FBCT.

Thermal or thermo-magnetic annealing can drastically change the magnetic properties of the raw materials used. The time and annealing heating rate are two key parameters that play a crucial role in the final result, enabling, for example, the fabrication of nanocrystalline materials. By adjusting these parameters the magnetic properties can be tuned in order to produce cores with different characteristics. Testing such cores as current transformers will then give an insight to the effect that different cores have on performance. By producing different cores and studying their behaviour it is possible to define the properties required to improve the characteristics of the instrument.

Content from this work may be used under the terms of the Creative Commons Attribution 3.0 licence. Any further distribution of this work must maintain attribution to the author(s) and the title of the work, journal citation and DOI.
Published under licence by IOP Publishing Ltd

IOP Conf. Series: Journal of Physics: Conf. Series **903** (2017) 012035 doi:10.1088/1742-6596/903/1/012035

The magnetic material used at CERN was specified for making the instrument as sensitive as possible. In 1998, Kottman studied the effects of noise on performance and resolution [4]. Simulations indicated that the shape of the magnetisation curve, coercivity and noise influenced the operation of the transformers. However, no measurements of these effects were done on the instruments nor the raw materials to verify these results.

For the DCCTs, soft ferromagnetic material with a relative complex permeability of more than 50 000 has been used up to now. Other characteristics sought include low Barkhausen Noise, a coercive field of around 1 A·m^{-1} and low magnetostriction. The coercivity limitation comes from the principle of operation, as the asymmetry in the B-H curve caused by the passing beam has to be detectable by the acquisition system. The magnetostriction effect produces changes in the size or shape of the core and can introduce noise if the cores are constrained [1].

2. Summary of materials used

For this study, the materials used were iron-based amorphous and nanocrystalline alloys and cobalt-based amorphous alloys. Two iron-based alloys were purchased from Qinhuangdao Yanqin Nano Science & Technology Co., Ltd., nanocrystalline 107A1 and amorphous 2065. Iron-based amorphous alloy FINEMET® FT-3 was bought from Hitachi Metals Europe GmbH. The amorphous cobalt-based materials were purchased from Nanostructured & Amorphous Materials (Nanoamor), Inc., VACUUMSCHMELZE GmbH & Co. KG as VAC 6025 G40 Z, and Hitachi Metals as alloy 2705M.

3. Material characterization

3.1. Permeability and B-H curve analysis

The relative complex permeability was calculated from the impedance measured with an Agilent Vector Impedance Analyser 4294 in the range of 40 Hz to 110 MHz. Cores of 40 mm of external diameter were used. The real and imaginary parts of permeability were derived from the in-series inductance and resistance of the core following [2]. The results are summarised in Table 1.

Table 1. Maximum relative complex permeability and coercivity for all samples analysed.

Sample	Permeability	Coercivity [A·m^{-1}]
Yanqin amorphous	1171	14.98
Yanqin nanocrystalline	697	19.43
FINEMET FT-3	1219	14.37
VAC 6025 G40 Z	165626	7.62
Nanoamor	158110	4.47
Metglas 2705M	250802	9.63

The high coercivity of the iron-based amorphous and nanocrystalline materials combined with their low permeability, practically rules them out as a suitable candidates for transformer cores, but does not mean that further annealing tests should not be performed on these materials to fully explore the potential magnetic properties that could be achieved by transforming the amorphous materials into nanocrystalline.

3.2. Barkhausen noise

The Barkhausen Effect is a physical phenomenon that manifests itself as a series of jumps in magnetization of ferromagnetic material when exposed to a varying magnetic field. Surrounding the sample by a secondary coil, the induced voltage can be transformed into acoustic noise, from which the term Barkhausen Noise (BN) derives [3]. This effect is often used as a non-destructive test to check changes in micro structure (grain boundaries, dislocations, inhomogeneities, etc.) and stress configurations of materials, and offers a good overview of changes in magnetic domains.

It is interesting to study the effect of thermal treatment on BN and on the final transformer response. BN tests were performed recreating the setup described in [3] for all the materials, looking at the voltage induced in the secondary coil surrounding the sample while it is subjected to a triangular excitation. The response of the FINEMET sample can be seen in Figure 1, where it is shown that the BN drops to almost zero when the sample is saturated.

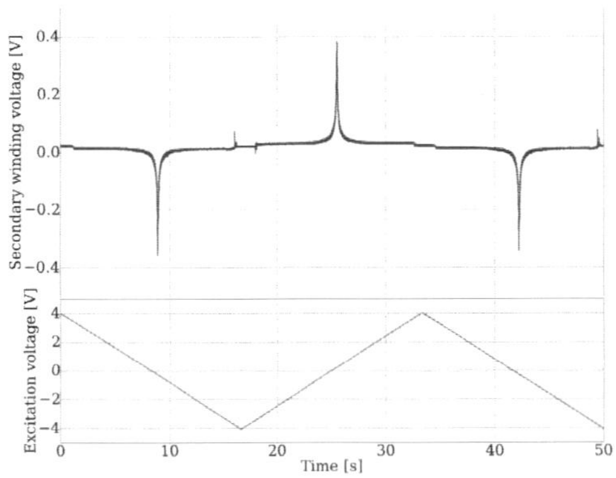

Figure 1. Barkhausen Noise response of the FINEMET sample (top) and excitation voltage (bottom).

Figures 2 to 7 show the BN (proportional to the secondary winding voltage) for all the materials. Once measured as cast, the samples where annealed for half an hour above their Curie temperatures before being re-measured. As can be seen, the BN decreases for Yanqin nanocrystalline and VAC 6025, while it increases for FINEMET. Yanquin amorphous, Metglas and Nanoamor show approximately the same noise level. Further XRD and magnetic domain visualisation studies of the annealed samples are being performed in order to understand these results, which may be caused by stress relaxation, dislocation displacement and crystal growth.

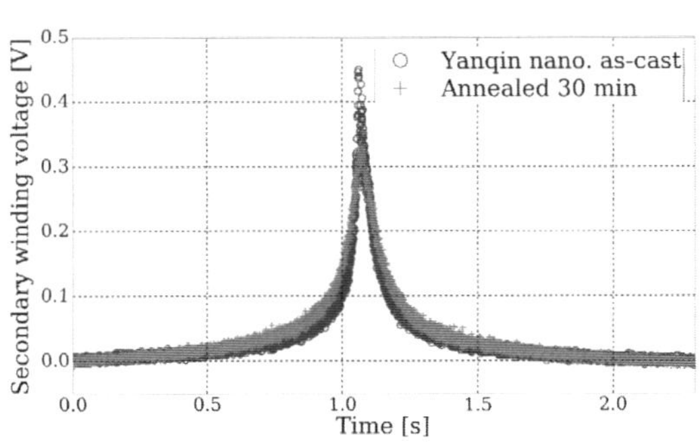

Figure 2. BN for Yanqin nanocrystalline before and after annealing.

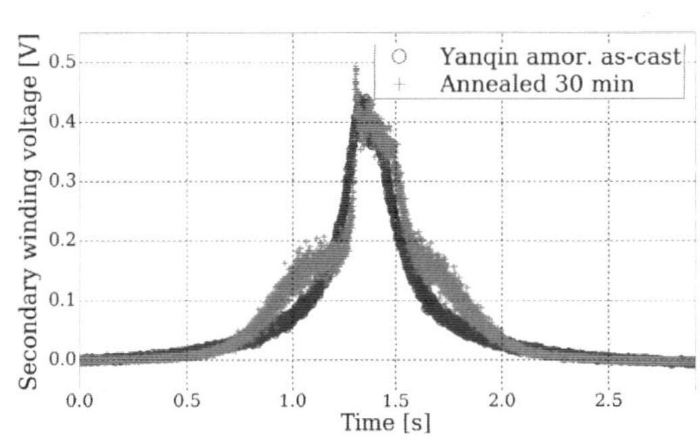

Figure 3. BN for Yanqin amorphous before and after annealing.

4. Conclusions

The cobalt-based materials (VAC, Nanoamor and Metglas) have the highest permeability and lowest coercivity in their amorphous states and are good base materials. However, the

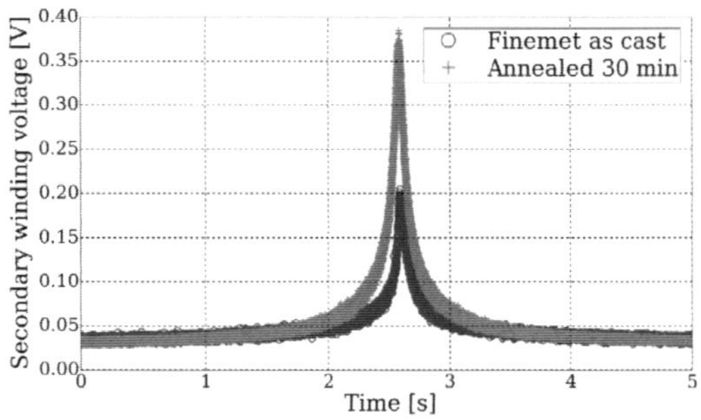

Figure 4. BN for FINEMET FT-3 before and after annealing.

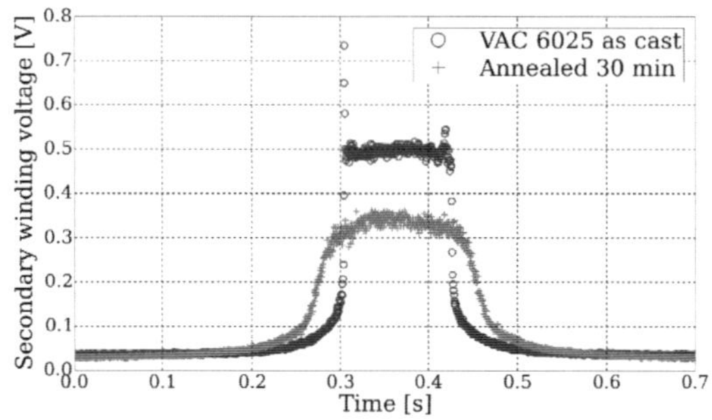

Figure 5. BN for VAC 6025 before and after annealing.

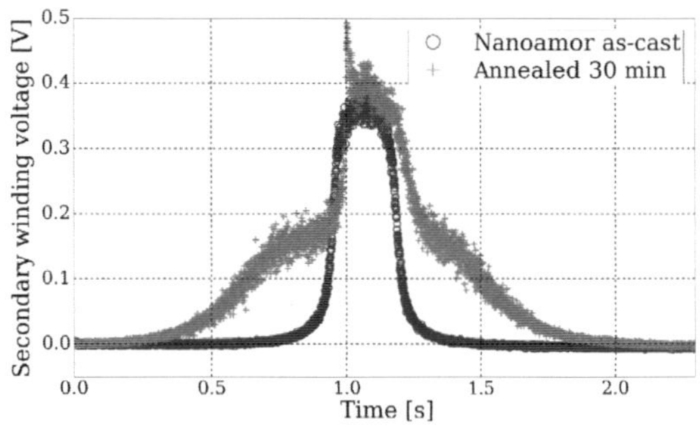

Figure 6. BN for Nanoamor before and after annealing.

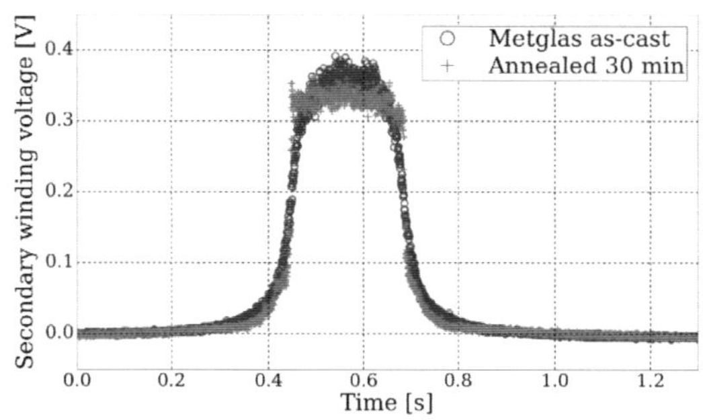

Figure 7. BN for Metglas before and after annealing.

potential properties of nanocrystalline materials should be taken into account and therefore, the FINEMET material, which has the lowest coercivity and highest permeability of the iron-based materials, will be studied further to analyse the properties of the material in its nanocrystalline form.

References

[1] Odier P 2004 *Proc. of Workshop on DC Current Transformers and Beam-Lifetime Evaluations (Lyon)* 3–5
[2] Agilent Technologies 2008 *Application Note 1369-1* p 17
[3] Spasojević D, Bukvić S, Milošević S and Stanley H 1996 *Phys. Rev. E* **54** no. 3 2531–46
[4] Kottman P 1998 *CERN PS-BD-Note-97-06-Rev.1*

IOP Conf. Series: Journal of Physics: Conf. Series **903** (2017) 012036 doi:10.1088/1742-6596/903/1/012036

Circularly polarized light interaction in topological insulators investigated by time-resolved ARPES

D Bugini[1,2,*], H Hedayat[1], F Boschini[1,5], H Yi[3], C Chen[3], X Zhou[3], C Manzoni[4], C Dallera[1], G Cerullo[1] and E Carpene[4]

[1] Dipartimento di Fisica, Politecnico di Milano, piazza Leonardo da Vinci 32, Milan, 20133, Italy

[2] Center for Nano Science and Technology@PoliMi, Istituto Italiano di Tecnologia, via Giovanni Pascoli 70/3, Milan 20133, Italy

[3] National Lab for Superconductivity, Institute of Physics, Chinese Academy of Science, Beijing, 100190, China

[4]IFN-CNR, Dipartimento di Fisica, Politecnico di Milano, 20133, Milan, Italy

E-mail: *davide.bugini@polimi.it

Abstract. Topological Insulators (TI) represent a hot-topic for both basic physics and promising applications because of the in-plane spin-polarized surface states (TSS) arising within the bulk insulating energy gap. The backscattering protection and the control of the spin polarization using ultrashort light pulses open new scenarios in the use of this class of materials for future opto-spintronic devices. Using time- and angle-resolved photoemission spectroscopy on $Sb_xBi_{(2-x)}Se_yTe_{(3-y)}$ class we studied the response of spin-polarized electrons to ultrashort circularly-polarized pulses. Here, we report for the first time the experimental evidence of a direct coupling between light and empty topological surface states (ESS) and the establishment of a flow of spin-polarized electrons in **k**-space i.e. a photon-induced spin-current.

1. Introduction

In the last years, topological insulators have become a hot-topic as benchmark for novel theoretical and experimental studies in spintronics. However, the electronic dynamics induced by ultrashort laser pulses is far from being completely unravelled in these materials. Because circularly-polarized light couples to the total angular momentum of electron (i.e. orbital momentum + spin), we performed time and angle-resolved photoemission spectroscopy (TR-ARPES) measurements using linear p-polarized pump beam and circularly-polarized probe beam. In this way we can reveal a spin-related parameter of the spin-polarized band structure. In Sec. 3.1 we present our circular dichroism (CD) TR-ARPES measurements on the topological insulator Bi_2Te_3. In Sec. 3.2 we show our measurements with circularly-polarized pump and linear p-polarized probe on the Bi_2Se_3 topological insulator. In this configuration the selective-coupling of helicities of light with spin leads to a spin-selective pumping.

[5] Present address: Department of Physics & Astronomy, University of British Columbia, Vancouver, BC V6T 1Z1, Canada.

Content from this work may be used under the terms of the Creative Commons Attribution 3.0 licence. Any further distribution of this work must maintain attribution to the author(s) and the title of the work, journal citation and DOI.
Published under licence by IOP Publishing Ltd

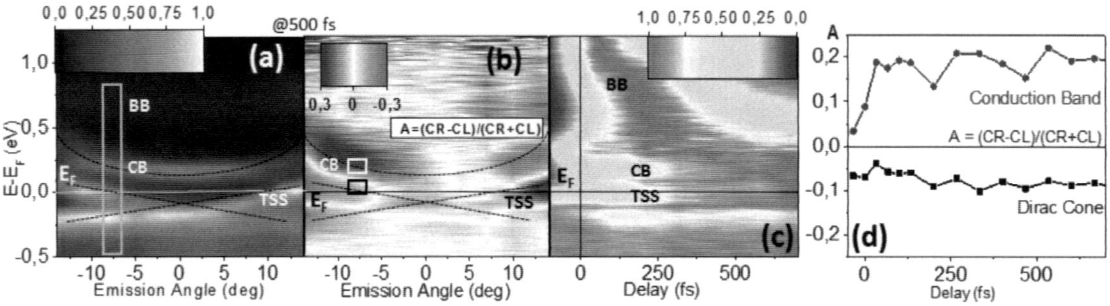

Figure 1. (a) Bi_2Te_3 ARPES map binding energy vs. angle of emission at 500 fs pump-probe delay. (b) Dichroic asymmetry $A = (CR - CL)/(CR + CL)$ map binding energy vs. angle of emission computed for (a). The colored squares represent the angles of emission and the energies at which the temporal evolutions of A shown in (d) are extracted. (c) Log intensity map binding energy vs pump-probe delay at the angle of emission marked by the green rectangle in (a). E_F:Fermi energy. Dashed black lines are a guide for eyes to follow linear dispersion of the Dirac cone and the parabolic dispersion of the conduction band.

2. Materials and methods

Our Yb-based TR-ARPES setup, described in detail in Ref. [1], provides pump pulses shorter than 30 fs at 1.85 eV and probe pulses at 6.02 eV with 65 fs time duration by means of a cascade of non-linear processes. The temporal resolution is ≈70 fs. Pump and probe beams impinge almost collinearly on the sample at an angle of about 45°. Photoemitted electrons are detected by means of a time-of-flight analyzer with energy and angular resolutions of ≈ 50 meV and 0.8°, respectively. High quality single crystals TIs were grown by the self-flux method. Powders of bismuth, selenium and tellurium were stoichiometrically mixed, put in alumina crucibles and sealed in a quartz tube under vacuum. The materials were heated to 1000° C for 12 hours and slowly cooled down to 500° C over 100 hours before reaching room temperature [2]. The resulting crystals, belonging to the R$\bar{3}m$ space group, were cleaved in situ at pressure <5 x 10^{-10} mbar for the TR-ARPES measurements. They naturally cleave along the (111) plane. All measurements here reported were performed at room temperature. Pump fluence was ≈ 0.5 $\frac{mJ}{cm^2}$.

3. Results

3.1. Circular Dichroism of probe on Bi_2Te_3

Fig. 1(a) shows ARPES map measured on Bi_2Te_3 for one helicity of the probe (circular right (CR)) and at 500 fs pump-probe delays. We see the parabolic-dispersing conduction band (CB) and the linear-dispersing topological surface state (TSS). In addition, a less-intense bulk band (BB) above the Fermi level at 0.8 eV binding energy can be observed. The binding energy vs. pump-probe delay map (Fig. 1(c)) for one emission angle (green rectangle in Fig. 1(a)) highlights the pump role and the dynamics of the electrons. The pump promotes electrons into high-energy bulk bands (BB) and after 100 fs they decay filling the lower energy empty bands i.e. the CB and the TSS. As already reported [3, 4], TSS is not directly populated by the pump beam. Computing the dichroic asymmetry $A = (CR - CL)/(CR + CL)$, we have a signature of the spin-polarization of the bands. As shown in Fig. 1(b), the two branches of the TSS present opposite dichroic signals passing throught the Γ-point. The asymmetry map also exhibits a dichroic contrast in the CB, always opposite to the signal in the upper part of the cone. We found a common trend of the dichroism in several compounds of $Sb_xBi_{(2-x)}Se_yTe_{(3-y)}$ class (not

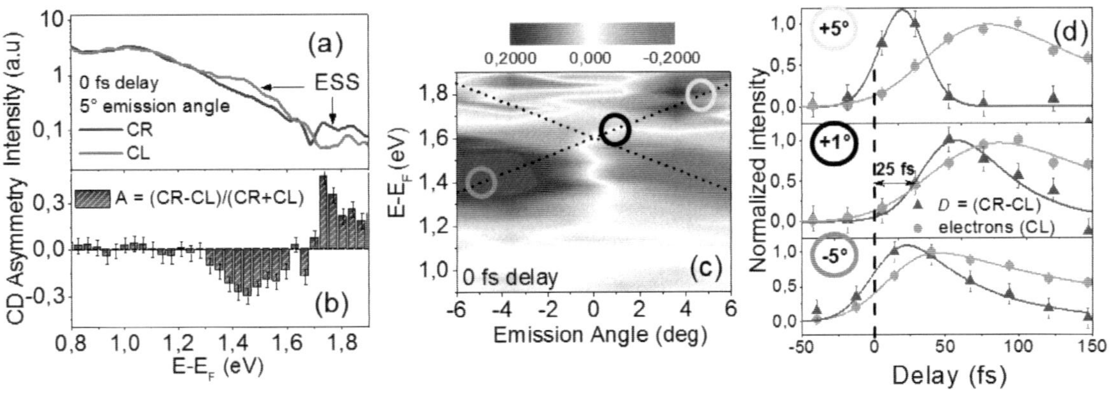

Figure 2. (a) Bi_2Se_3 energy distibution curves in the 0.8-1.9 eV binding energy range for two opposite pump helicities (circular right CR (blue) and circular left CL (red)) at 0 fs delay. (b) Dichroic asymmetry computed for (a). Dashed black lines are a guide for eyes to follow linear dispersion of the empty surface state (ESS). (c) Dichroic asymmetry map binding energy vs. angle of emission at 0 fs. The colored circles represent the angles of emission and the energies at which the CD difference and the related electronic background shown in (d) are calculated. (d) Comparison between dichroism (violet triangles) and the unpolarized electronic background induced by CL pump (orange circles) at three different angles of emission $+5°$ (top), $+1°$ (middle) and 5°(bottom) along the CR-pumped branch (blue branch in (c)).

shown) suggesting a universal property. A spin-polarized signature in the CB has already been reported as a surface-resonant state in Bi_2Se_3 [5] and in Sb_2Se_3 [6]. The dynamics of the dichroic signal of the two bands (Fig. 1(e)) suggests that the signal arising from TSS and from CB have different physical origin. For a spin-polarized band the dichroic asymmetry is expected to be constant in time. The dichroic signal of the TSS is constant at all delays confirming that it is a truly spin-polarized band. In contrast, the temporal evolution of the CB dichroism presents a rise around zero that resembles the relaxation-assisted population of the band. The electrons promoted in BB are unpolarized: when they decay, only the electrons matching the spin of the empty branch of TSS can fill it. The electrons of the opposite spin do not accumulate in BB due to the high energy and they decay into the first lower energy accessible band, i.e. the conduction band. Our data suggest the presence of a spin-dependent decay channel that leads to a polarization of the conduction band.

3.2. Selective circularly-polarized pumping of Bi_2Se_3

In TR-ARPES the main effect of the pump beam is to promote electrons into unoccupied bands by optical transitions. If the pump beam is circularly-polarized, it has been showed that this process becomes spin-selective [7]. In Fig. 2(a) the electron distribution curves (EDCs) of intrinsically n-doped Bi_2Se_3 for two different pump-helicities are reported. The unpolarized bulk bands are equally filled by the two opposite pump-helicities while, in the high-energy decay tails of these bands, two shoulders appear. Looking at the circular-asymmetry in Fig. 2(b), the two shoulders exhibit opposite character. In addition, they reverse their sign passing through the Γ-point (Fig. 2(c)) and a Dirac cone-like linear dispersion appears. This observation is consistent with previously reported 2PPE-measurements that disclosed an empty topological surface state (ESS) at the same binding energy [8, 9]. Focusing on the most intense band (blue one in Fig. 2(c)), we studied the dynamics of the spin-selectively promoted electrons by the circular-difference $D = C_R - C_L$. As shown in Fig. 2(a), only one helicity is able to promote electrons in the ESS while both are populating the underlying bulk bands. The circular-difference D then represents the spin-ordered electronic population promoted in ESS at the net of the bulk one. The dynamics for D and for the electrons of the bulks for three different angles (marked by different colored circles in Fig. 2(c)) are reported in Fig. 2(d). We see that for outer angles ($\pm 5°$), i.e. larger $\mathbf{k}_{||}$, the dynamics of the electrons promoted in ESS starts 25 fs earlier with respect to the region close to the Γ -point (i.e. 1°). This implies that the circular-pump is promoting a direct transition into the ESS only at outer angles starting from electrons occupying the valence and conduction bands. The spin-polarized electrons promoted into the ESS flow by electron-electron scattering along the linear branch of the Dirac cone without any spin-order dissipation and then, after 25 fs, the dichroic signal arises in the region close to the Γ-point. Our results highlight the capability to directly excite electrons in the ESS by circular light and, in addition, they show the establishment of a flow of spin-polarized electrons in **k**-space, i.e. a spin-current.

4. Conclusions

We have presented a circular dichroism time-resolved photoemission study on the Bi_2Te_3 topological insulator. We have observed a dichroic signal in the conduction band and studying the temporal evolution of this signal, we clarified that it originates from the presence of a spin-dependent decay channel for photon-excited electrons. Our results suggest the capability to generate a spin-polarized electron accumulation in the conduction band in addition to the spin-polarized topological surface state. We have presented also spin-selective pumping by time-resolved photoemission measurements on the topological insulator Bi_2Se_3. We reported a population of the empty topological surface state (ESS). The **k**-dependent temporal evolution of the electronic population in the ESS suggests an intraband-flow of spin-polarized electrons along the Dirac cone, i.e. a photon-induced spin-current.

5. References

[1] Boschini F, Hedayat H, Dallera C, Farinello P, Manzoni C, Magrez A, Berger H, Cerullo G and Carpene E 2014 *Rev. Sci. Instrum.* **85** 123903

[2] Chen C, He S, Weng H, Zhang W, Zhao L, Liu H, Jia X, Mou D, Liu S, He J *et al.* 2012 *Proc. Natl. Acad. Sci. U.S.A.* **109** 3694–3698

[3] Sobota J A, Yang S, Analytis J G, Chen Y L, Fisher I R, Kirchmann P S and Shen Z X 2012 *Phys. Rev. Lett.* **108**(11) 117403

[4] Hajlaoui M, Papalazarou E, Mauchain J, Lantz G, Moisan N, Boschetto D, Jiang Z, Miotkowski I, Chen Y, Taleb-Ibrahimi A *et al.* 2012 *Nano lett.* **12** 3532–3536

[5] Cacho C, Crepaldi A, Battiato M, Braun J, Cilento F, Zacchigna M, Richter M, Heckmann O, Springate E, Liu Y *et al.* 2015 *Phys. Rev. Lett.* **114** 097401

[6] Sessi P, Storz O, Bathon T, Wilfert S, Kokh K A, Tereshchenko O E, Bihlmayer G and Bode M 2016 *Phys. Rev. B* **93**(3) 035110

[7] Sánchez-Barriga J, Golias E, Varykhalov A, Braun J, Yashina L V, Schumann R, Minár J, Ebert H, Kornilov O and Rader O 2016 *Phys. Rev. B* **93**(15) 155426

[8] Sobota J A, Yang S L, Kemper A F, Lee J J, Schmitt F T, Li W, Moore R G, Analytis J G, Fisher I R, Kirchmann P S, Devereaux T P and Shen Z X 2013 *Phys. Rev. Lett.* **111**(13) 136802

[9] Niesner D, Fauster T, Eremeev S V, Menshchikova T V, Koroteev Y M, Protogenov A P, Chulkov E V, Tereshchenko O E, Kokh K A, Alekperov O, Nadjafov A and Mamedov N 2012 *Phys. Rev. B* **86**(20) 205403

Magnetotransport properties of nanogranular Bi and Co(c=0.2)Bi(1-c) thin films

P Athanasopoulos[1], Th Speliotis[2] and C Christides[1]
[1] Dept. of Computer Engineering & Informatics, University of Patras, 26504 Rion Patras, Greece.
[2] Institute of Nanoscience and Nanotechnology, NCSR "Demokritos", 15310 Athens, Greece

t.speliotis@inn.demokritos.gr

Abstract. A nanogranular thin film structure with concentration Co(c=0.2)Bi(1-c), and thickness 25 nm, exhibits: (i) a progressive crossing of Hall coefficient $R_H(B)$ curves from negative at lower fields to positive sign at higher fields between 5K and 200K, and (ii) the perpendicular magnetoresistance (PMR), $\rho_\perp(B,T)/\rho_\perp(B=0,T)$, curves are separated, exhibiting two different components above and below $|B|\approx 2T$. The low field PMR component can be described by the two band-model of bulk Bi. The effect of the low-field PMR component increases from 5K up to 300K by: $\Delta\rho_\perp(\Delta B=2T,300K)\approx 3\cdot\Delta\rho_\perp(\Delta B=2T,5K)$, following the same trend with $R_H(B,T)$ curves. Below 100K and for $|B|\geq 2T$ the second component in PMR curves can be simulated by a power law: $\rho_i(B)=b\cdot B^{\kappa}$, with an exponent κ: $1.3\leq\kappa<2$, which can be understood as a result of the two band-model equation. However, above 100K and for $|B|\geq 2T$ this second PMR component becomes linear in B: $\rho_i(B)=b\cdot B$, up to 300K. This behavior coincides with the temperature and field regions where the $R_H(B)$ curves reverse the sign of Hall coefficient. Such a linear MR with B can be related to an edge-resistance: $R_{edge}\sim B_\perp$, providing evidence for an *edge* conduction channel in grain boundaries of Co and Bi that competes the bulk component of bulk Bi, over the temperature and field regions.

1. Introduction

Granular electronic systems with metallic nanoparticles randomly dispersed in an insulating matrix can be made either a good metal or an insulator by tuning the coupling between granules. Topological insulators (TI) are materials that are insulating in their interior but can support the flow of electrons on their surface. These surface states are robust, maintained even in the presence of surface defects. A recent study [1] shows that the metallic surface states of Bi single-crystal nanoribbons dominate the transport properties in thinner sample (<40 nm) with insulating bulk. These metallic surface states were robust to the oxidation although the carrier density in the surface states are modified after the exposures. These results suggest that the metallic surface states in Bi nanoribbons should be topologically protected. Numerical simulations predicted [2] that, by introducing a layer of strong surface disorder, and patterning that layer's depth, a TI's surface conduction can be focused, directed along particular channels, and switched—all of the requirements for a *topological integrated circuit*. In this context, we performed experiments by introducing disorder on edge conduction channel of nanogranular Bismuth (Bi) thin films. Disorder is induced from grain boundaries between magnetic Co and Bi composite nanostructures. Our study focus on the magnetotransport properties of a composite Co(c = 0.2)Bi(1-c) thin film with granular nanostructure, deposited by magnetron sputtering [3] for a Co concentration-c=0.2. Experimental details are given in reference [3].

Content from this work may be used under the terms of the Creative Commons Attribution 3.0 licence. Any further distribution of this work must maintain attribution to the author(s) and the title of the work, journal citation and DOI.
Published under licence by IOP Publishing Ltd

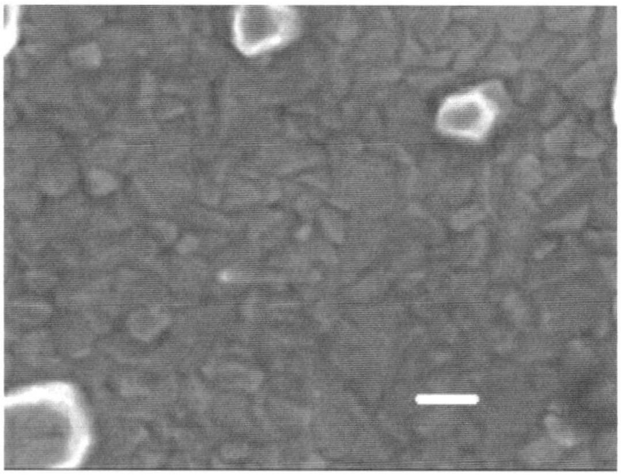

Figure 1. Surface topography of pure Bi film (c=0) with thickness 15 nm. Magnification 8×10^4. The scale bar is 100nm.

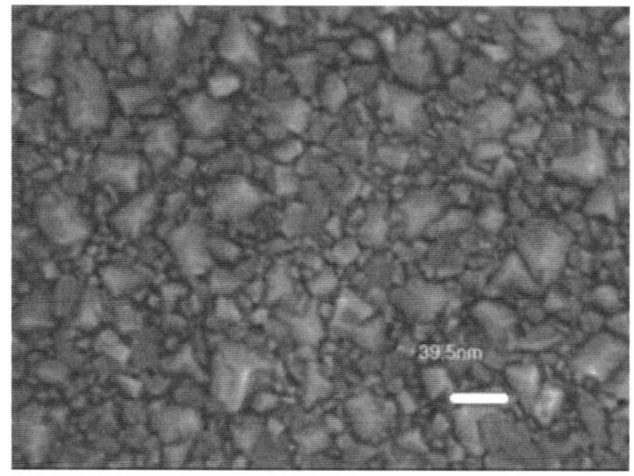

Figure 2. Surface topography of Co(c=0.2)Bi(1-c) with film thickness 25 nm. Magnification 8×10^4. The scale bar is 100nm.

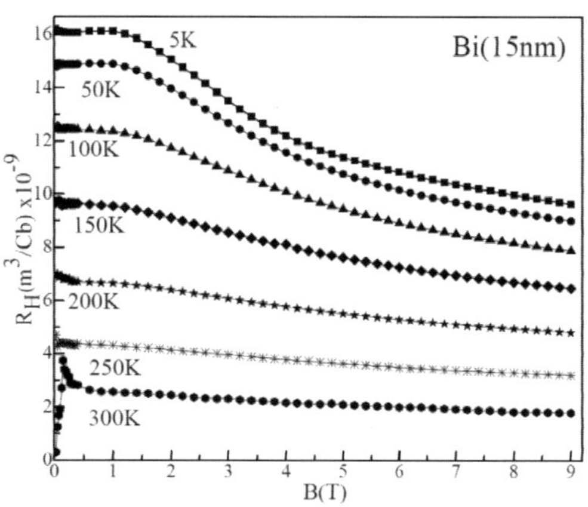

Figure 3. Hall coefficient $R_H(B)$ curves.

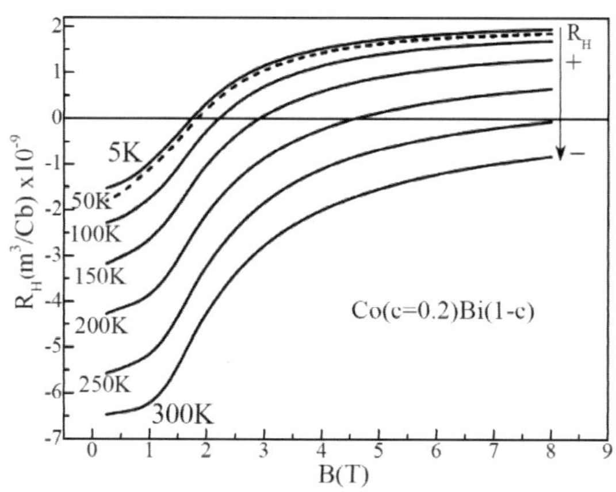

Figure 4. Hall coefficient $R_H(B)$ curves.

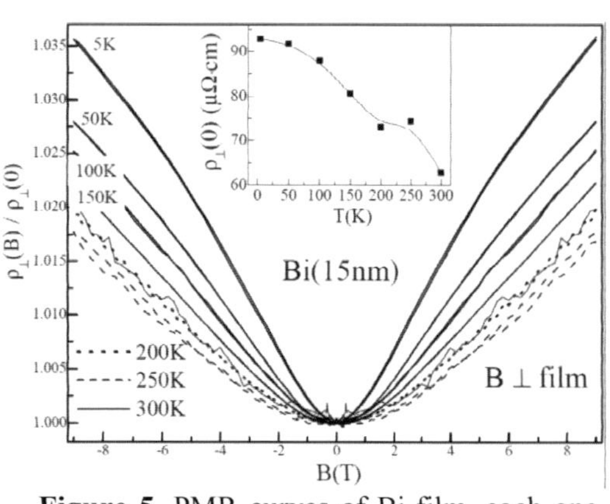

Figure 5. PMR curves of Bi film, each one normalized to resistivity at B=0 (inset).

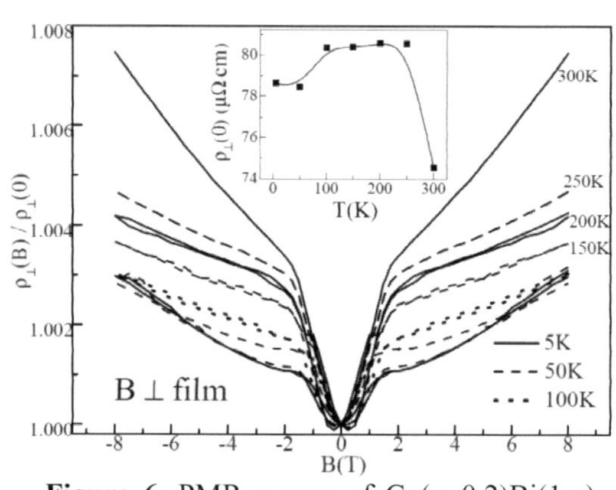

Figure 6. PMR curves of Co(c=0.2)Bi(1-c), each one normalized to resistivity at B=0 (inset).

2. Experimental results

Figures 1 and 2 show the surface topography of Bi and Co(c=0.2)Bi(1-c) films observed with a Field Emission Scanning Electron Microscope (FE-SEM), model Jeol JSM-7401F. Statistical analysis of grain size distributions from these images reveals that: (i) both films exhibit a log-normal distribution, (ii) in Bi film a mean value of grain sizes equals to D_{Bi}=71 nm and standard deviation σ_{Bi}=0.3 nm whereas in Co(c=0.2)Bi(1-c) film D_C=37 nm and σ_C=0.7 nm, evidencing that Co grains correspond to smaller sizes (almost half-size from those in Bi).

Magnetotransport measurements were performed with a horizontal rotator probe in a Physical Property Measurement System (Quantum design). Perpendicular magnetoresistance (PMR) and Hall coefficient (R_H) measurements were performed with the standard van der Pauw method. Figure 3 reveals that R_H(B) curves in Bi remain at positive values between 5 and 300K but, most important, the single slope from the ordinary Hall effect (OHE) at 250K and 300K exhibits below 250K a progressively increasing contribution from a second component at lower fields (B<2T). Other experimental results [4] provide evidence that this low field component can be due to a second conduction channel from surfaces states at the grain boundaries of Bi. Figure 4 shows that R_H(B) curves in Co(c=0.2)Bi(1-c) film exhibit negative values at 250K and 300K, whereas below 250K R_H(B) curves are crossing zero at intermediate field values, B_r, and become positive at higher fields, probing [3] the effective changes in conduction paths of charge carriers during Hall resistance measurements of composite films with ferromagnetic (Co) and semi-metallic (Bi) grain distributions. Thus, at lower temperatures the Hall effect displays 3 contributions: one from ordinary Hall effect, a second from the anomalous Hall effect that follows the magnetization, and a third one that, for $B>B_r$, reverses the sign of R_H(B) curves for T≤200K, and is reminiscent of the topological Hall effect observed [5] in $Mn_{1-x}Fe_xSi$.

Figures 5 and 6 show normalized PMR loops, $\rho_\perp(B,T)/\rho_\perp(B=0,T)$, of Bi and Co(c=0.2)Bi(1-c) films. Their insets indicate that zero field resistivity $\rho_\perp(B=0,T)$, in Bi and Co(c=0.2)Bi(1-c) films exhibit non-metallic temperature dependence. The PMR curves in figure 5 can be described [6] by the two band-model of bulk Bi, considering that:,

$$\rho_i(B) = a \cdot B^2 / (1 + g \cdot B^2)$$

where $g \sim (n_e - n_h)^2$ depends on charge density n_e (n_h) of electrons (holes) and the factor a depends on Hall mobility of charge carriers. Figure 6 shows that the PMR curves in Co(c=0.2)Bi(1-c) film can be separated in two different components above and below $|B|\approx 2T$. The effect of the low-field PMR component increases from 5K up to 300K by: $\Delta\rho_\perp^{Hall}(\Delta B=2T,300K)\approx 3\cdot\Delta\rho_\perp^{Hall}(\Delta B=2T,5K)$, following the observed behaviour of R_H(B,T) curves in figure 5. Below 100K and for $|B|\geq 2T$ the second component in PMR curves can be simulated by a power law: $\rho_i(B)=b\cdot B^\kappa$, with an exponent κ: $1.3 \leq \kappa < 2$, which is understood as a result of the two band-model equation in case of $n_e \neq n_h$ ($g>0$). Above 100K and for $|B|\geq 2T$ this second PMR component becomes linear in B: $\rho_i(B)=b\cdot B$, up to 300K. Such a linear MR with B can be related [7] to an edge-resistance: $R_{edge} \sim B_\perp$. This behaviour coincides with the temperature and field regions where the R_H(B) curves, in figure 4, change the sign of Hall coefficient. Both effects can be related with the predominance of an edge conduction channel, over the bulk Bi channel, at the grain boundaries of Co and Bi up to a Co concentration [3] of 0.3, indicating that edge conduction can create problems in the design [2] of *topological integrated circuits*.

References

[1] Ning W *et al*, *Nature Scientific Reports* | 4 : 7086 | DOI: 10.1038/srep07086
[2] Sacksteder V, Ohtsuki T, Kobayashi K, 2015 *Phys. Rev. Applied* **3** 064006
[3] Athanasopoulos P *et al*, 2013 *EPJ Web of Conferences* **40** 12002
[4] Steinberg H *et al*, 2010 *Nano Lett.* **10** 5032; Kong D *et al*, 2010 *Nano Lett.* **10** 329; Wang J *et al*, 2011 *Phys. Rev. B* **83** 245438; Yokota K *et al*, 2012 *Appl Phys Lett* **100** 251605.
[5] C. Franz *et al*, 2014 *Phys. Rev. Lett.* **112** 186601
[6] Gerber A *et al*, 2007 *Phys. Rev. Lett.* **99** 027201
[7] Alekseev P S *et al*, 2015 *Phys. Rev. Lett.* **114** 156601

IOP Conf. Series: Journal of Physics: Conf. Series **903** (2017) 012038 doi:10.1088/1742-6596/903/1/012038

Three-dimensional Dirac semimetal and magnetic quantum oscillations in Cd_3As_2

[1]**Filip Orbanić**, [1]**Mario Novak**, [2]**Nikola Biliškov**,[3]**Sanda Pleslić**, [1]**Ivan Kokanović**

[1]Department of Physics, Faculty of Science, Bijenička cesta 32, Zagreb, Croatia.
[2]Ruder Bošković Institute, Bijenička cesta 54, Zagreb, Croatia
[3]Department of Applied Physics, Faculty of Electrical Engineering and Computing, Unska 3, Zagreb, Croatia.

E-mail: forbanic@phy.hr

Abstract. Single crystals of topological Dirac semimetal Cd_3As_2 were synthesized using a modified vapour transport technique. The magnetization of Cd_3As_2 single crystals was measured in the temperature range from 4.2-300 K and magnetic field up to 5 T along [001] and [100] directions. In the low temperature region we observed the de Haas van Alphen (dHvA) oscillations which allow us to characterize the three dimensional Fermi surface by extracting its relevant parameters. The dHvA oscillations show the existence of two different Fermi surface cross sections and nontrivial Berry's phase, which is the signature of Dirac fermion in Cd_3As_2.

1. Introduction

Recently, three-dimensional (3D) topological Dirac semimetals (TDS) have been theoretically predicted [1, 2, 3] and a crystalline-symmetry-protected 3D bulk Dirac semimetal phase has been experimentally identified in a high-mobility Cd_3As_2 single crystal using the angle-resolved photoemission spectroscopy (ARPES) [4, 5]. Cd_3As_2 is a degenerate n-type semimetal of the II-V family with high mobility and low effective mass, which exhibits an inverted band structure ($E_g < 0$) [6, 7]. The conduction and valence bands of Cd_3As_2 touch at the Dirac point in the bulk band structure, giving rise to bulk Dirac fermions. Moreover, due to the lack of inversion symmetry this TDS phase raise a possibility of realizing the Weyl semimetal phase [8, 9].
Here we report measurements of magnetization in magnetic field up to 5 T parallel and perpendicular to the crystalline c-axis ([001] and [100] direction) for relatively large crystals of Cd_3As_2. The growth of the single crystals of Cd_3As_2 was carried out by a modified vapour transport technique. We observe a linear magnetization accompanied by the de Haas van Alphen (dHvA) quantum oscillations at low temperatures. The dHvA oscillations allow us to characterize the three dimensional Fermi surface. We analyzed the magnetization experimental data by using the Lifshitz-Kosevich (LK) theory which predicts the magnitude of the dHvA oscillations as well as their temperature dependence [10].

2. Experimental results and discussion

Single crystals of Cd_3As_2 were synthesized using the vapour transport technique. Starting polycrystalline material was prepared as described in [7]. Approximately 2 g of polycrystalline

Content from this work may be used under the terms of the Creative Commons Attribution 3.0 licence. Any further distribution of this work must maintain attribution to the author(s) and the title of the work, journal citation and DOI.
Published under licence by IOP Publishing Ltd

material is placed at one side of 16 mm in diameter and 10 cm long, vacuum sealed, quartz tube. The tube is then subjected to a temperature gradient in a modified two zone tube furnace. The empty side of the ampoule was kept at 565 °C while the temperature of the side with the material was slowly raised from 530 °C up to 585 °C with the rate of 1.7 °C /h. The tube was kept in this temperature gradient for 8 hours and then slowly cooled down to a room temperature. Two zones of the furnace are physically separated so there is a little drop in temperature at their junction when they are about the same temperature, Figure 1. Starting place of crystallization is restricted to the region of the initial lowest temperature. Without temperature kink material will be pushed to the end of ampoule and crystals will be compacted there. The grown crystals were granule like with dimensions from 1-5 mm.

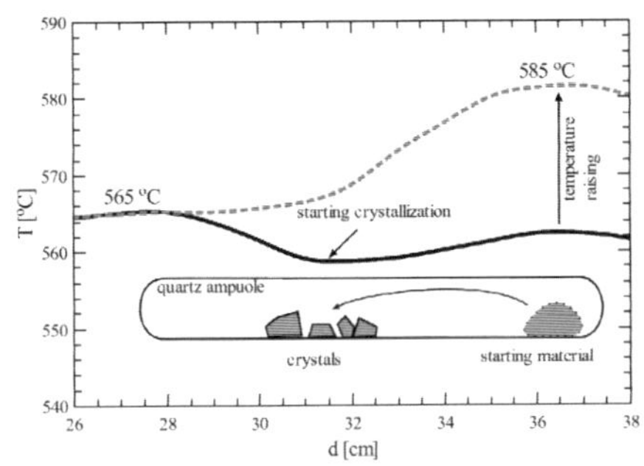

Figure 1. Shematic representation of crystal growth method.

The samples are examined by the x-ray diffraction. The crystal structure of the obtained Cd_3As_2 crystals has a tetragonal symmetry group $I4_1/acd$ with lattice parameters in agreement with previous studies [7].

Two samples of masses 82 mg (sample A) and 124 mg (sample B) were subjected to magnetization measurement. Magnetization was measured in the temperature range from 4.2-300 K and magnetic field up to 5 T using Quantum Design's Magnetic Property Measurement System which uses a SQUID amplifier as a sensitive magnetic field detector. The samples were first mounted and cooled in the SQUID magnetometer down to 4.2 K. Checks for the absence of possible ferromagnetic contamination were made by performing $M - B$ scans from 300 down to 5 K. Any intercepts at $B = 0$ T were less than $1.6 \cdot 10^{-5}$ emu for the measured samples. The samples are diamagnetic with magnetic susceptibilities that are only weakly temperature dependent down to about 60 K, below which an upturn can be observed. The quality of samples is also confirmed by measuring the temperature dependence of sample's resistance. Residual resistivity ratio (RRR) of nearly 6 is obtained for both samples.

In the low temperature region, from 25 K down to 4.2 K, we observe the dHvA oscillations. The oscillations can be traced down to field of 2 T. The magnetisation was measured along the [001] and [100] direction. By subtracting the diamagnetic background, pure oscillating part of magnetisation can be obtained. The dHvA oscillations of the sample B for magnetic field along [001] direction (a) and [100] direction (b) for different temperatures are shown in Figure 2.

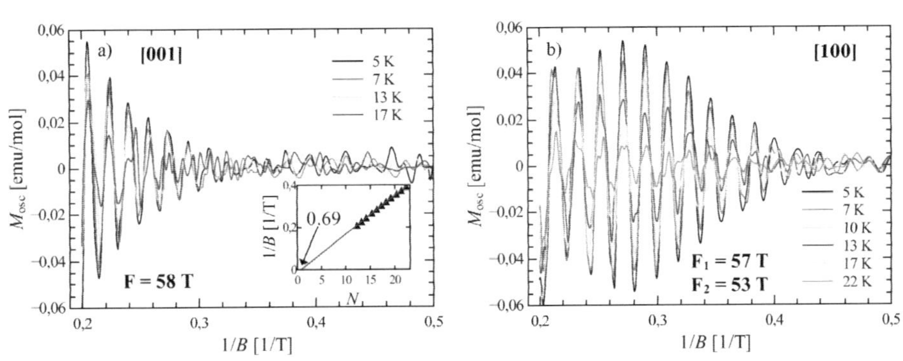

Figure 2. The dHvA oscillations for Cd_3As_2 sample (sample B) for magnetic field along [001] (a) and [100] direction (b).

According to the LK formula, for magnetic field along [001] direction a single frequency oscillation with $F = 55$ T (sample A) and $F = 58$ T (sample B) is obtained, Figure 2a. Oscillation frequency is determined by direct fitting of the LK formula for a single frequency and also by fast Fourier transform (FFT). Both methods give nearly the same values. Oscillation amplitude dependence on $1/B$ for magnetic field in [100] direction suggest that there is a contribution of at least two close frequencies, see Figure 2b. By FFT two frequencies can not been resolved. By fitting of the LK formula for two frequency contributions a good match can be achieved with frequencies $F_1 = 56$ T, $F_2 = 51$ T (sample A) and $F_1 = 57$ T, $F_2 = 53$ T (sample B). From temperature and $1/B$ dependence of single frequency oscillation amplitude the effective cyclotron mass of $m_c^* = 0.047m_e$ and Dingle temperature [10] of $T_D = 16.5$ K (related to quantum scattering time of $\tau_q = 7.4 \cdot 10^{-14}$ s) have been extracted. The quantum mobility of electrons is given by $\mu = e\tau_q/m_c^* = 0.28 \cdot 10^4$ cm^2/Vs. From oscillation frequency $F = \hbar k_F^2/2e = 55$ T one can obtain the Fermi wave vector $k_F = 0.041$ Å^{-1} which leads to the Fermi velocity $v_F = \hbar k_F/m_c* = 1.01 \cdot 10^6$ m/s. The charge carrier density has been obtained from $n = k_F^3/3\pi^2 = 2.3 \cdot 10^{18}$ cm^{-3}. Dirac dispersion of charge carriers can be confirmed by plotting the so called Landau level (LL) diagram [11] ($1/B$ vs. N-index corresponding to minima or maxima). For Dirac fermions the intercept of obtained line for $1/B = 0$ must be near 5/8 (for the minimal Fermi surface cross section), whereas for Scrödinger fermions the intercept is near 1/8. LL diagram for sample A for magnetic field along [001] direction gives intercept of 0.69 confirming the Dirac nature of charge carriers in Cd_3As_2, inset to Figure 2. Two different frequencies in dHvA oscillations, for magnetic field in given direction, indicate that there are two different extremal Fermi surface cross sections. Theoretically is predicted and experimentally confirmed that, below the Lifshitz transition, Fermi surface of Cd_3As_2 consist of two identical ellipsoids [2, 5], which can not explain two frequency contributions. The Fermi energy is also estimated, $E_F = m_c^* v_F^2 \approx 270$ meV, and according to [8, 12] E_F is above the Lifshitz point.

3. Conclusion

A very useful synthesis method for easy sublimating material, as Cd_3As_2, is described in detail. The synthesized samples were characterized by magnetization measurements and dHvA oscillations are observed at low temperatures. We obtained single frequency dHvA oscillations for magnetic field along [001] direction (not measured in [7]), whereas for field in [100] direction the contributions of two close frequencies are observed (authors in [7] report on just one frequency for a given direction). Modelling quantum oscillations by the LK formula the effective cyclotron mass, quantum scattering time, mobility, Fermi wave vector, Fermi velocity and density of carriers are estimated. From the LL diagram for single frequency quantum oscillations, a nontrivial π Berry phase is found, confirming the existence of the Dirac fermion in Cd_3As_2. Measurements like this will be helpful in further understanding of Cd_3As_2 Fermi surface which still remains unclear.

Acknowledgments

This work has been fully supported by Croatian Science Foundation under the project No. 6216.

References

[1] Young S M, Zaheer S, Teo J C Y, Kane C L, Mele E J and Rappe A M 2012 *Phys. Rev. B* **108** 140405
[2] Bohm J Y and Naoto N 2014 *Nat. Commun.* **5** 4898
[3] Wang Z, Sun Y, Chen X Q, Franchini C, Xu G, Weng H, Dai X and Fang Z 2012 *Phys. Rev B* **85** 195320
[4] Borisenko S *et al* 2014 *Phys. Rev. Lett.* **113** 027603
[5] Nupane M *et al* 2014 *Nat. Commun.* **5** 3786
[6] Liang T, Gibson Q, Ali M N, Liu M, Cava R J and Ong N P 2015 *Nat. Mat.* **14** 280-284
[7] Pariari A, Dutta P and Mandal P 2015 *Phys. Rev. B* **91** 155139
[8] Jeon S *et al* 2014 *Nat. Mat.* **13** 851-856
[9] Arnold F *et al* 2016 *Nat. Commun.* **7** 11615
[10] Shoenberg D 1984 *Magnetic Oscillations in Metals* (New York: Cambridge University Press)
[11] Xiong J, Luo Y, Khoo Y, Jia S, Cava R J and Ong N P 2012 *Phys. Rev. B* **86** 045314
[12] Feng J, Pang Y, Wu D, Wang Z, Weng H, Li J, Dai X, Fang Z, Shi Y and Lu L 2015 *Phys. Rev. B* **92** 081306

IOP Conf. Series: Journal of Physics: Conf. Series **903** (2017) 012039 doi:10.1088/1742-6596/903/1/012039

A semi-empirical approximation of static hysteresis for high flux densities in highly grain-oriented silicon iron

C Carrander, G Engdahl

Department of Electromagnetic Engineering, Royal Institute of Technology, Drottning Kristinas väg 6, 100 44 Stockholm, Sweden

claesca@kth.se

Abstract. In calculations and simulations regarding magnetic materials, it is important to have a have an accurate model of the hysteresis loop. The major loop, in particular, is used in many simulations. However, it is generally not possible to measure the true major loop, and it must therefore be approximated using a minor loop. There are several methods available for approximating magnetization curves, but they are primarily designed for paramagnetic materials, and are poorly suited to the highly grain-oriented steels used in modern transformers. Therefore, we propose two expressions for approximating the magnetization curves of grain-oriented silicon-iron steels. Both methods give close agreement with measurements and can be extrapolated to in order to describe the major loop.

1. Introduction

When modelling the magnetization of electromagnetic equipment such as transformers, it is important to have a hysteresis model that is accurate over the whole magnetization range. Over this range, the relative permeability varies from its nominal value, which in transformer steels is in the tens of thousands, to unity at saturation; a variation by more than four orders of magnitude.

There are several existing models for static magnetization, and many of the most common models [1] are based on the Brillouin function [2], with angular moments $m=-J,-J+1,\cdots,0,\cdots,J-1,J$:

$$B_J(x) = \frac{2J+1}{2J}\coth\left(\frac{2J+1}{2J}x\right) - \frac{1}{2J}\coth\left(\frac{1}{2J}x\right). \tag{1}$$

For large J (i.e. for a large sample with a large number of possible states), the ratio of the magnetization to the saturation magnetization M_S is the Langevin function with fitting parameter a:

$$\frac{M}{M_S} = \mathcal{L}\left(\frac{B}{a}\right) = \coth\left(\frac{B}{a}\right) - \frac{a}{B}. \tag{2}$$

The Brillouin function only holds for paramagnets, and does not account for hysteresis. This can partly be taken care of by replacing B in eq. (2) with an effective field $B_e = \mu_0(H + fM)$, where f is a constant and μ_0 is the permeability of free space. Still, the underlying physics hold only when the exchange energy is negligible compared to the Zeeman energy. As a result, using the Langevin function for ferromagnets tends not to give very good curve fits, as will be shown further on in this paper. In power transformers, it is often desirable to calculate the losses to within a few percent accuracy or better, and the Langevin approximation cannot deliver this level of accuracy. Additionally, it is impractical to have an expression where the magnetization appears in both the left and right hand terms.

Content from this work may be used under the terms of the Creative Commons Attribution 3.0 licence. Any further distribution of this work must maintain attribution to the author(s) and the title of the work, journal citation and DOI.
Published under licence by IOP Publishing Ltd

IOP Conf. Series: Journal of Physics: Conf. Series **903** (2017) 012039 doi:10.1088/1742-6596/903/1/012039

Finally, in engineering applications, it is often more useful to express *B*(*H*) than *M*(*B*). This can be done without much loss of generality since the derivation in [2] is based on minimization of energy. This minimization can be done, provided that the sample is large, using the per-volume energy *BH* instead of the per-electron magnetic energy $\mu_B B$, where μ_B is the Bohr magneton. In the rest of this paper, *B(H)* will be given. With this convention, the magnetization curve is made hysteretic by introducing the coercive force H_c, and $H - H_c$ thus fulfills the role of the effective field. In transformer steels, the two branches of the major loop do not have the same shape (see Figure 1) and must be expressed separately. Thus, by introducing fitting parameters $k_1, k_2, a_1,$ and a_2, Eq. (2) becomes:

$$\frac{B}{B_{sat}} - \mu_0 H = \begin{cases} k_1\left(\coth\left(\frac{H-H_c}{a_1}\right) - \frac{a_1}{H-H_c}\right), \text{for } B\frac{dB}{dH} < 0 \\ k_2\left(\coth\left(\frac{H-H_c}{a_2}\right) - \frac{a_2}{H-H_c}\right), \text{for } B\frac{dB}{dH} > 0 \end{cases} \quad (3)$$

where B_{sat} is the saturation flux density. It should be noted here that although k_2 should be equal to one, this does not, in our experience, give the best fit, and both k_1 and k_2 have therefore been left as variables. This modified Langevin function will be used as the benchmark for the other two magnetization expressions described in this paper.

2. Methods

The modified Langevin function in eq. (3) does not have the fat-tailed behavior of measured magnetization curves. This could be corrected by adding a constant slope, but doing so gives the wrong asymptotic behavior, as the slope of the *B-H* curve should tend toward μ_0 at saturation.

Instead, we propose two alternative methods. Both use $B_1(x)$, the Brillouin function with *J*=1. The choice of *J*=1 is arbitrary, but is not very significant. The first method applies the correct tail behavior directly in the Brillouin function and uses an exponential function to get the correct low-flux behavior:

$$\frac{B}{B_{sat}} - \mu_0 H = \begin{cases} -\left(1 - \exp\left(-\frac{H-H_c}{c_1}\right)\right) A_1 B_1\left(d_1(-H+H_c)^{b_1}\right), \text{for } B\frac{dB}{dH} < 0 \\ \left(1 - \exp\left(-\frac{H-H_c}{c_2}\right)\right) A_2 B_1\left(d_2(H-H_c)^{b_2}\right), \text{for } B\frac{dB}{dH} > 0 \end{cases} \quad (4)$$

Here, $A_1, A_2, b_1, b_2, c_1, c_2, d_1, and\ d_2$ are fitting parameters.

The second model uses a probability distribution function to get the correct tail behavior:

$$\begin{aligned} \frac{B}{B_{sat}} - \mu_0 H = {} & A_1 S_{CDF}(-H; \alpha, 1, \gamma_1, -H_c; 1) B_1\left(q_1(H-H_c)\right) \\ & + A_2 S_{CDF}(H; \alpha, 1, \gamma_2, H_c; 1) B_1\left(q_2(H-H_c)\right), \end{aligned} \quad (5)$$

where $A_1\ and\ A_2$ are amplitude factors, and $S_{CDF}(H; \alpha, \beta, \gamma, \delta; p)$ is a stable cumulative distribution function [3] with stability parameter α, skewness parameter β, shape parameter γ, location parameter δ, and parametrization *p*. In both methods, since the function must approach B_{sat} when *H* goes to infinity, A_2 must be equal to one, and A_1 can be deduced since the two branches must be equal at the reversal point. This gives a total of seven and six unknown fitting parameters, respectively.

3. Results and discussion

The static hysteresis curve of a highly grain-oriented electrical steel (27ZDKH) was measured at peak flux densities 1.7 T, 1.8 T, and 1.9 T.

The modified Langevin expression and the two alternative methods in Eqs. (4) and (5) were fitted to the measured curves. The best fit parameters are shown in Table 1. The table also shows the RMS deviation from the measured values. As can be seen in the table, the parameters vary only slightly depending on the peak flux density. Low variation means that the values are easy to extrapolate to higher peak flux densities. The large difference between the two branches of the hysteresis curve (i.e. between subscripts 1 and 2) also shows the need for the modified Langevin function as compared to the unmodified. The fat-tailed behavior of the magnetization curve is particularly marked close to the "knee" of the curve, as can be seen in Figure 1.

Table 1: Best fit parameters for the static hysteresis, using the two proposed models and the modified Langevin expression.

	Model 1				Model 2				Langevin		
	1.9 T	1.8 T	1.7 T		1.9 T	1.8 T	1.7 T		1.9 T	1.8 T	1.7 T
H_c	6.4	6.2	5.5	H_c	6.4	6.2	5.5	H_c	6.4	6.2	5.5
b_1	0.25	0.35	0.4	γ_1	0.35	0.35	0.5	k_1	0.92	0.89	0.86
b_2	0.2	0.2	0.2	γ_2	2	2	2	k_2	0.93	0.9	0.87
c_1	0.9	1.7	1.9	q_1	0.86	0.45	0.45	a_1	0.98	1.76	1.83
c_2	3.5	3.5	3.5	q_2	0.45	0.45	0.45	a_2	2.99	2.63	2.69
d_1	1.09	0.86	0.8	α	0.45	0.45	0.45				
d_2	0.82	0.82	0.82								
ΔB_{rms} [mT]	30	50	33		23	31	21		44	44	28

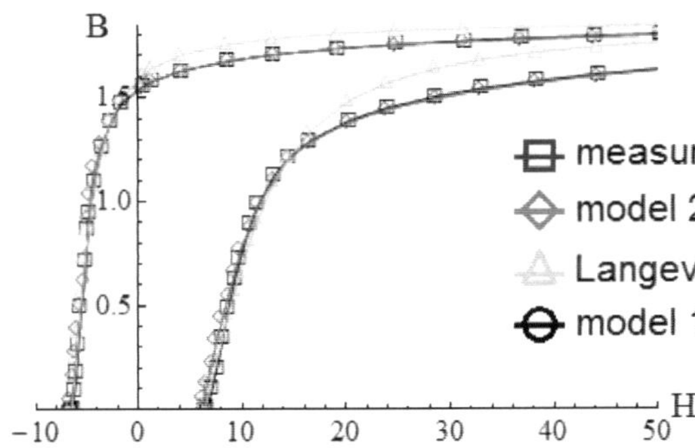

Figure 1: Measured static hysteresis at 1.9 T peak flux density and approximated curves using the modified Langevin expression and using model 2. Note that the two branches do not have the same shape.

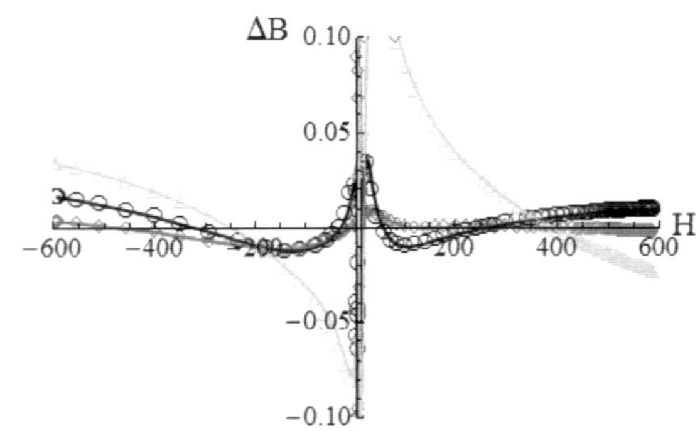

Figure 2: Deviation from measured static hysteresis at 1.9 T peak flux density for approximated curves using the modified Langevin expression and the proposed models.

The deviation of the fits from the measured values are shown in Figure 2. Here, it can be seen that the Langevin function gives an incorrect slope at high fields, while Model 2 agrees with the measurement to within 0.02 T for most values of H. Model 1 deviates more from the measured than Model 2 does, and the slope is not correct at high fields, as evidenced by the non-constant difference. All models give large errors close to the zero-crossing. This is because the slope is very large, and a small error in the model or in the measurement gives a large deviation.

It should be emphasized that both these methods, while based on valid theoretical models for paramagnets, are not intended to explain the true magnetization behavior of the steel, but merely to provide a highly accurate fit to measured values. In this sense, Model 2 is the most successful.

References

[1] David C. Jiles and David L. Atherton, 'Theory of ferromagnetic hysteresis', *J. Magn. Magn. Mater.*, vol. 61, no. 1–2, pp. 48–60, Sep. 1986.
[2] Leon Brillouin, 'Les moments de rotation et le magnétisme dans la mécanique ondulatoire', *J. Phys. Radium*, vol. 8, no. 2, pp. 74–84, 1927.
[3] John P. Nolan, *Stable Distributions - Models for Heavy Tailed Data.* (Boston: Birkhausen), 2015, In progress, Chapter 1 online at academic2.american.edu/$$jpnolan.

Development of High-Speed Switched Reluctance Motor for Electric Power Tools

Kenji Nakamura[1], Yuya Kumasaka[1], Osamu Ichinokura[1]

[1]Graduate School of Engineering, Tohoku University, 6-6-05 Aoba Aramaki Aoba-ku, Sendai 980-8579, Japan

E-mail: nakaken@ecei.tohoku.ac.jp

Abstract. This paper presents design and experimental evaluation of a switched reluctance (SR) motor used for electric power tools. First, characteristics of a previous designed 6/4-pole SR motor is shown and compared to a permanent magnet (PM) motor used in present electric power tools. Next, to further improve characteristics, a 12/8-pole SR motor is designed and evaluated in experiment. It is proved that the performance of the prototype 12/8-pole SR motor is almost comparable or superior to the present PM motor.

1. Introduction

Electric power tools including an electric screw driver and an electric impact driver *etc.* have generally employed less expensive universal motors. But recently, permanent magnet (PM) motors are being employed instead of the universal motors due to high-power, high-efficiency, and maintenance-free operation. However, the cost of rare-earth magnets used in the PM motors is high and its supply is unstable because of few producing countries. Therefore, it is expected to develop the rare-earth-free, high-power, high-efficiency, and maintenance-free motor for electric power tools.

A switched reluctance (SR) motor is simple and robust, less expensive due to no PMs, and maintenance-free. Thus, it is expected to be put into various fields [1]-[4], especially into high-speed applications like electric power tools. However, torque and efficiency of the SR motor are generally inferior to those of the PM motor.

In a previous paper [5], to overcome the above problem, we designed a 6/4-pole SR motor for electric power tools using a three-dimensional finite element method (3D-FEM), and indicated the possibility that the designed SR motor has almost the same performance as a PM motor used in present electric power tools. This paper describes further improvement of the SR motor and indicates experimental results of the prototype SR motor to demonstrate the performance of the prototype motor is almost comparable or superior to the present PM motor.

2. Characteristics of a previous 6/4-pole SR motor [5]

Figure 1(a) shows specifications of the PM motor used in present electric power tools. It is a three-phase, six-slots, and four-pole motor. The core material is non-oriented Si steel with a thickness of 0.35 mm, and the magnet material is Nd-Fe-B.

Figure 1(b) indicates the previous designed SR motor. The motor diameter and the axial length, and the core material are the same as those of the PM motor. On the other hand, the gap length is 0.2 mm, while the PM motor has 0.5 mm since the gap length greatly influences on the performance of the SR motor. In addition, the winding space factor is higher because the SR motor has an open slot structure.

Content from this work may be used under the terms of the Creative Commons Attribution 3.0 licence. Any further distribution of this work must maintain attribution to the author(s) and the title of the work, journal citation and DOI.
Published under licence by IOP Publishing Ltd

It was investigated from a geometrical viewpoint how many windings can be placed in the stator slot of the SR motor as shown in Figure 1(c). Thereby, the winding space factor of the designed SR motor was estimated about 47%, while the PM motor has about 24%.

Figure 1(d) indicates current density versus torque characteristics calculated by the 3D-FEM. It is understood that the torque of the designed SR motor is almost the same as that of the PM motor, however that the torque increasing becomes gentle under heavy load since magnetic saturation occurs in the stator salient pole.

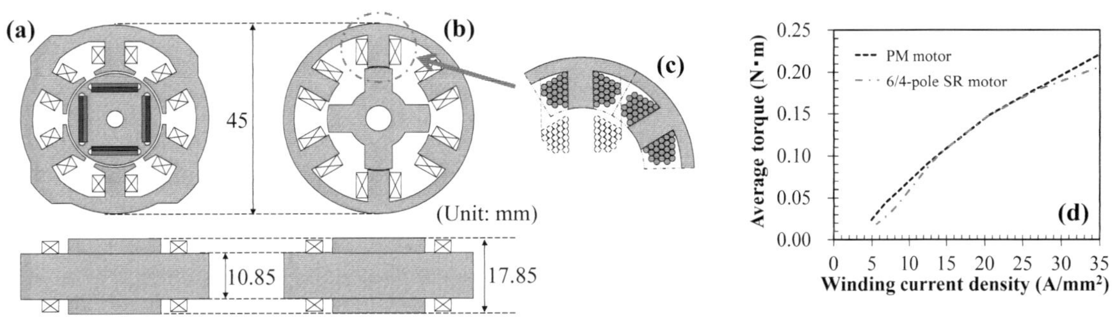

Figure 1. (a) Present PM motor used in electric power tools, (b) Designed 6/4-pole SR motor, (c) Winding arrangement of the 6/4-pole SR motor, and (d) Calculated current density versus torque characteristics.

3. Design and experimental evaluation of a 12/8-pole SR motor

To improve the magnetic saturation, the number of the poles is increased twice, namely, a 12/8-pole SR motor is designed as follows: let the winding diameter and the number of windings per phase of the 12/8-pole SR motor be the same as those of the 6/4-pole one. Accordingly, the number of windings per pole of the 12/8-pole SR motor is half the 6/4-pole one. Under the above conditions, the slot shape and dimensions are optimized from a geometrical viewpoint in the same manner shown in Figure 1(c). As a result, the widths of the stator pole and yoke of the 12/8-pole SR motor are about 60% of the 6/4-pole one, namely, it is expected to improve the magnetic saturation because the relative widths of the stator pole and yoke of the 12/8-pole SR motor are wider than those of the 6/4-pole one. Figure 2 indicates a comparison of contour diagrams of flux density of the 6/4-pole and 12/8-pole SR motors. The figure reveals that the magnetic saturation is relieved in the 12/8-pole SR motor.

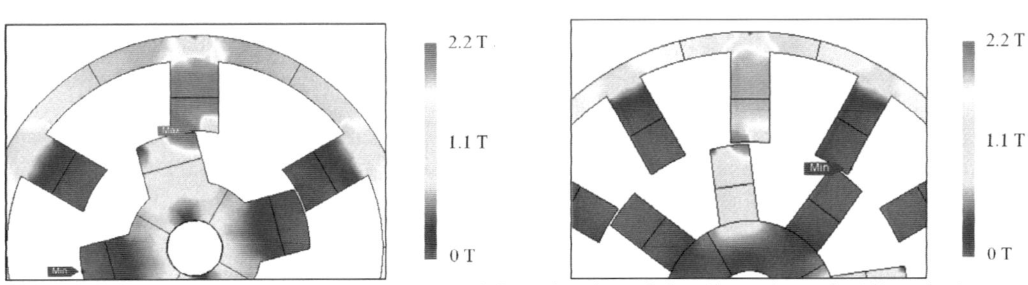

Figure 2. Comparison of contour diagrams of flux density of the 6/4-pole and 12/8-pole SR motors.

In addition, the length of the coil-end can be shortened since the number of windings per pole becomes half. Thereby, the stack length of the stator core of the 12/8-pole SR motor can be increased to improve the torque. Figure 3(a) shows shape and dimensions of the prototype 12/8-pole SR motor which is designed based on the above conditions. As shown in the figure, the stack length of the stator core is increased from 10.85 mm to 13.45 mm. Figure 3(b) indicates an appearance of the prototype SR motor, and (c) illustrates an experimental system.

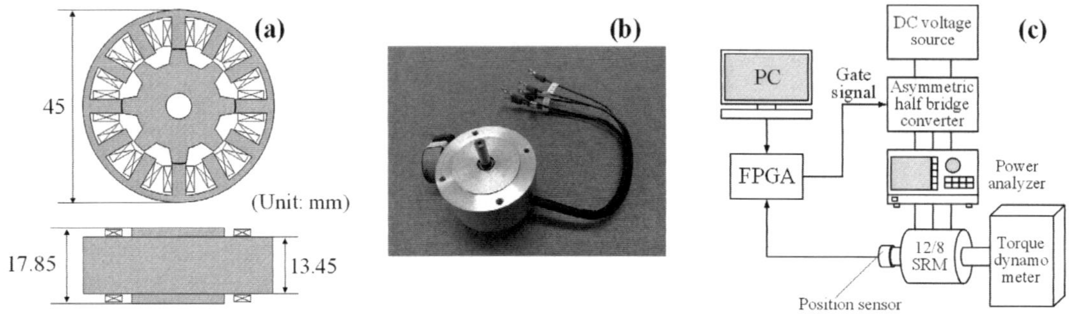

Figure 3. (a) Shape and dimensions of the prototype 12/8-pole SR motor, (b) Appearance of the prototype SR motor, and (c) Experimental system.

Figure 4(a) shows comparison of the current density versus torque characteristics. It is understood that the maximum torque of the prototype SR motor is larger than that of the present PM motor, and the measured values are agree well with the calculated ones. Figure 4(b) indicates the torque versus speed characteristics. The prototype SR motor exhibits the good performance as designed. Figure 4(c) shows the efficiency characteristics. From the figure, it is understood that the maximum efficiency of the prototype 12/8-pole SR motor is about 80%.

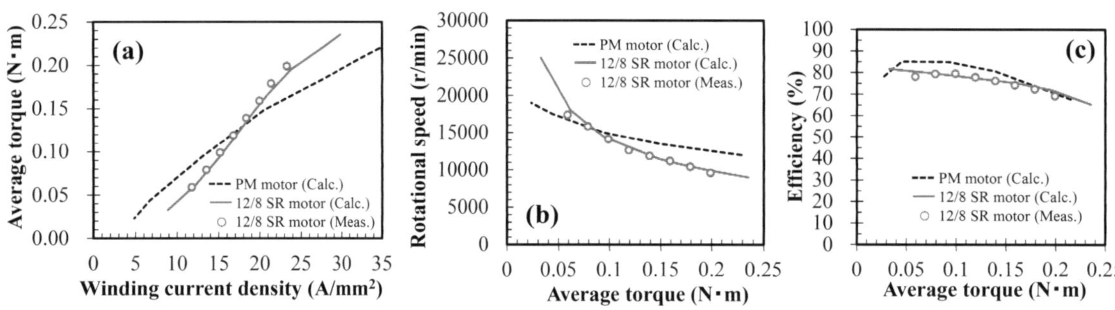

Figure 4. Comparison of the motor characteristics.

4. Conclusion

This paper presented the design and experimental evaluation of the SR motor used for electric power tools. First, characteristics of the previous designed 6/4-pole SR motor was indicated and compared to the PM motor used in present electric power tools. Next, to further improve characteristics, the 12/8-pole SR motor was designed. As a result, the relative cross sections of the stator pole and yoke of the 12/8-pole SR motor were larger than those of the 6/4-pole one, and thereby the magnetic saturation was relieved. Finally, the prototype 12/8-pole SR motor was evaluated experimentally. It was proved that the prototype SR motor has good performance as designed, and that the maximum torque of the prototype SR motor is larger than that of the present PM motor.

References

[1] Becerra R C, Ehsani M, and Miller T J E 1993 *IEEE Trans. Power Electron.* **8** 257-263
[2] Ferreira C A, Jones S R, Drager B T and Heglund W S 1995 *IEEE Trans. Power Electron.*, **10** 55-61
[3] Rahman K M, Fahimi B, Suresh G, Rajarathnam A V and Ehsani M 2000 *IEEE Trans. Ind. Applicat.* **36** 111-121
[4] Nakamura K, Suzuki Y, Goto H and Ichinokura O 2005 *Journal of Magnetism and Magnetic Materials* **290-291** 1334-1337
[5] Isobe K, Nakamura K and Ichinokura O 2014 *Journal of the Magnetics Society of Japan* **38** 194-198

IOP Conf. Series: Journal of Physics: Conf. Series **903** (2017) 012041 doi:10.1088/1742-6596/903/1/012041

Synthesis of magnetocaloric $LaFe_{11.6}Si_{1.4}$ alloy by spark plasma sintering

P. Shamba[1], G. Kerridge[1] N.A. Morley[1], I. M. Reaney[1] and W. M. Rainforth[1]
[1]Department of Materials Science and Engineering, University of Sheffield, Sheffield, S1 3JD, UK

n.a.morley@sheffield.ac.uk

Abstract. $LaFe_{11.6}Si_{1.4}$ alloys have been successfully fabricated by spark plasma sintering (SPS). An annealing study of the SPS $LaFe_{11.6}Si_{1.4}$ alloys at different temperatures ranging from 1373 to 1523 K for annealing times from 30 minutes to 72 hours was carried out. This annealing study showed that $LaFe_{11.6}Si_{1.4}$ samples annealed at 1473K (for annealing times between 30 minutes and 6 hours) have a significantly higher amount of the $NaZn_{13}$-type phase compared to samples annealed at other temperatures. Thus the critical annealing temperature which enhances the formation of the $NaZn_{13}$-type phase in SPS $LaFe_{11.6}Si_{1.4}$ compounds is 1473K. A second study investigated the effect of different particle sizes of the starting powders on the formation of the $NaZn_{13}$-type phase. This study found that the samples synthesized using larger sized powder particles exhibited a significantly higher amount of the $NaZn_{13}$-type phase compared to samples synthesized using smaller sized powder particles, for the same heat treatment.

1. Introduction

Magnetocaloric (MC) materials are candidate materials for solid state refrigeration, due to their theoretical cooling efficiency being 60% of the Carnot engine. Prototype devices have been demonstrated including the Haier wine cooler, launched in January 2015 [1], used MnFePSiGe for the MC component. One of the industrially preferred MC material is $LaFe_{11.6}Si_{1.4}$ ($La(Fe,Si)_{13}$) in the $NaZn_{13}$-type phase. This is because the change in the magnetic entropy ($-\Delta S_M$) is 20-25 $Jkg^{-1}K^{-1}$ for $La(Fe,Si)_{13}$, which is one of the highest of the MC materials with the Curie Temperature (T_c) near room temperature [2], plus it is non-toxic and consists of relatively cheap elements. $La(Fe,Si)_{13}$ alloys disadvantages include the T_c~187K and the difficulty in achieving the $NaZn_{13}$-type phase. The first is solved by chemically doping with either Co or H, which shifts T_c up to ~270K and maintains $-\Delta S_M$ at 10-20 $Jkg^{-1}K^{-1}$, as required for applications [2,3].

For the second, the main method of fabricating $La(Fe,Si)_{13}$ involves using conventional casting methods (arc-melting and induction melting) [3] followed by annealing at 1373K for up to 14 days, to allow the correct $NaZn_{13}$-type phase to be achieved. One method of producing MC materials in a shorter time frame is melt-spun ribbons. Lyubina et al [4] investigated melt-spun $La(Fe,Si)_{13}$ ribbons, and determined that only a short annealing time of 1 hr at 1273K was required to achieve 90% 1:13 phase within the samples. The $La(Fe,Si)_{13}$ ribbons had no thermal hysteresis as a function of magnetic field, but still had a first order transition. The main disadvantage of melt-spun $La(Fe,Si)_{13}$ ribbons was that they were brittle, so tended to break under magnetic cycling. Thus further investigations into different synthesis routes are required, so to reduce the production time, while maintaining the large $-\Delta S_M$.

Spark plasma sintering (SPS) is becoming a popular method for fast synthesis of a range of materials from metals to ceramics [5]. The process involves applying a high pressure at the same time as an electric

Content from this work may be used under the terms of the Creative Commons Attribution 3.0 licence. Any further distribution of this work must maintain attribution to the author(s) and the title of the work, journal citation and DOI.
Published under licence by IOP Publishing Ltd

current to a powder sample within a die, at a holding temperature for a short period of time (5 – 30 mins) [5]. This produces dense samples as well as control of the grain size. Further annealing can take place after the SPS synthesis. Both Patisser et al. [6] and Shamba et al. [7] have used SPS to successfully sinter the $NaZn_{13}$-type phase in $La(Fe,Si)_{13}$. In this paper, we present two studies of how the annealing temperature and particle size influences the amount of $NaZn_{13}$-type phase present in $La(Fe,Si)_{13}$ synthesized using SPS.

2. Experimental Procedure

5g of high purity (99.9% or higher) La, Fe and Si elemental powders were mixed in the required stoichiometric ratio using a speedmixer (Speedmixer DAC 800 FVZ) for 20 minutes. The thoroughly mixed powders were introduced into a 20mm graphite die under argon atmosphere and consolidated using spark plasma sintering (FCT Systeme GmBH SPS system, type HPD 1050). The variation of the sintering parameters for the 12 samples studied are given in Table 1. The sintering parameters varied included the particle sizes of the starting materials from 5 - 800μm, the applied pressure from 41-51MPa, and the holding time from 5-15 minutes. After the SPS process the samples were polished to remove the graphite coatings, wrapped in tantalum foil and place in evacuated quartz tubes, then annealed at different temperatures ranging from 1373-1523K.

Analysis of the samples crystal structure was carried out using x-ray diffraction (XRD) with Cu K_α radiation using a Siemens D5000 diffractometer. A scanning electron microscope (SEM), (FEI Inspect F (FEI, Netherlands)) was used to investigate the microstructures of the SPS $La(Fe,Si)_{13}$ alloys including elemental mapping using energy dispersive spectroscopy (EDS). The magnetization measurements were carried out using a Quantum Design Physical Property Measurement system (PPMS) at 0.02T, varying the temperature from 100 – 350K.

Table 1. The fabrication and annealing parameters for the SPS $La(Fe,Si)_{13}$ samples

	Starting particle size (μm)	Applied pressure (MPa)	Holding time (mins)	Holding Temp (K)	Annealing temp (K)	Annealing time (hrs)
Sample 1	45 - 800	41	5	1423	1373	72
Sample 2	45 - 800	41	10	1423	1373	72
Sample 3	45 - 800	41	15	1423	1373	72
Sample 4	45 - 800	45	5	1423	1373	72
Sample 5	45 - 800	51	5	1423	1373	72
Sample 6	45 - 800	41	5	1423	1473	0.5
Sample 7	45 - 74	41	5	1423	1523	6
Sample 8	45 - 74	41	5	1423	1473	0.5
Sample 9	45 - 74	41	5	1423	1473	6
Sample 10	45 - 74	41	5	1423	1373	72
Sample 11	45 - 74	41	5	1423	1373	72
Sample 12	5 - 74	41	5	1423	1373	72

3. Results and Discussion

In this work, we investigated the effects of various high-temperature annealing times from 1373-1523K on the phase formation of the SPS $La(Fe,Si)_{13}$ compounds. For the $La(Fe,Si)_{13}$ material system, the $NaZn_{13}$-type structure phase (known as the 1:13 phase), whose three strong diffraction peaks are at $2\theta = 27$, 35 and 38.5° is the phase that exhibits promising MC properties. The α-Fe phase has strong diffraction peaks at $2\theta = 45$ and 65° (Fig. 1a) and is detrimental to the MC properties. From the XRD patterns of the SPS $La(Fe,Si)_{13}$ samples, it is clearly observed that samples 6, 8 and 9 exhibit a significant amount of the 1:13 phase as compared to the rest of the samples, as the 1:13 phase XRD peaks have a higher intensity suggesting an increase in its volume fraction. Table 1 shows that samples 6 and 8 were annealed at 1473K for 30 minutes, whilst sample 9 was annealed at 1473K for 6 hours. The differing

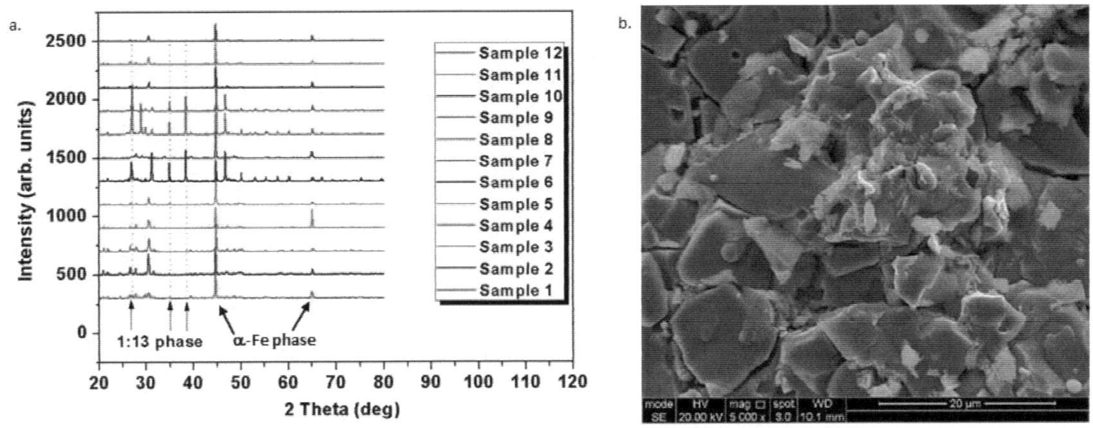

Figure 1a. XRD of the different SPS $La(Fe,Si)_{13}$ alloys. **1b**. Backscattered SEM micrograph of $La(Fe,Si)_{13}$ sample 6.

amounts of the 1:13 phase in samples 6, 8 and 9 can be attributed to the different particle sizes of the starting powders (Table 1). Sample 6 which was synthesized using larger La powder particles (800μm) exhibited a significantly higher amount of $NaZn_{13}$-type phase compared to samples 8 and 9, which were synthesized using smaller sized La powder particles (74μm), but for different annealing times.

It is also observed in Fig. 1a that samples 1-5 and samples 10-12, which were annealed at 1373K for 72 hours consisted mainly of the α-Fe phase suggesting that annealing $La(Fe,Si)_{13}$ compounds at lower temperatures for longer times independent of particle size does not significantly increase the formation of the 1:13 phase. While sample 7 was annealed at 1523K for 6 hours, it also consisted mainly of the α-Fe phase, which suggests that annealing $La(Fe,Si)_{13}$ compounds at temperatures above 1500K for longer times also does not increase the formation of the 1:13 phase. Thus the XRD confirms that short time annealing at 1473K causes homogenization of the $La(Fe,Si)_{13}$ alloys by enhancing the formation of the 1:13 phase, thereby suggesting that 1473K is the critical temperature for obtaining the 1:13 phase.

Fig 1b shows the backscattered SEM micrograph of the SPS $La(Fe,Si)_{13}$ sample 6 alloy in the unpolished condition. The different phases in these alloys were identified by EDS (Fig. 2a). Fig. 1b shows that the SPS $La(Fe,Si)_{13}$ alloy has a texture with grains of various diameters ranging from 5 to 17μm. The main grey phase within the grains contains La, Fe and Si, which corresponds to the 1:13

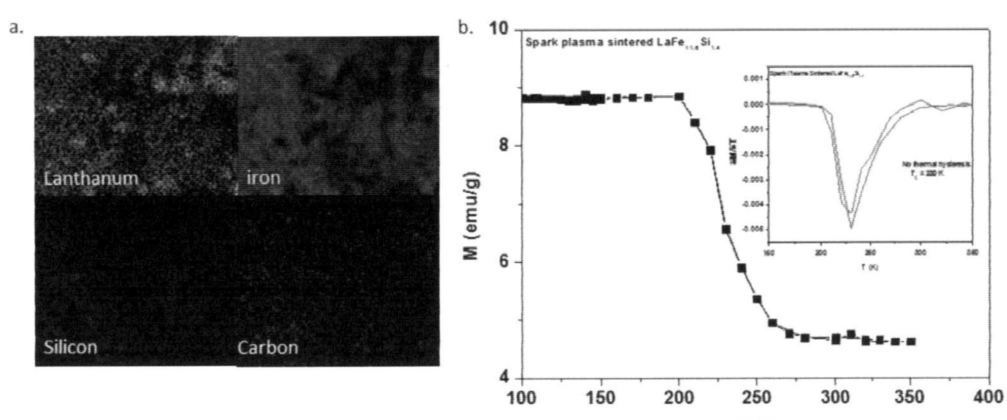

Figure 2a. EDS elemental mapping of the $La(Fe,Si)_{13}$ sample 6. **2b**. Magnetisation as a function of temperature of sample 6 $\mu_0H = 0.02$ T. **Inset:** *dM/dT* of $La(Fe,Si)_{13}$ sample 6.

phase. The presence of a cracked appearance of the SPS La(Fe,Si)$_{13}$ alloy can be attributed to the short annealing time (30 minutes) which prevented full densification. The composition distribution in these La(Fe,Si)$_{13}$ alloys was investigated by SEM and EDS experiments. Fig. 2a shows the element mappings of the La(Fe,Si)$_{13}$ sample 6, in which the La, Fe, Si C elements were scanned to enable distributions of these elements to be obtained. The element mappings do not show significant inhomogeneity. As expected, carbon is observed in the entire sample, which is due to contamination from the SPS graphite die. Fu et al [8] investigated arc-melted La(Fe,Si)$_{13}$ doped with carbon, and found that the carbon diffused through the microstructure. They also found that the carbon lowered the optimum annealing temperature and accelerated the formation of the 1:13 phase by improving the nucleation rate. Thus the present of the die carbon, could be beneficial to the synthesis process and explain why a shorter annealing time gives the preferential phase.

Fig 2b shows the temperature dependence of the magnetization for the La(Fe,Si)$_{13}$ sample 6 measured in the zero field cooled (ZFC) and field-cooled (FC) process under a magnetic field of 0.02T over a temperature range from 100K to 350K. T_c is defined as the temperature at which the *dM/dT* (fig. 2b inset) of the heating *M-T* curves is at a minimum. For the SPS La(Fe,Si)$_{13}$ sample 6, T_c = 230K, compared to an arc-melted La(Fe,Si)$_{13}$ sample T_c = 186K. The SPS sample T_c is higher, due to a larger amount of residual α-Fe impurity phase. This α-Fe impurity phase was confirmed by the residual magnetization to be ~ 4.5emu/g after the phase transition (Fig. 2b). Fig 2b, also shows the broadened FM-PM transition of the SPS La(Fe,Si)$_{13}$ alloy with no thermal hysteresis. This broadening of the *M-T* cuvre is an indication of a 2nd order magnetic transition of this compound. The MC properties as well as other magnetic properties of the SPS La(Fe,Si)$_{13}$ alloy are discussed in details elsewhere [7].

4. Conclusions

The optimisation of SPS conditions for the synthesis of La(Fe,Si)$_{13}$ alloys has been investigated in detail by varying sintering parameters such as the applied pressure and holding times. An annealing study has shown that the optimum annealing temperature which results in the highest yield of the NaZn$_{13}$- type phase is 1473K. Results from this work show that the particle size of the starting powders play an important role in optimising the yield of NaZn$_{13}$-type phase. This study found that samples synthesised using larger La sized powder particles exhibited a significantly higher amount of the NaZn$_{13}$-type phase compared to samples synthesised using smaller La sized powder particles after undergoing the same heat treatment.

Acknowledgements

This research was funded from the Engineering and Physical Sciences Research Council (EPSRC), Grant Number L107563

References

[1] Haier, BASF and Astronautics, Jan 2015, News Release
[2] Gutfleisch O, Willard M A, Bruck E, Chen C H, Sankar S G and Liu J P, 2011, *Adv. Mater.* **23**, 821-842
[3] Liu J, Moore J D, Skokov K P, Krautz M, Lowe K, Barcza A, Katter M and Gutfleisch O, 2012, *Scripta Mater.* **67**, 584-589
[4] Lyubina J, Gutfleisch O, Kuz'min M D and Richter M, 2008, *J. Magn. Magn. Mat*, **320**, 2252-2258
[5] Wang L, Zhang J and Jiang W, 2013, *Int. J. Refract. Met. Hard Mats*, **39**, 103-112
[6] Patissier A and Paul-Boncour V, 2015, *J. Alloys Comp*, **645**, 143-150
[7] Shamba P, Morley N A, Cespedes O, Reaney I M and Rainforth W M, 2016, *Appl Phys A*, **122**, 8, 1-9
[8] Fu S, Long Yi, Sun Y and Hu J, 2013, *Intermetallics*, **39**, 79-83

IOP Conf. Series: Journal of Physics: Conf. Series **903** (2017) 012042 doi:10.1088/1742-6596/903/1/012042

Thermal stability of $Sm_2Fe_{17}N_3$ magnet powders

N Imaoka[1,2], K Takagi[1], K Ozaki[1], T Iriyama[3] and H Kato[4]

[1]National Institute of Advanced Industrial Science and Technology, Nagoya, 463-8560, Japan
[2]Asahi Kasei Corporation, Fuji, 416-8501 Japan
[3]Daido Steel Co., Ltd., Nagoya, 457-8545 Japan
[4]Yamagata University, Yonezawa, 992-8510 Japan

E-mail: n-imaoka@aist.go.jp

Abstract. The time dependence of coercivity of fine $Sm_2Fe_{17}N_3$ powder compacts exposed in air and in He gas (oxygen-free atmosphere) was investigated at 110°C and 150°C. At 110°C in air, coercivity rapidly decreased within 5 h, reached a minimum at around 5h, and then gradually increased over 1000 h. Similar behavior was observed in the He gas atmosphere, although the minimum coercivity value was higher than that in air exposure. Since the oxygen content was confirmed to be constant, the coercivity minimum behavior is independent of oxidation. The temperature dependence of the anisotropy field, coercivity and saturation magnetization of fine $Sm_2Fe_{17}N_3$ powder compacts was measured in air at 110°C and analyzed by using Kronmüller's formula.

1. Introduction

$Sm_2Fe_{17}N_3$ compounds are good candidates for high performance magnets because of their high saturation magnetization and strong uniaxial anisotropy [1,2]. However, it is fairly difficult to obtain fully dense magnets, because $Sm_2Fe_{17}N_3$ compounds decompose into SmN and α-Fe when heated above 600°C, and in particular, the coercivity of fine $Sm_2Fe_{17}N_3$ powder (about 2 μm) is unstable even at 200°C [3]. For this reason, application of $Sm_2Fe_{17}N_3$ compounds is currently limited to resin bonded magnets. Recently, several researchers have reported $Sm_2Fe_{17}N_3$ powders having high coercivity of 10-23 kOe [4-7]. These works demonstrate the potential for application not only to the bonded magnets, but also to excellent bulk magnets rivalling the Nd-Fe-B magnets. Surface modification and consolidation methods for maintaining large coercivity even at high temperatures are the key techniques for realizing bulk magnets with $Sm_2Fe_{17}N_3$ powders. It is also important to know the fundamental properties of the powder at high temperatures. Therefore, in this work, we investigated the time dependence of coercivity of fine $Sm_2Fe_{17}N_3$ powder compacts exposed in air and in He gas at 110°C and 150°C.

2. Experiment

Coarse $Sm_2Fe_{17}N_3$ powders (particle size: 30 μm) were prepared by nitrogenation of Sm_2Fe_{17} alloy powders in a mixed gas of ammonia and hydrogen [2], and were then pulverized into fine powders (particle size: 2 μm) by jet milling. The fine $Sm_2Fe_{17}N_3$ powders were compacted at a pressure of 12 tonf/cm^2 under an alignment field of 15 kOe. Some $Sm_2Fe_{17}N_3$ compacts were exposed for 0 h (initial sample) to 2500 h at 110°C. The other $Sm_2Fe_{17}N_3$ compacts were encapsulated in Pyrex glass tubes with 0.3 atm helium gas, after the inner gas in these tubes had been replaced three times with helium gas through a liquid nitrogen cold trap. Subsequently, the glass tubes containing the $Sm_2Fe_{17}N_3$ compacts

Content from this work may be used under the terms of the Creative Commons Attribution 3.0 licence. Any further distribution of this work must maintain attribution to the author(s) and the title of the work, journal citation and DOI.
Published under licence by IOP Publishing Ltd

were annealed at 110ºC and 150ºC. The coercivities of the compacts were measured by a vibrating sample magnetometer (VSM) after magnetization with a pulse field of 60 kOe. The temperature dependence of magnetic properties from the temperature of liquid helium to room temperature was measured by VSM after magnetization with a static field of 150 kOe. The oxygen contents in the compacts were analyzed by an inert gas fusion method.

3. Results and Discussion

The time dependence of coercivity of the $Sm_2Fe_{17}N_3$ compacts at 110ºC is shown in Fig. 1. The coercivity of $Sm_2Fe_{17}N_3$ decreased to about 50% (4.8 kOe) of the initial value of 9.4 kOe when the sample was exposed in air for around 5 h, then increased gradually and recovered to 70% of the initial value at 1000 h. Similar behavior was observed in the helium atmosphere, although the minimum coercivity value (6.9 kOe after 5 h exposure) was larger than that in air exposure (4.8 kOe). From this result, it does not seem that the coercivity decrease-increase process occurs with the progress of oxidation. Fig. 2 shows the time dependences of the coercivity and oxygen contents of $Sm_2Fe_{17}N_3$ at 150ºC in the helium atmosphere. The significant difference between Fig. 1 and 2 is the time of minimum coercivity, which occurred at 5 hours at 110ºC and 30 minutes at 150ºC. The oxygen contents of all the compacts showed approximately constant values of 1.3 wt% from 0 to 8 hours. This indicates that the coercivity minimum behavior is independent of the oxygen content.

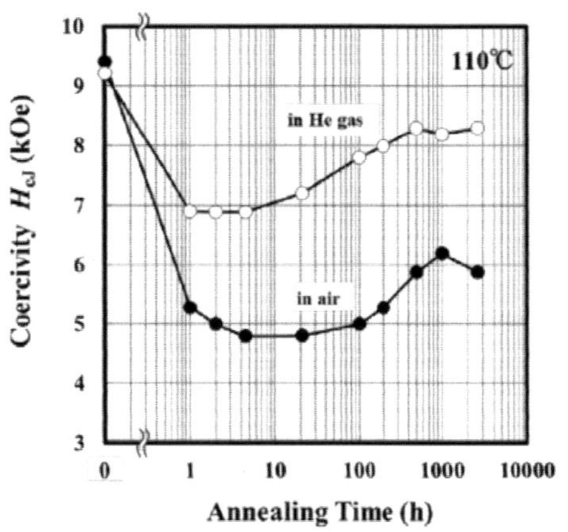

Figure 1. Time dependences of coercivity of $Sm_2Fe_{17}N_3$ compacts at 110ºC in air (closed circles) and in He gas (open circles).

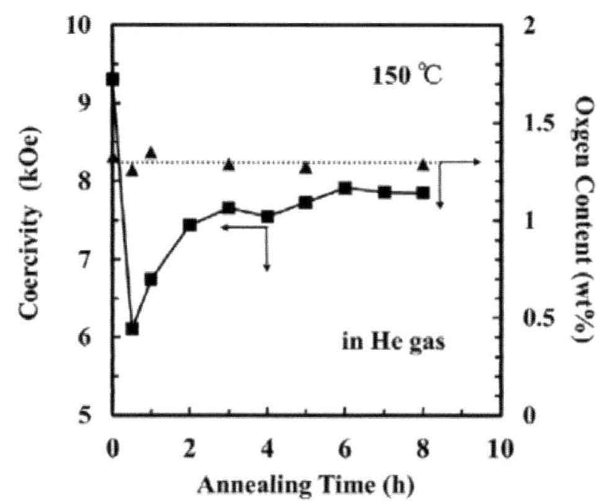

Figure 2. Time dependences of coercivity (closed squares) and oxygen content (closed triangles) of $Sm_2Fe_{17}N_3$ compacts at 150ºC in He gas.

In order to understand the mechanism of the coercivity decrease-increase process, we measured the temperature dependence of the anisotropy field $H_A^* = 2(K_1 + K_2) / M_s$ [8], where $H_A^* M_s/2$ equals the anisotropy energy, coercivity H_{cJ} and saturation magnetization M_s of fine $Sm_2Fe_{17}N_3$ powder compacts which were exposed in air at 110ºC for A) 0 h (non-exposure), B) 100 h, and C) 1000 h. Fig. 3 shows the temperature dependence of the coercivity values of samples A)-C). The coercivity values of all the samples monotonically increased with decreasing temperature. Since the coercivity of fine $Sm_2Fe_{17}N_3$ powders is controlled by the nucleation process [8, 9], their coercivity can be expressed as $H_{cJ} = cH_A^* - NM_s$ [8]. This formula is adapted from Ref. 10 by using H_A^* instead of $2K_1/M_s$, which is another approach shown in Ref. 9. All the samples A)-C) exhibited linear relationships between H_A^* / M_s and H_{cJ} / M_s except at the temperature of liquid helium (Fig. 4). Consequently, the coefficients c and N can be derived from these results in the temperature range from −173°C to 27°C (Table 1). The initial value of c (A: 0.080) decreased to 0.064 at 100 h (B) and then was constant to 1000 h (C). The N value decreased monotonically (A: 7.3, B: 5.9, C: 5.2). Therefore, the initial decrease of coercivity would be due to a reduction of the *c* value, which suggests easier nucleation of the reversed domains in this sample. On the other hand, the gradual increase of coercivity in the 1000 h sample (C) can be attributed to a decrease in the local demagnetization factor value *N*. Such a change in the *N* value would be related to the sharpness of the edges and corners [6, 11] of the fine $Sm_2Fe_{17}N_3$ powders. Fig. 5 shows a SEM image of fine $Sm_2Fe_{17}N_3$ powder immediately after jet milling in this work. Sharp edges and corners can be observed in all the powder particles, and these portions of the powder presumably have very high *N* values. We speculate that these sharp edges and corners become rounder as annealing time increases, probably because of endogenous diffusion or reaction of oxygen at the surface amorphous layers of the powder [3, 12].

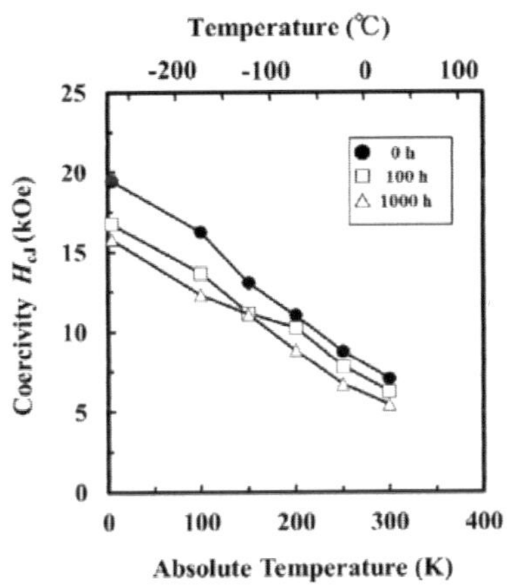

Figure 3. Temperature dependence of coercivity for $Sm_2Fe_{17}N_3$ compacts exposed at 110°C for 0, 100, 1000 h in air.

Table 1. *c* and *N* values for $Sm_2Fe_{17}N_3$ compacts exposed in air at 110°C for 0, 100 and 1000 h.

Annealing time (h)	*C*	*N*
0	0.080	7.3
100	0.064	5.9
1000	0.064	5.2

Figure 5. SEM image of fine $Sm_2Fe_{17}N_3$ powder prepared by jet milling.

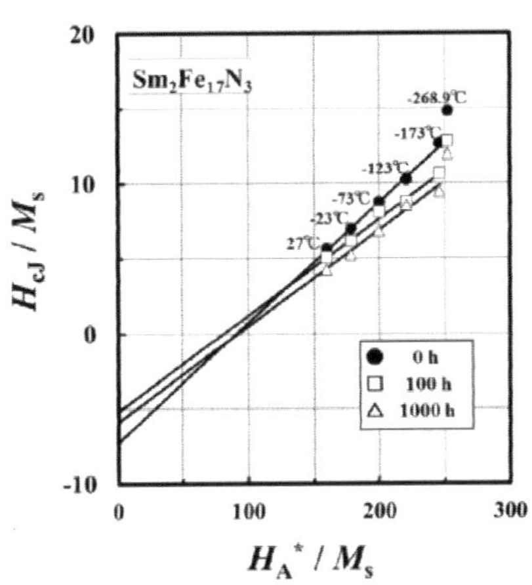

Figure 4. Relationship between H_A^* / M_s and H_{cJ} / M_s of $Sm_2Fe_{17}N_3$ compacts exposed at 110°C for 0, 100 and 1000 h in air.

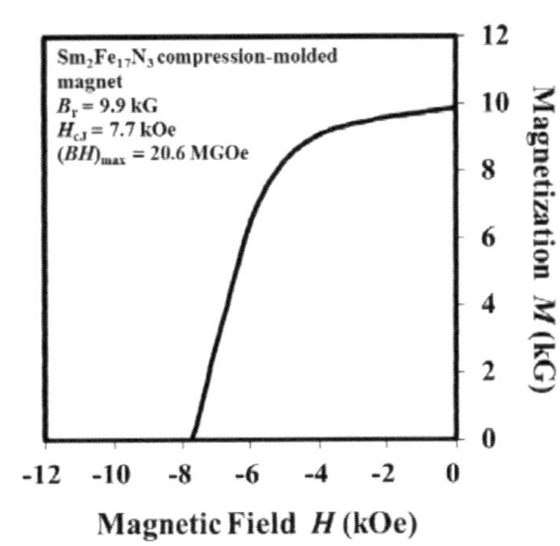

Figure 6. Demagnetization curve of $Sm_2Fe_{17}N_3$ compression-molded magnet.

We produced $Sm_2Fe_{17}N_3$ compression-molded magnets based on this work by using the following conditions. Fine $Sm_2Fe_{17}N_3$ powders, which had been surface treated with a silane coupling agent at 150°C in a vacuum, were mixed with 3 wt% of lubricant and epoxy-resin, and subsequently compressed at a pressure of 14 tonf/cm^2 under an alignment field of 15 kOe. The magnets were then cured in a vacuum at 150°C. The total annealing time at 150°C in an oxygen-free atmosphere was several hours; this was just the coercivity recovery time (see Fig. 2). Fig. 6 shows an example of the demagnetization curve of the $Sm_2Fe_{17}N_3$ compression-molded magnets. The coercivity of the compression-molded magnets recovered to 7.7 kOe as a result of annealing, but nevertheless did not reach the initial value of 9.3 kOe. Amorphous oxide layers (thickness: 10 nm) were observed on the surfaces of the $Sm_2Fe_{17}N_3$ powder particles [12]. Takagi et al. [3] reported that the coercivity of fine $Sm_2Fe_{17}N_3$ powders was deteriorated by sintering for 1 minute at 350°C, because a soft magnetic α-Fe phase appeared in the surface amorphous oxide layers. We consider the origin of the rapid decrease in the *c* value observed in this work to be related to a similar appearance of the soft magnetic phase. Therefore, when using fine $Sm_2Fe_{17}N_3$ powders, a lower temperature and an oxygen-free process (e.g., explosive consolidation using water [13]) are preferable conditions for fabrication of consolidated magnets. Alternatively, techniques to remove the surface amorphous oxide layers on the fine $Sm_2Fe_{17}N_3$ powder particles, such as acid etching, coating with a stable phase (e.g., iron-ferrite [14, 15]), or surface cleaning by reaction (e.g., Zn metal coating [16]), can be proposed as possible methods for obtaining excellent consolidated magnets.

4. Conclusion

The coercivity of fine $Sm_2Fe_{17}N_3$ powders decreased rapidly and subsequently increased gradually during annealing at 110°C and 150°C. According to an analysis by Kronmüller's formula, the initial decrease and the subsequent increase in coercivity were related to the rapid reduction in the *c* value and the gradual decrease in the *N* value, respectively. We speculate that the former arises from easier nucleation of reversed domains, while the latter is related to the change of sharpness in the edges and corners of the fine $Sm_2Fe_{17}N_3$ powders with increasing annealing time. In addition, sample compression-molded magnets were produced, and methods for consolidating fine $Sm_2Fe_{17}N_3$ powders while maintaining their high coercivity values were discussed.

References

[1] Coey J M D and Sun H 1990 *J. Magn. Magn. Mater.* **87** L251
[2] Iriyama T, Kobayashi K, Imaoka N, Fukuda T, Kato H and Nakagawa Y 1992 *IEEE Trans. Magn.* **28** 2326
[3] Takagi K, Nakayama H and Ozaki K 2012 *J. Magn. Magn. Mater.* **324** 2326
[4] Isikawa T, Yokosawa K, Watanabe K and Ohmori K 2011 *J. Phys. Conf. Ser.* **266** 012033
[5] Tada S, Tomimoto T and Kume M 2012 *Proc. the 22nd Int.l Workshop on Rare-Earth Permanent Magnets and Their Applications (Nagasaki)* p 48
[6] Hirayama Y, Panda A K, Ohkubo T and Hono K 2016 *Scripta Mater.* **120** 27
[7] Okada S, Takagi K and Ozaki K 2016 *J. Alloys Compd.* **663** 872
[8] Iriyama T, Okamoto A, Imaoka N, Kato H, Shiomi J, Nakagawa Y, Katori A K and Goto T 1994 *J. Magn. Soc. Jpn.* **18** 782 in Japanease
[9] Kou X C, Qiang W J, Kronmüller H and Schultz L 1993 *J. Appl. Phys.* **74** 6791
[10] Kronmüller H 1987 *Phys. Stat. Sol.(b)* **144** 385
[11] Li W F, Hu X C, Cui B Z, Yang J B, Han J Z and Hadjipanayis G C 2013 *J. Magn. Magn. Mater.* **339** 71
[12] Kobayashi K, Iriyama T, Yamaguchi T, Kato H and Nakagawa Y 1993 *J. Alloys Comp.* **193** 235
[13] Kakimoto E, Dohke K, Shibasaki I, Imaoka N and Chiba A 2008 *U.S. patent* 7,364,628
[14] Imaoka N, Koyama Y, Nakao T, Nakaoka S, Yamaguchi T, Kakimoto E, Tada M, Nakagawa T and Abe M 2008 *J. Appl. Phys.* **103** 07E129
[15] Imaoka N, Kakimoto E, Takagi K, Ozaki K, Tada M, Nakagawa T and Abe M 2016 *AIP advances* **6** 056022
[16] Noguchi K, Machida K, Yamamoto K, Nishimura M and Adachi G 1999 *Appl. Phys. Lett.* **75** 1601

IOP Conf. Series: Journal of Physics: Conf. Series **903** (2017) 012043 doi:10.1088/1742-6596/903/1/012043

First-order antiferromagnetic to ferromagnetic transition in Mn(Co,Fe)P probed by x-ray absorption experiments

F. Guillou, A. Hen, F. Wilhelm and A. Rogalev

ESRF, European Synchrotron Research Facility, 71 Av. des Martyrs, F-38043 Grenoble Cedex 09, France

E-mail: fguillou@ameslab.gov

Abstract. The first-order antiferromagnetic to ferromagnetic transition taking place near to room temperature in $MnCo_{0.53}Fe_{0.47}P$ has been investigated by x-ray absorption techniques: in the near edge region (XANES), in the extended energy range (EXAFS) and by using magnetic circular dichroism (XMCD). The absence of significant changes in the electronic and crystal structures across the transition are at odds with recent results obtained for the first-order magnetic transitions of the closely related Mn alloys, MnFe(P,Si,B) and MnCoGe. This points out that the microscopic mechanism responsible for the discontinuous antiferromagnetic to ferromagnetic transition in the Mn(Co,Fe)P system is intrinsically different from these latter prototypical examples.

1. Introduction

Mn alloys with chemical compositions deriving from MnMX, where M is a $3d$ transition metal (TM) and X is a pnictide or a metalloid (P, Si, As, Ge..), generally belong to one of the four following crystal-type: Fe_2P ($P_{\bar{6}2m}$), Fe_2As ($P_{4/nmm}$), TiNiSi (P_{nma}), and Ni_2In ($P_{63/mmc}$).[1] These crystal structures actually share many similarities, as they are build from the same two types of crystallographic blocks with specific local symetries: a TM in the center of square-pyramid formed by metalloid atoms, and a tetrahedron site. In addition, the distances between TM and non-metals are quite similar irrespectively of the crystal-type. Despite their crystallographic similarities, MnMX alloys present a large diversity of magnetic and structural transitions. As they are based on widely available $3d$ transition metals and metalloids, these Mn alloys are particularly promising as functional magnetic materials.[2] One can note a great interest currently paid on their magnetocaloric and magnetostrictive properties. For further improvement of these properties, it is desirable to understand the microscopic mechanisms at the origin of the various first-order magnetic transitions (FOMT). This primarily concerns the interplay between the discontinuous changes in electronic structure, crystal lattice and magnetism at the FOMT. In MnFe(P,Si,B) alloys with Fe_2P-type structure, pronounced changes in the x-ray absorption spectra (XANES) at the K-edge of P are fingerprinting the magneto-elastic ferromagnetic transition. It indicates that an electronic reconstruction on the metalloid site is a key element of this FOMT.[3] In MnCoGe alloys presenting a ferromagnetic TiNiSi to paramagnetic Ni_2In magneto-structural transition, pronounced changes were observed on the

Content from this work may be used under the terms of the Creative Commons Attribution 3.0 licence. Any further distribution of this work must maintain attribution to the author(s) and the title of the work, journal citation and DOI.
Published under licence by IOP Publishing Ltd

x-ray absorption spectra of all elements, but the most significant FOMT feature is an instability of the local Mn moments revealed by magnetic circular dichroism experiments (XMCD).[4]

Another type of FOMT in MnMX materials having attracted attention is the antiferromagnetic to ferromagnetic transition in orthorombic (P_{nma}) materials. This transition is very specific in many aspects. $MnCo_{1-x}Fe_xP$ compounds with 0.53$\leq x \leq$0.2 as well as MnCoSi constitute archetypical examples of such a behavior.[5, 6, 7] At intermediate magnetic field (B= 0.1 T), $MnCo_{0.53}Fe_{0.47}P$ presents an antiferromagnetic to ferromagnetic transition (T_N) at room temperature with very narrow thermal hysteresis, and a Curie temperature around 350 K. No significant discontinuities in the temperature dependence of the lattice parameters could be detected at the magnetic transitions. In addition, the difficulty to detect a latent heat in differential scanning calorimetry experiments and the large dT_N/dB suggest a very weak first-order transition. This picture however strongly evolves at higher magnetic field (B$\geq$1 T), where a large thermal hysteresis starts to develop. This conversion from weak to strong FOMT in large magnetic field is particularly noticeable in MnCoSi, and manifests itself by the appearance of a giant-magnetostriction.[8, 9]. The origin of the strengthening of the FOMT in large magnetic field is not yet fully understood. It is the aim of the present work to investigate the properties of $MnCo_{0.53}Fe_{0.47}P$ at high magnetic field with a probe bringing simultaneous informations about electronic, crystal structures and magnetism .

2. Experimental details

The x-ray absorption experiments were carried out at the ID12 beamline of the ESRF.[10] The experimental station was the 17 T superconducting magnet equipped with a Si photodiode fluorescence detector in backscattering geometry. The HU-52 helical undulator was used as source of polarized x-ray beam to cover the energy range from 2.05 to 9 keV. The isotropic x-ray absorption spectra (XANES) and the extended x-ray absorption fine structure (EXAFS) correspond to the average of spectra with opposite helicities. The magnetic circular dichroism spectra (XMCD) were measured by reverting the helicity as well as the magnetic field. The spectra presented hereafter were corrected for self-absorption. Typical probing depths are $\approx 1\mu$m at P K-edge, and $\approx 10\mu$m for the transition metals. The bulk $MnCo_{0.53}Fe_{0.75}P$ polycrystalline sample originates from a batch previously studied for its magnetocaloric effect.[7] The sample surface was hand-polished prior the mounting in high-vacuum chamber.

3. Results and Discussion

In order to study the FOMT above the tricritical point where a large thermal or magnetic hysteresis appears, all x-ray absorption experiments were performed at 255 K, i.e. about 30 K below the antiferromagnetic to ferromagnetic transition temperature. At T= 255 K, the critical field for which the FOMT develops is $H_C \approx 1.3$ T, and the ferromagnetic phase is fully induced by a magnetic field of 9 T. Due to the selection rules governing the x-ray absorption process, the K-edge absorption spectra for $3d$ transition metals and pnictides mostly originate from the excitation of a $1s$ core state to p unoccupied densities of states, $4p$ and $3p$, respectively. As a consequence, if significant changes in the densities of states occur at the transition, they are likely to be reflected in the XANES spectra. Fig. 1 presents the isotropic x-ray absorption spectra at the K-edge of all elements in $MnCo_{0.53}Fe_{0.47}P$ across the FOMT. The most striking feature is the almost perfect overlap of the antiferromagnetic and ferromagnetic XANES spectra for all elements. This result is in stark contrast with the evolutions observed at the FOMT of closely related MnMX alloys, MnFe(P,Si,B) and MnCoGe.[3, 4] The absence of changes in the conduction band are also astonishing in regards to theoretical studies which were suggesting an increase of the density of state near the Fermi level between the antiferromagnetic and ferromagnetic phases.[6] It is interesting to note that the P K-edge in $MnCo_{0.53}Fe_{0.47}P$ presents an overall spectral shape, in particular a secondary feature located about 8 eV above the edge,

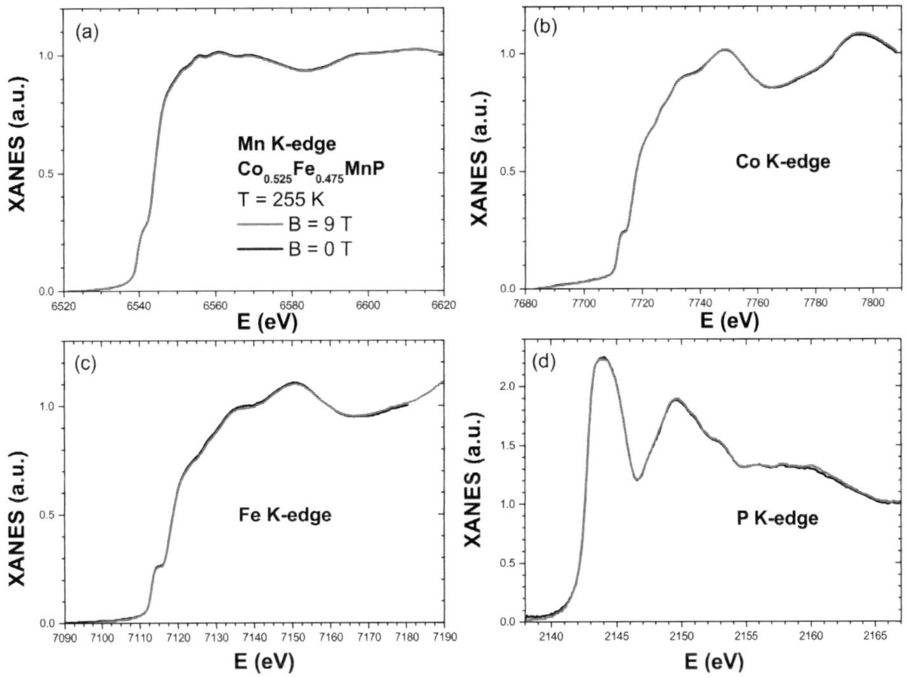

Figure 1. Isotropic x-ray absorption spectra (XANES) at the K-edge of all elements in $MnCo_{0.53}Fe_{0.47}P$ in zero and 9 T magnetic fields.

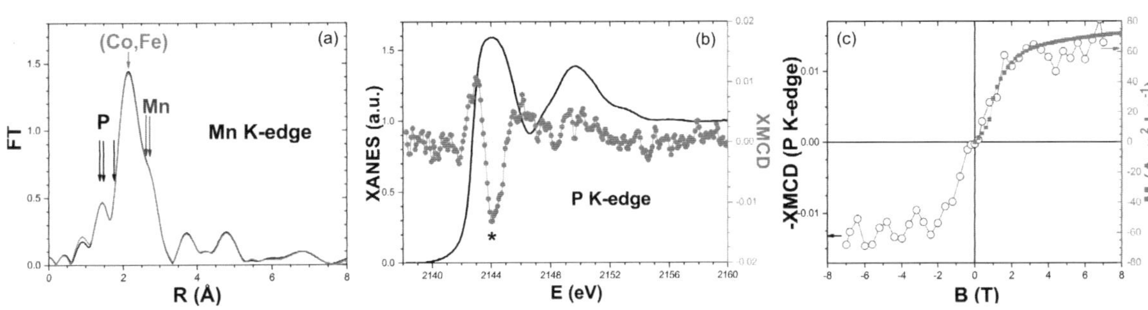

Figure 2. $MnCo_{0.53}Fe_{0.47}P$. (a) Fourrier transform of the EXAFS at the Mn K-edge in zero and 9 T magnetic fields. (b) XANES and XMCD spectra at the K-edge of phosphorus at 255 K and 9 T, the asterisk marks the energy for panel (c). (c) Field dependence of the XMCD at the P K-edge and bulk magnetization at 255 K.

comparable to that observed in $MnFeP_{2/3}Si_{1/3}$.[3] This highlights the structural similarities between these series of Mn alloys, in particular the TM-pnictide distances lead to analogous band hybridization between TM and phosphorus in these materials.

Fig 2(a) shows the experimental EXAFS spectra for Mn in the antiferromagnetic and ferromagnetic states. The EXAFS data were analysed using the ATHENA code to obtain pseudoradial distribution function, the presented Fourrier transforms are k^3 weighted.[11] In order to ascribe the features to different scattering paths, the EXAFS spectra were calculated using the experimental powder x-ray diffraction data for $MnCo_{0.53}Fe_{0.47}P$. The main features reflect the shortest distances Mn-P, then Mn-Co and Mn-Mn (with offets compared to the XRD data). Crossing the FOMT by application of a large magnetic field does not lead to a significant evolution of the EXAFS at the Mn K-edge. The estimate of the change in lattice parameters by fitting of the EXAFS data falls within the error barrs. One can nevertheless note that the magnestriction in $MnCo_{0.53}Fe_{0.47}P$ is smaller than the $\Delta a/a \approx -6.10^{-3}$ observed in MnCoSi.[8] Fig 2(b) presents the isotropic x-ray absorption (XANES) and the magnetic circular dichroism (XMCD) spectra at the phosphorus K-edge. A relatively large XMCD maximum of 1.2% is observed. This is a direct experimental confirmation of the development of induced magnetic moments on phosphorus atoms due to hybridization with TM predicted by KKR-CPA calculations.[5] Interesting to note that the amplitude of the phosphorus XMCD in $MnCo_{0.53}Fe_{0.47}P$ is smaller than for $MnFeP_{2/3}Si_{1/3}$ ($\approx$1.6%), which is in line with their predicted magnetic moments, -0.07 and -0.11 μ_B, respectively. Bulk magnetization measurements at 255 K are shown in Fig. 2(c). A metamagnetic transition is clearly noticeable around 1.3 T. The agreement between the field dependence of the XMCD at P K-edge and the macroscopic magnetization measurements clearly illustrates that all x-ray absorption experiments as a function of the magnetic field were carried out across the antiferromagnetic-ferromagnetic FOMT, and are representative of bulk properties.

4. Conclusion

The x-ray absorption study of $MnCo_{0.53}Fe_{0.47}P$ allowed us to clarify a few points in this class of orthorhombic MnMX alloys. The observation of a sizeable XMCD signal on phosphorus demonstrates the development of an induced magnetization, in line with former theoretical reports. However, in contrast to theoretical predictions, no significant electronic reconstructions are experimentally observed across the FOMT. The analysis of the magnetostriction through the EXAFS data as a function of the magnetic field suggests that the changes in lattice parameters are weaker in $MnCo_{0.53}Fe_{0.47}P$ than in the MnCoSi system. The present x-ray absorption study thus indicates that the first-order magnetic transition has little consequences on the electronic and crystalline structures. This weaker interplay between the different degrees of freedom in Mn(Co,Fe)P compared to the closely related MnFe(P,Si,B) and MnCoGe series explains the lower magnetocaloric effect of the former. It most probably originates from the specific mechanism of the FOMT, which is often described as an exchange interaction inversion driven by tiny changes in Mn-Mn distances.

5. Acknowledgements

The authors would like to thank P. Voisin for his technical support. The European Synchrotron Research facility is acknowledged for provision of 1 shift of beamtime. The samples were prepared at TU Delft, during a project supported by FOM and BASF New Business.

References

[1] Krumbügel-Nylund A, Boursier D, Rouault A, Senateur J P and Fruchart R 1974 *Mat. Res. Bull.* **9** 21
[2] Tegus O, Brück E, Zhang L, Dagula, Buschow K and de Boer F 2002 *Physica B: Condens. Matter* **319** 174
[3] Guillou F, Ollefs K, Wilhelm F, Rogalev A, Yaresko A, Yibole H, van Dijk N and Brück E 2015 *Phys. Rev. B* **92** 224427
[4] Guillou F, Wilhelm F, Tegus O and Rogalev A 2016 *Appl. Phys. Lett.* **108** 122405
[5] Sredniawa B, Zach R, Fornal P, Duraj R, Bombik A, Tobola J, Kaprzyk S, Niziol S, Fruchart D, Bacmann M, Fruchart R and Stanek J 2001 *J. Alloys Compd.* **317** 266
[6] Zach R, Tobola J, Sredniawa B, Kaprzyk S, Guillot M, Fruchart D and Wolfers P 2007 *J. Phys.:Condens. Matter* **19** 376201
[7] Guillou F and Brück E 2013 *J. Appl. Phys.* **114** 143903
[8] Barcza A, Gercsi Z, Knight K and Sandeman K 2010 *Phys. Rev. Lett.* **104** 247202
[9] Barcza A, Gercsi Z, Michor H, Suzuki K, Kockelmann W, Knight K and Sandeman K 2013 *Phys. Rev. B* **87** 064410
[10] Rogalev A and Wilhelm F 2015 *Phys. Met. Metallography (USSR)* **116** 1285
[11] Ravel B and Newville M 2005 *J. Synchrotron Rad.* **12** 537

IOP Conf. Series: Journal of Physics: Conf. Series **903** (2017) 012044 doi:10.1088/1742-6596/903/1/012044

Influence of La substitutions on the magnetic after-effect in hexaferrites and garnets

P Hernández-Gómez[1], X Liu[2], J M Muñoz[1], C Torres[1] and O Alejos[1]

[1]Univ. Valladolid, Dpto. Electricidad y Electrónica, 47071 Valladolid, Spain

[2]Univ. Anhui, School of Physics and Materials Science, Hefei 230039, PR China

E-mail: pabloher@ee.uva.es

Abstract. Magnetic relaxation has been measured in La-Co substituted M-type Sr hexaferrites and La substituted YIG polycrystalline samples prepared by means of standard ceramic techniques. In the temperature range between 80 and 420 K, the magnetic disaccommodation has been measured. The isochronal disaccommodation spectra show the presence of different relaxation processes. In hexaferrites, the La doping distort the lattice size and the arrangement of cations within the structure, changing the strenght of relaxation processes regarding unsubstituted samples. In garnets, substitution rate only modify lattice size, whereas with CO_2 atmosphere the formation of secondary perovskite phase is detectable with this technique.

1. Introduction

Hexagonal ferrites have been widely used as permanent magnets owing to their appropriate magnetic properties: big saturation magnetization, large uniaxial anisotropy, and high Curie temperature. In addition, they show promising properties in microwave devices and recording media [1]. There are different stable phases derived from the M-type $SrFe_{12}O_{19}$. Physical properties needed for different applications can be tailored by choosing the right phase, carrying out adequate cation substitutions and improving the microstructure, by using different synthesis methods. Among the differentt cation substitutions the addition of rare-earth La^{3+} and transition Co^{2+} improve the performance of the permanent magnetic hexagonal ferrites (H_c increase from 275 kA/m to 355 kA/m, i.e. about 25% higher) [2]. YIG are key ferrimagnetic materials for use at high frequencies and their properties can also be tailored. Yttrium substitution for a light RE is known for years, but now the interest is renewed, as La doped YIG exhibit a strong magnetoelectric effect [3]

Regarding magnetic properties, the magnetic after-effect processes have to be taken into account in order to minimize the losses. In addition, the study of this kind of processes provides information about the underlying mechanisms governing the dynamic behavior of Bloch walls [4]. Among the different techniques available, the magnetic disaccommodation measurements are a powerful tool in the detection of this kind of phenomena. It consists in the time variation of the mobility of domain walls after a magnetic shock, and is shown by a temporal evolution of the magnetic permeability after a demagnetization stage, and the relaxation time which characterizes each relaxation process is strongly temperature-dependent. This relaxation phenomenon has been observed firstly in spinel ferrites [4], but is also present in hexaferrites and garnets [5-6]. In this paper we analyze the effect of La substitution on the magnetic disaccommodation in SrM hexaferrites and yttrium iron garnets.

Content from this work may be used under the terms of the Creative Commons Attribution 3.0 licence. Any further distribution of this work must maintain attribution to the author(s) and the title of the work, journal citation and DOI.
Published under licence by IOP Publishing Ltd

2. Experimental

We have prepared two series of polycrystalline samples of La-Co substituted M-type hexaferrites (details can be found in [7]): one of them with equal substitution of La and Co, i.e. $Sr_{1-x}La_xFe_{12-x}Co_xO_{19}$ (x=0.09 to 0.18), and other with different amount of them: $Sr_{1-x}La_xFe_{12-y}Co_yO_{19}$ (x=0.09, y=0.08; x=0.13, y=0.11; x=0.16, y=0.13). Pressed samples were sintered at 1350º C in air atmosphere, keeping the top temperature for 4 hours, then they were quenched. La doped YIG series $Y_{3-x}La_xFe_5Co_yO_{12}$ (0<x<0.6) [6] were prepared in air and CO_2 sintering atmospheres at 1420 ºC for 5 h .

Magnetic disaccommodation measurements were carried out with a computer aided system based on a LCR bridge, in the 80 K < T < 420 K temperature range. The results have been represented as isochronal curves, i. e, the relative variation of the initial permeability after sample demagnetization between an initial time t_1= 2 s and different window times t_2 = 4, 8, 16, 32, 64 and 128 s in the form

$$\frac{\mu(t_1,T)-\mu(t_2,T)}{\mu(t_1,T)}(\%) \qquad (1)$$

When the time window (t_2-t_1) is of the same order of magnitude that the relaxation time, this curve exhibits a maximum. So, isochronal spectra disclose different after-effect processes in the temperature range tested, and characteristic parameters of the relaxation process, as activation energy, are obtained.

3. Results and discussion

Representative results of the two series of La-Co substituted SrM hexaferrites are presented in Figure 1. Several relaxation processes emerge at temperatures 380 K (A peak), 300 K (B), and 165 K (D) like in pure SrM hexaferrites [5]: B process, with activation energies of 0.84 eV, is caused by anisotropic reorientation of the local symmetry axis due to the jump of a ferrous cation towards a vacancy in the octahedral site 2a located in spinel S blocks within the hexagonal lattice. The A peak has a similar origin with increased activation energies (1 eV) due to the presence of the Sr ion. On the other hand, the D peak is attributed to similar process in the octahedral $4f_2$ sites located in R blocks. These are face-sharing sites so that the energy involved in this process, 0.5 eV, is lower than the corresponding to the B peak. The increase in doping rate in $Sr_{1-x}La_xFe_{12-x}Co_xO_{19}$ causes a progressive diminution of relaxation processes, especially A and B peaks, and a shift of D process peak temperature to higher values (from 165 K to 180 K). Mössbauer investigations of La^{3+}-Co^{2+} M-type ferrites revealed that most of the Co^{2+} cations enter in the octahedral 2a sites [8], then causing the observed diminution of B peak amplitude, and making energetically unfavourable the A process. In addition, local neutrality favors the presence of the divalent ferrous cation in the vicinity of La^{3+} thus enhancing the D process which takes place in face-sharing octahedral $4f_2$ sites. On the other hand, substitution of part of Sr^{2+} (ionic radius 1.27 Å) with La^{3+} (1.16 Å), and Co^{2+} instead Fe^{3+} (0.65 Å and 0.55 Å resp) distort the hexaferrite structure. With this in mind, increased doping rate almost suppresses B peak, due to the small amount of Fe^{2+} in 2a sites occupied by Co, and also modifies the D peak in two ways: increasing its activation energy due to lattice distortion and lowering its amplitude, because the unit cell is smaller and then the presence of localized ferrous cations in $4f_2$ sites is less probable in favour of diffusion of cations via lattice vacancies, thus increasing the high temperature process (over 400 K).

Concerning $Sr_{1-x}La_xFe_{12-y}Co_yO_{19}$ samples, the effect observed with doping rate is similar, i.e. diminution of relaxation amplitudes with increasing doping. We can see that D peak is prominent, and B peak is strongly diminished, even for the lower substitution rate analyzed. As the amount of Co^{2+} is lower than La^{3+} substitution, a higher amount of Fe^{3+} reduces to Fe^{2+} to maintain electrical neutrality. The preference of Co^{2+} for octahedral 2a sites promotes the preferential occupation of ferrous cations in octahedral sites in R blocks near La^{3+}, thus enhancing the D process and almost suppressing the B process regarding $Sr_{1-x}La_xFe_{12-x}Co_xO_{19}$ samples. The increased amount of ferrous cations can not enter completely in the $4f_2$ sites due to higher ionic radius of Fe^{2+} (0.61 Å) regarding Fe^{3+} hence increasing also the diffusion processes.

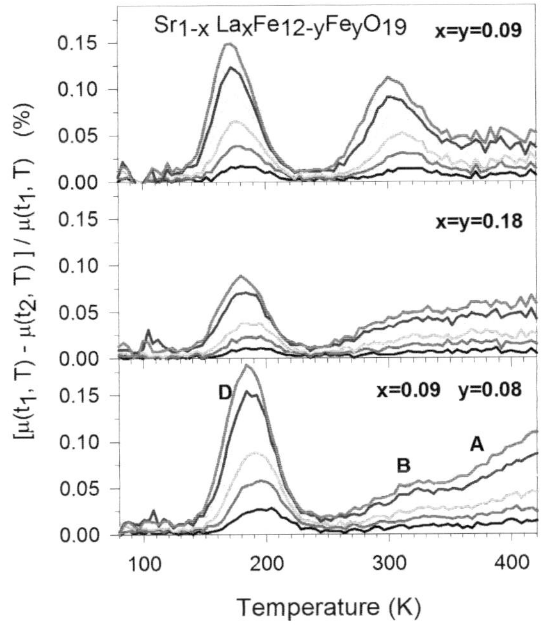

Figure 1. Representative disaccommodation spectra of $Sr_{1-x}La_xFe_{12-x}Co_xO_{19}$ and $Sr_{1-x}La_xFe_{12-y}Co_yO_{19}$ hexaferrites sintered in air at 1340 ºC.

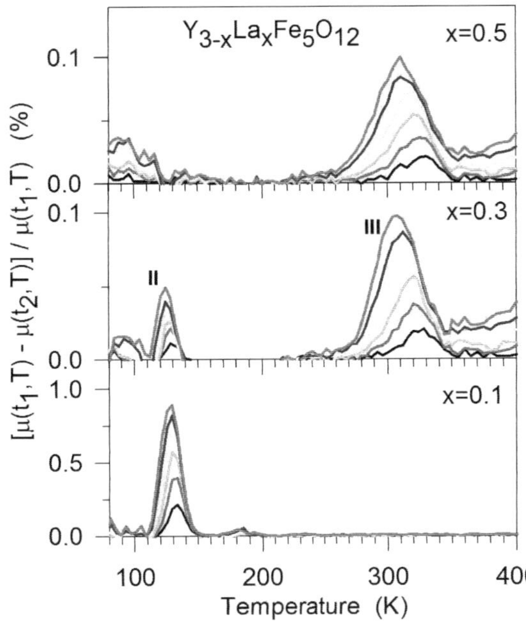

Figure 2. Representative disaccommodation spectra of $Y_{3-x}La_xFe_5Co_yO_{12}$. sintered in CO_2

In YIG, characteristic disaccommodation peak II process, centred at 130 K (0.35 eV), is associated with Fe^{2+}-Fe^{3+} electronic transfer in octahedral sites of the garnet lattice. Ferrous ion is necessary to maintain electrical neutrality due to the existence of oxygen vacancies. The temperature range of this peak, and hence the activation energy lowers with increasing La doping, from 130 K to 120 K, due to the larger size of the unit cell caused by the introduction of the dopant (12.376 Å with x=0 to 12.423 Å with x=0.5). For La doped YIG sintered in CO_2 (Figure 2), the garnet-perovskite transition takes place at a doping rate x=0.3. This effect is clearly visible in disaccommodation results for the suppression of YIG process due to the formation of antiferromagnetic perovskite and emergence of III process of magnetite, that appears as secondary phase. However, garnet-perovskite transition does not take place when sintering in air. Thermal dependence of magnetic permeability (not shown) support these conclusions. Low doped samples behaves similary to undoped YIG, and a progressive decrease with increasing substitution rate is observed. Once the garnet-perovskite transition has been reached, the permeability decreases strongly due to the high amount of antiferromagnetic perovskite in the sample. As a conclusion, La doping affects relaxation due to different ionic size in both systems, but has additional characteristic effects in hexaferrites and garnets.

Finnancial support of Univ. Valladolid is acknowledged

References

[1] M. P. Sharrock, 1989 *IEEE Trans Magn.* **25**(6) 4374
[2] F. Kools, et al., 2002 *J. Magn. Magn. Mater.* **242–245** 1270
[3] H. Wu, et al. 2015 *J. App. Phys.* **117** 144101
[4] F.Walz, V.A.M.Braber, S.Chikazumi, H.Kronmüller,M.O.Rigo 1982 *Phys. Stat. Sol. B*, **110** 470
[5] P. Hernández-Gómez, et al. 2004 *J. Magn. Magn. Mater.* **272–276** e1843
[6] C.Torres, O.Alejos, J.M.Muñoz, P.Hernández-Gómez, C.Francisco 2002 *Phys. RevB* **66** 024410
[7] X. Liu et al. 2006 *J. Magn. Magn. Mat*, **305** 524
[8] L. Lechevallier, J. M. Le Breton, 2005, *J. Magn. Magn. Mater.* **290-291** 1237

Ni-Mn-Sn Heusler: milling and annealing effect on structural and magnetic properties

Florin Popa[1], Traian Florin Marinca[1], Horea Florin Chicinaş[1], Olivier Isnard[2], Ionel Chicinaş[1]

[1]Department of Materials Science and Engineering, Technical University of Cluj-Napoca, 103-105 Muncii Avenue, 400641 Cluj-Napoca, Romania

[2]Institut Néel, CNRS/ University J. Fourier, BP166, 38042 Grenoble, Cédex 9, France

florin.popa@stm.utcluj.ro

Abstract. Nanocrystalline $Ni_{51}Mn_{19}Sn_{30}$ Heusler alloy was prepared in the form of powder by solid state reaction in a planetary ball mill under argon atmosphere. After 10 h of milling the samples exhibit a mixture of two phases: disordered Heusler structure and half-Heusler structure. The stability of the phases was studied and a transformation of the disordered Heusler phase into the ordered Heusler and Ni_3Sn_2 phase was observed after heat treatment. Magnetic properties strongly depend on the phases promoted during milling and annealing. The phase change after annealing leads to the increase of the sample's magnetisation.

1. Introduction

Heusler alloys are materials having the X_2YZ formula with a complex crystalline structure ($L2_1$) composed of four face centred cubic sub lattices [1]. Generally, Heusler alloys are prepared by high temperature arc melting, with a careful composition control [2, 3]. Among Heusler alloys, Ni-Mn-Sn represents an interesting case, presenting inverse magnetocaloric effect [4]. For the stoichiometric composition, Ni_2MnSn alloy has no martensitic transition, however, non-stoichiometric alloys with higher Mn content exhibit the austenite-martensitic transition and the magnetic order can be antiferromagnetic [5]. A method for obtaining nanocrystalline alloys is mechanical alloying, consisting in elemental powder mixture processing in high-energy ball mills, in order to obtain a solid state reaction. By this method, nanocrystalline alloys, amorphous phases, metastable alloys, out of equilibrium structures/microstructures, extended solid solution, quasicrystals can be obtained [6]. In the Ni-Mn-Sn system, several studies were performed on samples obtained by milling [7] or by melt spinning [8], showing the influence of the nanocrystalline state and of annealing on the structure and transition temperature of the martensitic transformation. The present study is focused on the phase formation of the $Ni_{51}Mn_{19}Sn_{30}$ non-stoichiometric alloy by mechanical alloying and the evolution of its magnetic and thermomagnetic properties during milling and annealing of the samples.

2. Experimental

$Ni_{51}Mn_{19}Sn_{30}$ (at. %) alloy was prepared in a planetary ball mill, Fritch-Pulverisette 6, starting from a mixture of elemental nickel, manganese and tin powder, weighed in the proper ratio. The mixture was first homogenized for 15 min in a Turbula type apparatus, and then milled for up to 10 hours in argon

Content from this work may be used under the terms of the Creative Commons Attribution 3.0 licence. Any further distribution of this work must maintain attribution to the author(s) and the title of the work, journal citation and DOI.
Published under licence by IOP Publishing Ltd

atmosphere. The ball/powder mass ratio was 9.2:1 at a filling factor of 40%. Several samples were annealed in a Setaram Labsys differential scanning calorimetry furnace at 400 °C with a heating rate of 10 °C/min and maintaining at 400 °C for 60s. The alloy formation was investigated by X-ray diffraction with an INEL 3000 Equinox diffractometer, operating with Co - Kα radiation. The magnetization curves were recorded at room temperature by extraction method technique in a continuous field of up to 2 T. The Curie temperatures have been determined by thermomagnetic measurements with a Faraday balance in small fields (less than 0.1 T).

3. Results and discussion

$Ni_{51}Mn_{19}Sn_{30}$ alloy formation was studied by X-ray diffraction indicating the phase evolution during milling, as presented in figure 1.

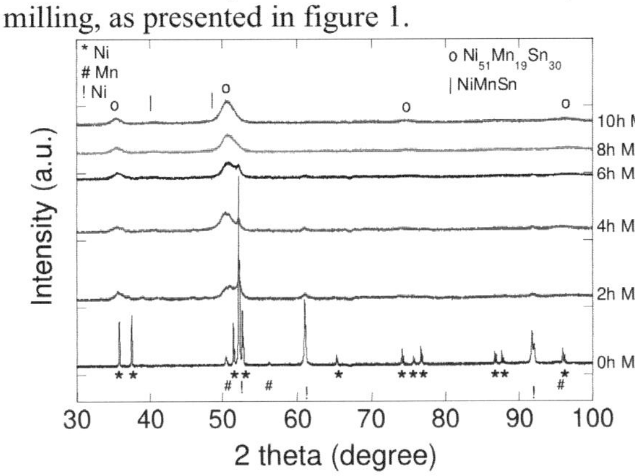

Figure 1. X-ray diffraction patterns for the as-milled samples.

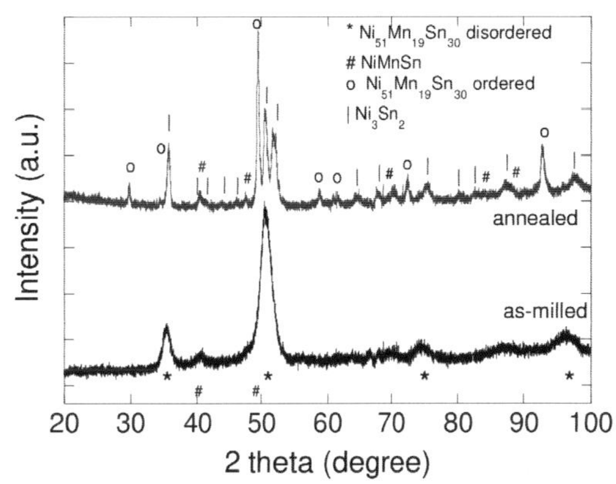

Figure 2. X-ray diffraction patterns for annealed and as-milled 10 h samples.

From Figure 1, the formation of the alloy can be described by the disappearance of the elemental peaks of Ni, Mn and Sn (recorded for the starting sample) and the emergence of a new set of peaks (Heusler structure) for prolonged milling. The Bragg peaks of the Sn and Mn, are no longer observed for samples milled for more than 4 h. In opposition, the Ni peaks are clearly visible up to 6 h of milling. In the sample milled for 8 h the presence of Ni is related to the broad peak asymmetry at its position compared with the starting sample. The Ni-Mn-Sn phase is the majority phase for the samples milled for 10 h, having nanocrystalline structure, with mean crystallite size of 15 ± 2 nm. A closer analysis of the Ni-Mn-Sn phase is presented in Figure 2. Probably due to the non-stoichiometry of the sample and to the high amounts of the stresses induced by milling, the obtained phase is the disordered Ni_2MnSn. The disordered Heusler phase is accompanied by a second phase, identified as being the half-Heusler compounds (NiMnSn). By heating at low temperature (400 °C), the disordered Heusler phase exhibits a decomposition process (Figure 2). Under the influence of temperature, the powder's initial phase transforms into three components: a Ni_2MnSn ordered Heusler phase, a half-Heusler phase (NiMnSn) and Ni_3Sn_2 phase. This phase transformation has important consequences on the magnetic properties, as presented in Figure 3.

The magnetisation of the milled samples (Figure 3) decreases as the milling is prolonged, from 21.5 Am^2/kg (0 h MA) down to 2 Am^2/kg (10 h MA). The low magnetization value is connected to the decrease of the Ni content, the Heusler phase being antiferromagnetic having low magnetization [9]. However, the annealing of the sample leads to an important magnetization increase, associated with the ordering of the Heusler phase [9] and probably with the apparition of the Ni_3Sn_2 phase (with Curie temperature below room temperature). By analysing the 4 h milled sample, when the solid state reaction is initiated (the powders are mechanically activated), it can be observed that a subsequent annealing leads to the same phases - ordered Heusler phase and Ni_3Sn_2 phase - as in the case of the sample milled for 10 h, which leads to the complete formation of the alloy by milling. Both 4 h and 10 h milled and annealed samples possess almost the same magnetization (15.9 Am^2/kg).

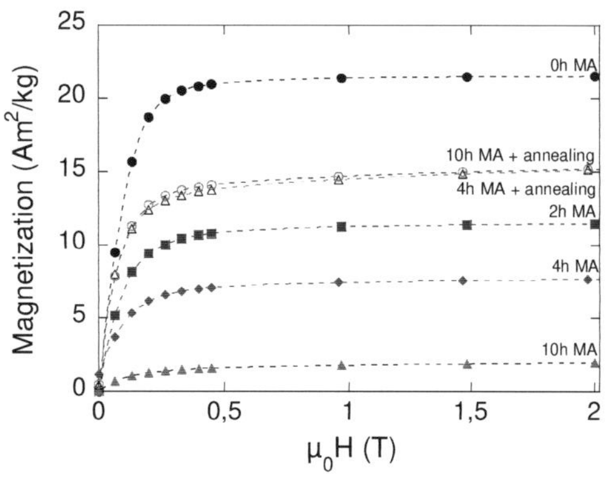

Figure 3. Room temperature magnetization curves recorded for the as-milled samples. For comparison annealed sample are also presented.

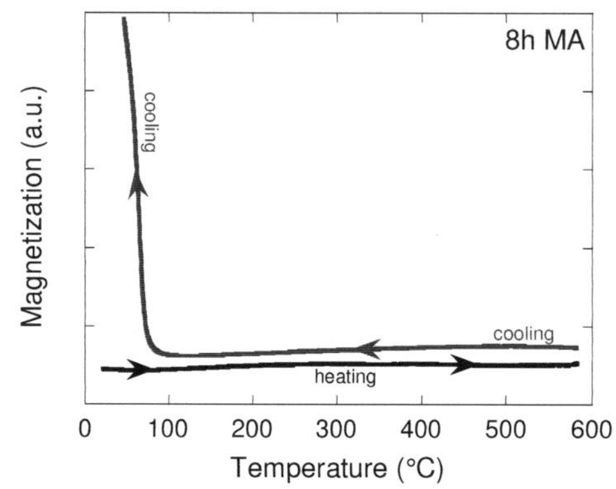

Figure 4. Thermomagnetic curve of the 8h milled sample.

The higher magnetization values observed after annealing can be better understood by performing thermomagnetic measurements (figure 4), where heating curves show that the magnetization remains almost constant up to 600 °C. After compound ordering, the cooling curves show a sudden increase of magnetization recorded at 65 °C, which is expected for the Curie temperature of the Ni_2MnSn ordered compound.

4. Conclusions

Milling up to 10 h of the elemental powders in the $Ni_{51}Mn_{19}Sn_{30}$ ratio, leads to the formation of the Ni_2MnSn disordered Heusler phase accompanied by a minor Half-Heuser phase. By annealing the samples at 400 - 600 °C, the powder's phase modifies to an ordered Heusler phase (Ni_2MnSn), a Ni_3Sn_2 compound and a Half-Heusler phase (NiMnSn). The magnetization decreases as the disordered Ni-Mn-Sn phase forms, but strongly increases when the phase becomes ordered. The Curie temperature of the ordered phase is 65 °C. The formation of the $Ni_{51}Mn_{19}Sn_{30}$ alloy by solid state reaction is found to be complex and further studies are in progress.

5. Acknowledgment

This work was supported by the UTCN Internal competition grant CI 6/1.2/2015 and Romania-France bilateral program (PHC Brancusi 2013/2014 - 28847YK).

References

[1] Handbook of magnetic materials, Ed. K.H. J. Buschow, Vol. 21, North-Holland - Elsevier, 2013
[2] T. Graf, F. Casper, J. Winterlik, B. Balke, G.H. Fecher, C. Felser , Z. Anorg. Allg. Chem. 635 (2009) 976 – 981
[3] W.M. Yuhasz, D.L. Schlagel, Q. Xing, R.W. McCallum, T.A. Lograsso, J. All. Compd. 492 (2010) 681-684
[4] T. Krenke, E. Duman, M. Acet, E. F. Wassermann, X. Moya, L. Manosa, A. Planes, Nature mater. 4 (2005) 450-454
[5] L.H. Yang, H. Zhang, F.X. Hu, J.R. Sun, L.Q. Pan, B.G. Shen, J. All. Compd. 588 (2014) 46-48
[6] C. Suryanarayana, Progress Mater. Sci. 46 (2001) 1-184
[7] A.G. Varzaneh, P. Kameli, V.R. Zahedi, F. Karimzadeh, H. Salamati, Met. Mater. Int. 21 (2015) 758-764
[8] H.C. Xuan, Y. Deng, D.H. Wang, C. L. Zhang, Z. D. Han, Y.W. Du, J. Phys. D: Appl. Phys. 41 (2008) 215002
[9] P. Lazpita, J. M. Barandiarán, V.A. Chernenko, B. Valle García, E. Díaz Tajada, T. Lograsso, D.L. Schlagel, J. All. Compd. 594 (2014) 171 – 174

IOP Conf. Series: Journal of Physics: Conf. Series **903** (2017) 012046 doi:10.1088/1742-6596/903/1/012046

Kinetics of heat flux avalanches at the first order transition in La(Fe-Mn-Si)$_{13}$-H$_{1.65}$ compounds

M Piazzi[1], **C Bennati**[1,2] **and V Basso**[1]

[1] Istituto Nazionale di Ricerca Metrologica, Strada delle Cacce 91, 10135 – Torino, Italy
[2] Dipartimento di Scienza Applicata e Tecnologia, Politecnico di Torino, C.so Duca degli Abruzzi 24, 10129 – Torino, Italy

E-mail: m.piazzi@inrim.it

Abstract. We study heat flux avalanches occurring at the first order transition in La(Fe-Mn-Si)$_{13}$-H$_{1.65}$ magnetocaloric material. As the transition is associated to the phase boundaries motion that gives rise to the latent heat, we develop a non equilibrium thermodynamic model. By comparing the model with experimental calorimetry data available for Mn=0.18, we find the values of the intrinsic kinetic parameter R_L, expressing the damping for the moving boundary interface, at different magnetic fields. We conclude that by increasing field, thus approaching the critical point, the avalanches increase in number and their kinetics is slowed down.

1. Introduction

The kinetics underlying the first order ferromagnetic (FM) to paramagnetic (PM) transitions of La(Fe-Si)$_{13}$ and related magnetocaloric materials is still an open issue [1–4]. Novel aspects about this feature can be learnt by analyzing heat flux signals obtained through temperature scans at very low rates [5, 6]. Indeed, the low scan rate allows to distinguish single heat flux avalanches associated to the microscopic individual processes occurring during the phase transition. Then, the characteristic times governing the behaviour of the avalanches are also those determining the kinetics of the transition. In particular, these indications may help to further improve the working frequency of magnetic refrigerators [1].

We develop a model, based on the non equilibrium thermodynamic theory of linear systems [7], in which the kinetics of first order transitions is related to the FM–PM phase boundaries motion. The model allows to describe single heat flux avalanches through an intrinsic kinetic parameter R_L relating the change in latent heat to the difference between the transition temperature T_t and the sample temperature T_s (Sec. 2). We use experimental data provided by Peltier calorimetry under magnetic field performed at 1 mK/s rate on LaFe$_{11.60}$Mn$_{0.18}$Si$_{1.22}$-H$_{1.65}$ compounds [8]. In Sec. 3 we present the results of the comparison between the model and the experimental data, while in Sec. 4 we discuss them.

2. Thermodynamic theory of heat flux avalanches

We will describe the first order phase transition of the system as due to the motion of the boundaries between the FM and PM states. We assume these boundaries to be ideally thin interfaces characterized by a position $x(t)$, changing in time, and a surface area A. When the transition starts, the nucleation of the interface occurs at a temperature different from the

Content from this work may be used under the terms of the Creative Commons Attribution 3.0 licence. Any further distribution of this work must maintain attribution to the author(s) and the title of the work, journal citation and DOI.
Published under licence by IOP Publishing Ltd

equilibrium transition temperature T_t because of the presence of local defects. Therefore, when transforming in heating from the FM to the PM phase, or viceversa in cooling from PM to FM, the nucleation event will be characterized by a critical temperature value T_h higher or T_c lower than T_t, respectively. The qualitative behaviour of the resulting path representing the sample temperature T_s close to the transition region is sketched in Figure 1.

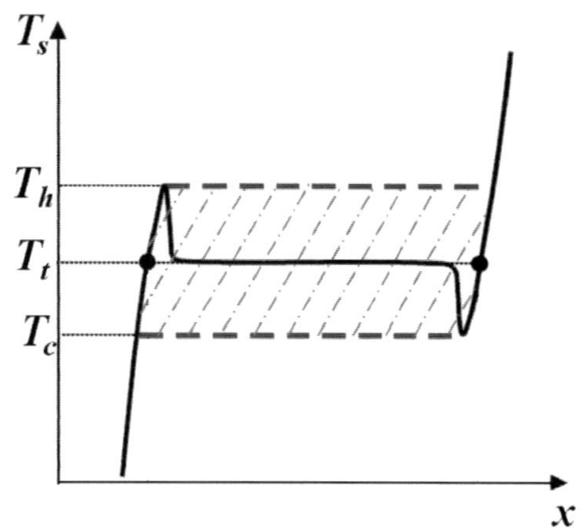

Figure 1: Sketched behaviour of temperature T_s in systems with first order transition and moving phase boundary position x. T_h, T_c, T_t are the heating, cooling and equilibrium temperatures, respectively.

$\mu_0 H$ (T)	τ_s (s)	ΔT_{hys} (K)	$\Delta T_{\text{hys}}/(2\tau_s R_L)$ ($\times 10^{-3}$ W/s)	R_L (K/W)
0	1.22	0.9	6.11	60
0.5	2.95	0.6	1.10	134
1	2.78	0.4	0.56	129
1.4	2.44	0.3	0.28	215

Table 1: Values of the parameters τ_s, $\Delta T_{\text{hys}}/(2\tau_s R_L)$ and R_L, appearing in Eq. (8), at different magnetic fields $\mu_0 H$. Experimental values of ΔT_{hys} obtained from Ref. [8].

With the above assumptions, the enthalpy change rate $\mathrm{d}U_{\mathrm{L}}/\,\mathrm{d}t$, due to the motion of the full phase boundary interface, is given by

$$\frac{\mathrm{d}U_{\mathrm{L}}}{\mathrm{d}t} = \Delta u_{\mathrm{L}} A \frac{\mathrm{d}x}{\mathrm{d}t} \tag{1}$$

where Δu_{L} is the enthalpy change per unit volume, representing the latent heat of the system.

To relate the velocity of the phase boundary interface $\mathrm{d}x/\,\mathrm{d}t$, appearing in Eq. (1), to the thermodynamic forces bringing the system back to equilibrium we employ the non equilibrium thermodynamic theory of linear systems [7]. The fact that the transition may occur at temperature $T_s \neq T_t$ implies that the enthalpy variation $\mathrm{d}U_{\mathrm{L}}$ is written as $\mathrm{d}U_{\mathrm{L}} = T_s\,(\mathrm{d}S - \sigma_s\,\mathrm{d}t)$, where σ_s is the entropy production rate, which is a definite positive term. Since the enthalpy change is also related to the entropy change $\mathrm{d}S$ by $\mathrm{d}U_{\mathrm{L}} = T_t\,\mathrm{d}S$, we obtain the following expression for σ_s:

$$\sigma_s = \left(\frac{1}{T_t} - \frac{1}{T_s}\right)\frac{\mathrm{d}U_{\mathrm{L}}}{\mathrm{d}t}. \tag{2}$$

Eq. (2) states that the entropy production rate is the product of a distance-like term, i.e. $1/T_t - 1/T_s$, and of a velocity-like term, i.e. $\mathrm{d}U_{\mathrm{L}}/\,\mathrm{d}t$. The former represents how far from equilibrium the system is, while the latter describes the velocity of the process. Since we are considering that T_s is close enough to T_t, we can safely assume that the displacement and the velocity terms are linked by a linear relation, obtaining this way the following equation:

$$\frac{\mathrm{d}U_{\mathrm{L}}}{\mathrm{d}t} = \alpha_s A T_t^2 \left(\frac{1}{T_t} - \frac{1}{T_s}\right) \simeq \alpha_s A\,(T_s - T_t) = \frac{1}{R_L}\,(T_s - T_t). \tag{3}$$

In Eq. (3), the last equality is justified by the assumption $T_s \simeq T_t$, the proportionality coefficient α_s is expressed in units $\mathrm{W\,K^{-1}\,m^{-2}}$ and the intrinsic kinetic parameter $R_L = (\alpha_s A)^{-1}$, which has the units of a thermal resistance, expresses the presence of damping processes for the phase boundary interface. By comparing Eq. (3) with Eq. (1) we can derive an expression for the velocity of the interface as: $\mathrm{d}x/\,\mathrm{d}t = \alpha_s T_t^2 \left(1/T_t - 1/T_s\right)/\Delta u_\mathrm{L} \simeq \alpha_s \left(T_s - T_t\right)/\Delta u_\mathrm{L}$.

The total enthalpy of the magnetocaloric material is the sum of the contributions from the latent heat, U_L, and from the specific heat, U_C: $U_e = U_\mathrm{C} + U_\mathrm{L}$. The contribution from the specific heat is taken as:

$$\frac{\mathrm{d}U_\mathrm{C}}{\mathrm{d}t} = C_s \frac{\mathrm{d}T_s}{\mathrm{d}t} \tag{4}$$

where C_s is the heat capacity of the region over which the transition takes place.

By means of Eq. (3) and Eq. (4), we have an expression of the heat flux q_s exchanged with the surroundings: $q_s = \mathrm{d}U_\mathrm{C}/\,\mathrm{d}t + \mathrm{d}U_\mathrm{L}/\,\mathrm{d}t$. The time behaviour of q_s is obtained by adding the effect of the measuring setup which is in contact with the material. A quasi-isothermal setup can be thought as a thermal bath at temperature T_b in contact with the sample at temperature T_s by a thermal resistance R_s. The heat $q_s(t)$ is then given by

$$q_s(t) = \frac{T_b(t) - T_s(t)}{R_s}. \tag{5}$$

By defining t_0 as the time at which the transition starts and t_1 as the time at which it ends, from Eqs. (3), (4), (5) we obtain the following couple of differential equations for $T_s(t)$:

$$\begin{cases} C_s \dfrac{\mathrm{d}T_s}{\mathrm{d}t} = \dfrac{T_b(t) - T_s(t)}{R_s} - \dfrac{T_s(t) - T_t}{R_L} & \text{for } t_0 \leq t \leq t_1 \\ C_s \dfrac{\mathrm{d}T_s}{\mathrm{d}t} = \dfrac{T_b(t) - T_s(t)}{R_s} & \text{for } t > t_1. \end{cases} \tag{6}$$

For an heating process, the $t_0 \leq t \leq t_1$ regime describes the behaviour of the sample when it is absorbing the latent heat due to the phase boundary motion and the temperature T_s decreases with respect to T_b. The $t > t_1$ regime describes instead what happens at the system when the transition is ended and T_s relaxes towards T_b.

We have solved Eq. (6) in the case of a constant T_b, which is an assumption justified when dealing with temperature scans performed at very low rates. By assuming $T_b = T_s$ for $t \leq t_0$ and since at $t = t_0$, when the transition begins, the sample temperature T_s reaches the threshold T_h, the constant T_b value for an heating process is $T_b = T_b(t_0) = T_h$. Moreover, we have that also the initial conditions $T_0 = T_s(t_0) = T_h$ and $T_1 = T_s(t_1)$ hold. The solution $q_s(t)$ for an individual avalanche is:

$$q_s(t) = \begin{cases} \dfrac{T_h - T_t}{R_s + R_L}\left(1 - e^{-\frac{t-t_0}{\tau_{SL}}}\right) & \text{for } t_0 \leq t \leq t_1 \\ q_{s1} e^{-\frac{t-t_1}{\tau_s}} & \text{for } t > t_1 \end{cases} \tag{7}$$

where $q_{s1} = q_s(t_1)$ is a constant factor. The time constants appearing in Eq. (7) are defined as $\tau_{SL} = \left(1/\tau_S + 1/\tau_L\right)^{-1}$, $\tau_S = C_s R_s$ and $\tau_L = C_s R_L$. It is worth noting that Figure 2 shows that the growth part of an individual avalanche, being linear, is completed in a time $t_1 - t_0$ much shorter than the time constant τ_{SL}. Hence, we can approximate Eq. (7) by using the expansion $1 - \exp\left[-\left(t - t_0\right)/\tau_{SL}\right] \simeq \left(t - t_0\right)/\tau_{SL}$, valid for $t - t_0 \leq t_1 - t_0 \ll \tau_{SL}$. Furthermore, we observe that the time constant τ_{SL}, being the parallel of the time constants τ_S, τ_L, is dominated by the smaller one between τ_S and τ_L.

3. Comparison with experiments

Figure 2 shows the heat flux q_s measured by Peltier calorimetry on a $LaFe_{11.60}Mn_{0.18}Si_{1.22}$-$H_{1.65}$ fragment of mass $m = 4.79$ mg, close to the transition, as a function of time t. The experimental data refer to heating scans performed at 1 mK/s rate at increasing applied magnetic field H. It is clear the presence of subsequent heat flux avalanches increasing in number by increasing H. Each avalanche is characterized by an initial linear growth and an exponential decay whose slope and time constant change with the field. In particular, as soon as H is getting closer to the critical field $\mu_0 H_c \sim 2.2$ T (μ_0 being the vacuum permeability) the transition becomes more and more of second order type [8] and the avalanches start to be less clearly distinguishable from the background signal. The avalanches are related to the presence of latent heat, while the background is due to the reversible part of the specific heat of the system.

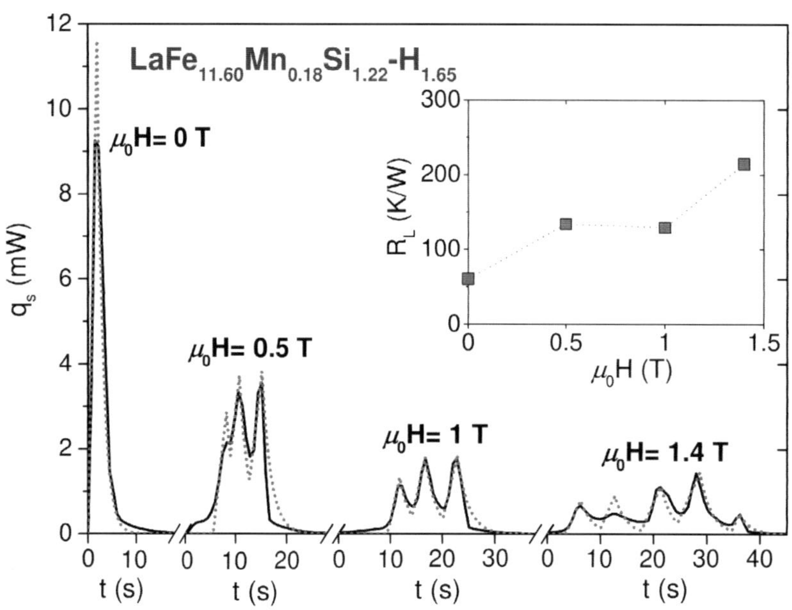

Figure 2: Experimental (black solid lines) and modelled (dotted red lines) behaviour of heat flux q_s vs. t at different applied fields H, close to the first order transition. Measurements performed in heating, at 1 mK/s rate, on a sample of mass $m = 4.79$ mg through Peltier calorimetry under magnetic field. Model curves obtained through Eq. (8) with the parameters values reported in Table 1. Inset shows the intrinsic kinetic parameter R_L of Eq. (3) at various H.

From the values of the resistivity of the silver varnish employed to contact the sample with the Peltier cell (2.5×10^3 W K^{-1} m^{-2}) and of the contact area surface (~ 1 mm^2), we have $R_s \simeq 400$ K/W. By comparing the model with the experimental data we have found $R_L < R_s$ (see Table 1), so we can safely assume $\tau_L < \tau_S$, so that $\tau_{SL} \simeq \tau_L$ and $R_s + R_L \simeq R_s$.

By using the above observations, Eq. (7) takes the simpler form:

$$q_s(t) = \begin{cases} \dfrac{T_h - T_t}{\tau_L R_s}(t - t_0) = \dfrac{\Delta T_{\text{hys}}}{2\tau_s R_L}(t - t_0) & \text{for } t_0 \leq t \leq t_1 \\ q_{s1} e^{-\frac{t-t_1}{\tau_s}} & \text{for } t > t_1 \end{cases} \tag{8}$$

where we have used the approximation $T_h - T_t \approx \Delta T_{\text{hys}}/2$, with $\Delta T_{\text{hys}} = T_h - T_c$.

Eq. (8) contains the free parameters τ_s and $\Delta T_{\text{hys}}/(2\tau_s R_L)$ that we have determined by comparison with the experimental data, trying to reproduce the qualitative behaviour of the avalanches. The result of the comparison is shown in Figure 2. The key observation is that the initial growth and decay times, t_0 and t_1, have been chosen by concatenating subsequent individual avalanches characterized, for every fixed value of H, by the same slope $\Delta T_{\rm hys}/\left(2\tau_s R_L\right)$ and the same time constant τ_s. The values of these parameters, together with those of $\Delta T_{\rm hys}(H)$ and R_L, are reported in Table 1. R_L values for different H are shown also in the inset of Figure 2.

4. Conclusions

We have applied the non equilibrium thermodynamic theory of linear systems to describe the heat flux avalanches observed by Peltier calorimetry temperature scans at low rates around the transition temperature of $LaFe_{11.60}Mn_{0.18}Si_{1.22}$-$H_{1.65}$ magnetocaloric material. We have found that a single avalanche shows an initial linear growth, describing the magnetic sample during the transition, and a subsequent exponential decay, corresponding to the system relaxation towards equilibrium. The decay behaviour is governed by a time constant $\tau_s \simeq 1$–$2\,$s increasing with the applied field H and related only to the contact between the measuring setup and the sample. The growth depends instead on the intrinsic kinetic parameter R_L introduced to describe the damping of the phase boundaries motion. Although we have analyzed only few experimental avalanches, the values of R_L show indeed that the kinetic coefficient increases with H, indicating that the kinetics is slowed down by approaching the critical point of the transition. A detailed estimate of the value of the time constant τ_L associated to R_L strictly depends on the typical length scales involving the sample region undergoing the transition and will be the subject of future work. Future investigations shall be also dedicated to understand how these outcomes may influence the operational frequency of magnetocaloric-based devices.

References

[1] Kuz'min M D 2007 *Appl. Phys. Lett.* **90** 251916
[2] Lovell E, Pereira A M, Caplin A D, Lyubina J and Cohen L F 2015 *Adv. Energy Mater.* **5** 1401639
[3] Fujita A and Yako H 2013 *J. Alloy. Compd.* **577S** S48
[4] Kuepferling M, Sasso C P and Basso V 2013 *EPJ Web of Conferences* **40** 06010
[5] Kuepferling M, Bennati C, Laviano F, Ghigo G and Basso V 2014 *J. Appl. Phys.* **115** 17A925
[6] Bennati C, Gozzelino G, Olivetti E S and Basso V 2016 *in preparation*
[7] Callen H B 1985 *Thermodynamics and an Introduction to Thermostatistics* 2nd ed (New York: J. Wiley & Sons)
[8] Basso V, Küpferling M, Curcio C, Bennati C, Barcza A, Katter M, Bratko M, Lovell E, Turcaud J and Cohen L F 2015 *J. Appl. Phys.* **118** 053907

IOP Conf. Series: Journal of Physics: Conf. Series **903** (2017) 012047 doi:10.1088/1742-6596/903/1/012047

Magnetic Circuit Model combined with Play Model Obtained from Landau-Lifshitz-Gilbert Equation

H Tanaka[1], K Nakamura[1] and O Ichinokura[1]

[1]Graduate School of Engineering, Tohoku University, 6-6-05 Aoba Aramaki Aoba-ku, Sendai 980-8579, Japan

E-mail: hideaki.tanaka.s7@dc.tohoku.ac.jp

Abstract. This paper presents a novel magnetic circuit model considering magnetic hysteresis behavior, in which dc hysteresis is expressed by a play model while classical and anomalous eddy current losses are calculated by magnetic circuit elements. We describe an efficient method for obtaining the play model from the minimum measured *B-H* loops. It is proved that the proposed magnetic circuit model can calculate hysteresis loops under Pulse Width Modulation (PWM) excitation with high accuracy in a short time.

1. Introduction

In recent years, establishment of a practical and simple method for calculating iron loss including magnetic hysteresis behavior is required for developing high-efficient electric machines with short-term. A magnetic circuit model is one of the potential candidates because it is simple model with fast computation time. In our previous paper [1], a magnetic circuit model considering the magnetic hysteresis was proposed. In this model [1], dc hysteresis is expressed by a simplified micromagnetic model using Landau-Lifshitz-Gilbert (LLG) equation [2], while classical and anomalous eddy current losses are calculated by magnetic circuit elements. Although the simplified micromagnetic model is derived from a few measured dc hysteresis loops, the magnetic circuit model [1] is able to simulate dc hysteresis loops in various maximum flux densities with high accuracy. However, this model takes relatively long time for calculation since the LLG equation requires convergence calculation. Meanwhile, a Play model [3] is a practical method from a viewpoint of the calculation time. However, a large amount of measured hysteresis loops is required to obtain the play model. To overcome these problems, we present an efficient method for obtaining the play model from the simplified micromagnetic model which can be derived from the minimum measured dc hysteresis loops.

2. Play model obtained from simplified micromagnetic model

The motion of the magnetizations in a magnetic substance can be represented by the following LLG equation:

$$\frac{d\boldsymbol{m}_i}{dt} = -\frac{|\gamma|}{1+\alpha^2}\{(\boldsymbol{m}_i \times \boldsymbol{H}_{eff}) + \alpha(\boldsymbol{m}_i \times (\boldsymbol{m}_i \times \boldsymbol{H}_{eff}))\} \tag{1}$$

where the magnetization vector is $\boldsymbol{M}_i = M_s\boldsymbol{m}_i$, the spontaneous magnetization is M_s, the normalized magnetization vector is $\boldsymbol{m}_i$, the gyromagnetic ratio of electron is γ, the damping constant is α, and the effective field is $\boldsymbol{H}_{eff}$, respectively.

Content from this work may be used under the terms of the Creative Commons Attribution 3.0 licence. Any further distribution of this work must maintain attribution to the author(s) and the title of the work, journal citation and DOI.
Published under licence by IOP Publishing Ltd

IOP Conf. Series: Journal of Physics: Conf. Series **903** (2017) 012047 doi:10.1088/1742-6596/903/1/012047

The simplified micromagnetic model [2] is derived based on several assumptions: each crystal grain has single domain particle, and the domain wall motion is neglected, *etc.* Thereby, the effective field $\boldsymbol{H}_{eff}$ in (1) is given by

$$\begin{aligned}\boldsymbol{H}_{eff} &= \boldsymbol{H}_{app} + \boldsymbol{H}_{ani} + \boldsymbol{H}_{ela} \\ &= H_{app} - \frac{1}{M_s}\frac{\partial}{\partial \boldsymbol{m}_i}(E_{ani} + E_{ela}),\end{aligned} \tag{2}$$

where the applied field is $\boldsymbol{H}_{app}$, the anisotropy field is $\boldsymbol{H}_{ani}$, the field generated by magnetoelastic energy is $\boldsymbol{H}_{ela}$, the magnetic anisotropy energy is E_{ani}, and the magnetoelastic energy is E_{ela}, respectively. The magnetic anisotropy energy E_{ani} and the magnetoelastic energy E_{ela} are defined as

$$E_{ani} = \frac{h_{ani} M_s}{2}(a_1^2 a_2^2 + a_2^2 a_3^2 + a_3^2 a_1^2), \tag{3}$$

$$E_{ela} = b_2 \overline{m}^2 + b_4 \overline{m}^4 + b_6 \overline{m}^6, \tag{4}$$

where the direction cosines of magnetization vectors with respect to x, y, and z axes (easy axes) of each grains are a_1, a_2, and a_3, the coefficient of magnetic anisotropy is h_{ani}, and the parameters of Taylor expansion are b_2, b_4, and b_6, mean magnetization is $\overline{m}$, respectively. It is considered that the magnetoelastic energy is symmetric with respect to the direction of the mean magnetization. Therefore, it is assumed that the magnetoelastic energy is expressed by Taylor expansion. In addition, although the LLG equation is generally used for expressing the dynamic hysteresis behavior, only steady-state value of the LLG equation is used for expressing dc hysteresis. Figure 1 shows dc hysteresis loops calculated by the simplified micromagnetic model. All the parameters of the simplified micromagneti model are determined to fit the dc hysteresis loops of the maximum flux density B_m = 1.2T in Figure 1. As shown in the figure, the model can calculate dc hysteresis loops in various flux densities with high accuracy using the same parameters. However, the calculation time of the simplified micromagnetic model is relatively long because the LLG equation requires convergence calculation.

Meanwhile, a play model [3], which needs no convergence calculation, is efficient for simulating hysteresis behavior from a viewpoint of calculation time. However, the play model requires a large amount of measured symmetric dc hysteresis loops. To overcome these problems, we present an efficient method for obtaining the play model from the simplified micromagnetic model. Figure 2 indicates the symmetric dc hysteresis loops in various maximum flux densities at intervals of 0.04 T which is calculated by the simplified micromagnetic model. By using the simplified micromagnetic model, the play model can be obtained without a large amount of measurement.

3. Proposed magnetic circuit model and simulation results

To prove the usefulness of the proposed method, the dynamic analysis of a transformer shown in figure 3 was conducted. As shown in the proposed magnetic circuit model, dc hysteresis is expressed by the play model which is determined by the simplified micromagnetic model [2], while classical and anomalous eddy current losses are calculated by magnetic circuit elements. Figure 4 shows the measured and calculated hysteresis loops under Pulse Width Modulation excitation. It is clear that the proposed model is able to calculate the hysteresis loops under PWM excitation with high accuracy including minor loops. Table 1 indicates the comparison of the calculation time between previous [1] and proposed models. It is understood that the proposed magnetic circuit model reduced remarkably the calculation time.

4. Conclusion

This paper presented the novel magnetic circuit model which is combined with the play model. The play model can be obtained from the simplified micromagnetic model using LLG equation without a large amount of measurement. The proposed model can calculate the hysteresis loops under PWM excitation with high accuracy and remarkably reduced the calculation time in comparison with the previous model.

This work was supported by Grant-in Aid for JSPS Fellows (26-5193).

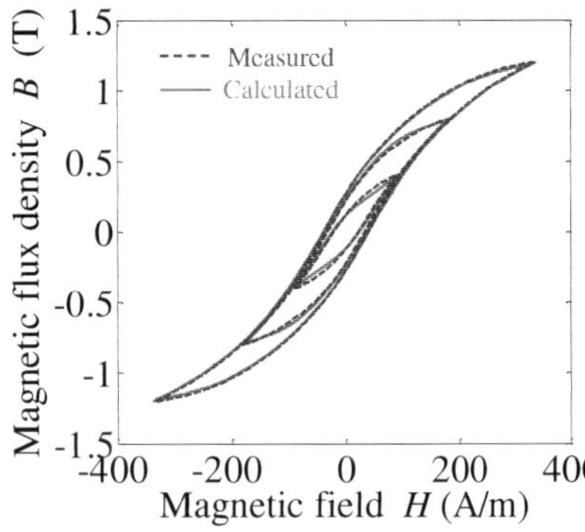

Figure 1. Measured and calculated dc hysteresis loops of non-oriented silicon steel at the maximum flux density B_m = 0.4 T, 0.8 T, 1.2 T.

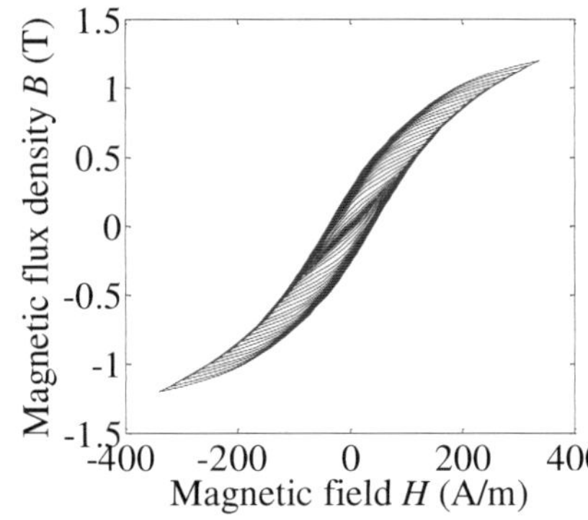

Figure 2. Calculated dc hysteresis loops from B_m = 0.04 T to 1.2 T at intervals of 0.04 T.

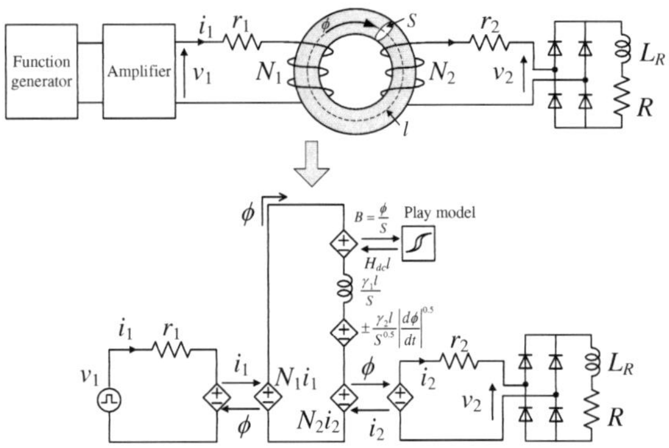

Figure 3. Experimental circuit and the proposed magnetic circuit model.

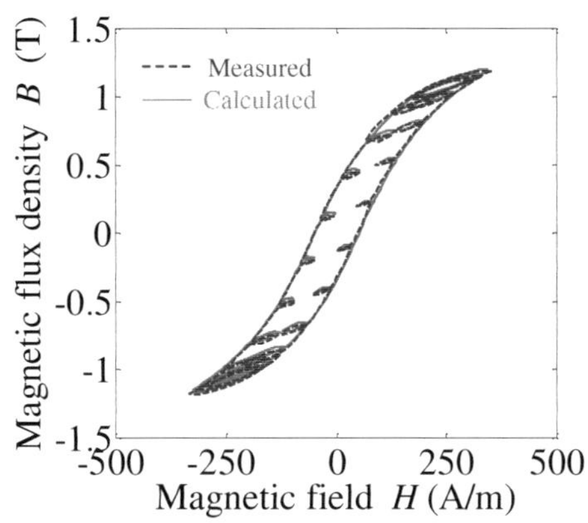

Figure 4. Measured and calculated hysteresis loops under PWM excitation.

Table 1. Comparison of the calculation time between previous [1] and proposed models.

	Previous model [1]	Proposed model
Number of steps per 1 cycle	400	400
Calculation time of 1 cycles	85 min.	0.15 min.

References

[1] Tanaka H, Nakamura K, and Ichinokura O 2015 *J. Magn. Soc. Jpn.* **39** 65-70

[2] Oshima H, Uehara Y, Shimizu K, Inagaki K, Furuya A, Fujisaki J, Suzuki M, Kawano K, Mifune T, Matsuo T, Watanabe K, and Igarashi H 2014 *J. Jpn. Soc. Powder Metallurgy* **61** S238-S241

[3] Bobbio S, Milano G, Serpico C, and Visone C 1997 *IEEE Trans. Magn.* **33** 4417-4426

IOP Conf. Series: Journal of Physics: Conf. Series **903** (2017) 012048 doi:10.1088/1742-6596/903/1/012048

Accuracy Improvement of Magnetic Hysteresis Calculated by LLG Equation

H Tanaka[1], K Nakamura[1] and O Ichinokura[1]

[1]Graduate School of Engineering, Tohoku University, 6-6-05 Aoba Aramaki Aoba-ku, Sendai 980-8579, Japan

E-mail: hideaki.tanaka.s7@dc.tohoku.ac.jp

Abstract. Quantitative estimation of iron loss including magnetic hysteresis behavior is essential to the development of high-efficient electrical machines. A simplified micromagnetic model using Landau-Lifshitz-Gilbert (LLG) equation is one of the useful models for calculating the hysteresis behavior. However, further improvement of the calculation accuracy under magnetic saturation is required. This paper presents the accuracy improvement of the magnetic hysteresis calculated by the LLG equation.

1. Introduction

Quantitative analysis of iron loss including magnetic hysteresis behavior is necessary to the development of high-efficient electrical machines. A lot of methods for calculating the hysteresis behavior have been proposed. Among them, a simplified micromagnetic model using Landau-Lifshitz-Gilbert (LLG) equation [1] is one of the useful models for calculating the hysteresis behavior. In a previous paper [2], the magnetic circuit model combined with the above-mentioned LLG equation was proposed, in which dc hysteresis is expressed by the above-mentioned LLG equation, while classical and anomalous eddy current losses are calculated by magnetic circuit elements. The proposed model can calculate the hysteresis loops of various materials under various exciting voltage waveforms. However, further improvement of the calculation accuracy under magnetic saturation is required.

To calculate the hysteresis loops with high accuracy, first, the equation expressing the magnetoelastic energy in [1] is improved to express the strong magnetic nonlinearity. Second, the equation for the magnetic anisotropy field in [1] is also improved to express the shape of the hysteresis loop with high accuracy.

2. Accuracy improvement of magnetic hysteresis calculated by LLG equation

The motion of the magnetizations in a magnetic substance can be represented by the following LLG equation:

$$\frac{d\boldsymbol{m}_i}{dt} = -\frac{|\gamma|}{1+\alpha^2}\{(\boldsymbol{m}_i \times \boldsymbol{H}_{eff}) + \alpha(\boldsymbol{m}_i \times (\boldsymbol{m}_i \times \boldsymbol{H}_{eff}))\} \ , \tag{1}$$

where the magnetization vector is $\boldsymbol{M}_i = M_s\boldsymbol{m}_i$, the spontaneous magnetization is M_s, the normalized magnetization vector is $\boldsymbol{m}_i$, the gyromagnetic ratio of electron is γ, the damping constant is α, and the effective field is $\boldsymbol{H}_{eff}$, respectively.

In the reference [1], the effective field $\boldsymbol{H}_{eff}$ in equation (1) is given by

Content from this work may be used under the terms of the Creative Commons Attribution 3.0 licence. Any further distribution of this work must maintain attribution to the author(s) and the title of the work, journal citation and DOI.
Published under licence by IOP Publishing Ltd

$$\boldsymbol{H}_{eff} = \boldsymbol{H}_{app} + \boldsymbol{H}_{ela} + \boldsymbol{H}_{ani}\,, \tag{2}$$

$$\boldsymbol{H}_{ela} = -\frac{1}{M_s}\frac{\partial E_{ela}}{\partial \boldsymbol{m}_i},\quad E_{ela} = b_2\overline{m}^{2} + b_4\overline{m}^{4} + b_6\overline{m}^{6}\,, \tag{3}$$

$$\boldsymbol{H}_{ani} = -\frac{1}{M_s}\frac{\partial E_{ani}}{\partial \boldsymbol{m}_i},\quad E_{ani} = \frac{h_{ani}M_s}{2}(a_1^2a_2^2 + a_2^2a_3^2 + a_3^2a_1^2)\,, \tag{4}$$

where the applied field is $\boldsymbol{H}_{app}$, the field generated by magnetoelastic energy is $\boldsymbol{H}_{ela}$, the anisotropy field is $\boldsymbol{H}_{ani}$, the magnetoelastic energy is E_{ela}, the parameters of Taylor expansion are b_2, b_4, and b_6, the magnetic anisotropy energy is E_{ani}, the coefficient of magnetic anisotropy is h_{ani}, and the direction cosines of magnetization vectors with respect to x, y, and z axes (easy axes) of each grains are a_1, a_2, and a_3, respectively. In the simplified micromagnetic model [1], all the parameters above mentioned are determined by fitting the measured dc hysteresis loops. In addition, although the LLG equation is generally used for expressing the dynamic hysteresis behavior, only steady-state value of the LLG equation is used for expressing dc hysteresis.

Figure 1 shows the measured and calculated dc hysteresis loops. It is found that the previous model [1] cannot simulate the strong nonlinearity. To calculate the hysteresis loops under the strong nonlinearity, the order of equation (3) is increased as follows:

$$E_{ela} = \sum_{j=1}^{10} b_{2j}\overline{m}^{2j}\,. \tag{5}$$

Figure 2 shows the measured and calculated hysteresis loops using equation (5). Figure 3 is an enlarged view of figure 2. It is clear that the strong nonlinearity can be expressed by using equation (5). However, the shape of the hysteresis loops is still different from the measured loops, especially under magnetic saturation.

To express the shape of the hysteresis loop with high accuracy, the magnetic anisotropy field in [1] is also improved as the following equation:

$$\boldsymbol{H}_{ani} = -k_{ani}\frac{1}{M_s}\frac{\partial E_{ani}}{\partial \boldsymbol{m}_i}\,, \tag{6}$$

$$k_{ani} = 1 + k_1 \min(1, \exp(k_2(|\overline{\boldsymbol{m}} \cdot \boldsymbol{m}_i| - k_3)))\,, \tag{7}$$

where the k_{ani} is the newly defined coefficient. To restrict the maximum value of k_{ani}, "min" is used in equation (7). Figure 4 shows the coefficient k_{ani} in equation (7) at $k_1 = 2$, $k_2 = 5$, and $k_3 = 0.7$ which are determined by fitting the measured dc hysteresis loops.

Figure 5 indicates the measured and calculated dc hysteresis loops using the proposed method. Figure 6 is an enlarged view of figure 5. It is understood that the shape of calculated hysteresis loops is improved by using equation (6).

Figure 7 illustrates the magnetic circuit model combined with the improved LLG equation. Using the improved model, current waveforms are calculated when the sinusoidal voltage waveform corresponding to the maximum flux density B_m = 1.8 T is applied. Figure 8 and 9 denote the current waveforms calculated by the previous [2] and improved models. It is clear that the calculation accuracy is remarkably improved.

3. Summary

This paper presented the accuracy improvement of the magnetic hysteresis calculated by the LLG equation. First, the equation expressing the magnetoelastic energy in [1] was improved to simulate the strong magnetic nonlinearity. Second, the equation for the magnetic anisotropy field in [1] was also improved to express the shape of the hysteresis loop with high accuracy. It was demonstrated that the proposed method can improve the accuracy of the magnetic nonlinearity and hysteresis behavior.

This work was supported by Grant-in Aid for JSPS Fellows (26-5193).

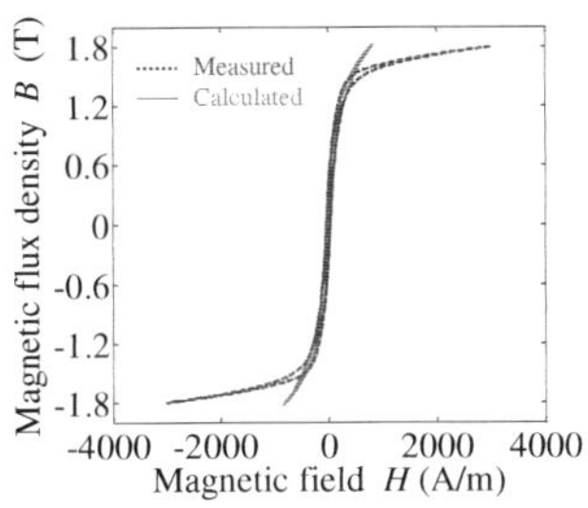

Figure 1. Measured and calculated dc hysteresis of non-oriented Si steel.

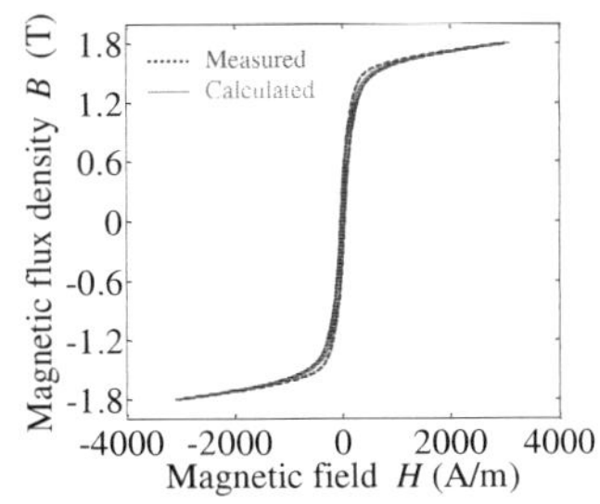

Figure 2. The dc hysteresis of non-oriented Si steel calculated by using equation (5).

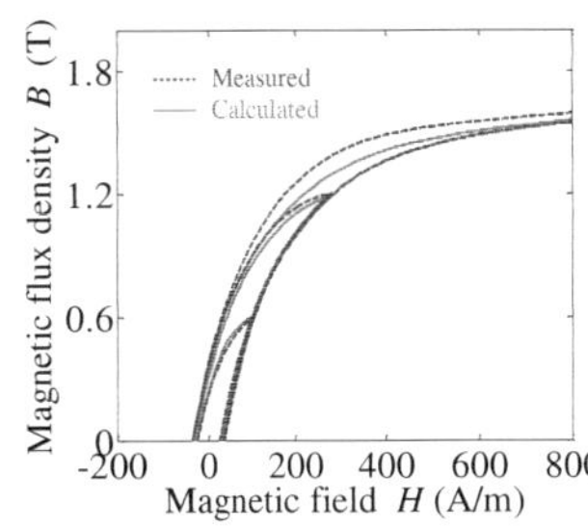

Figure 3. Enlarged view of figure 2.

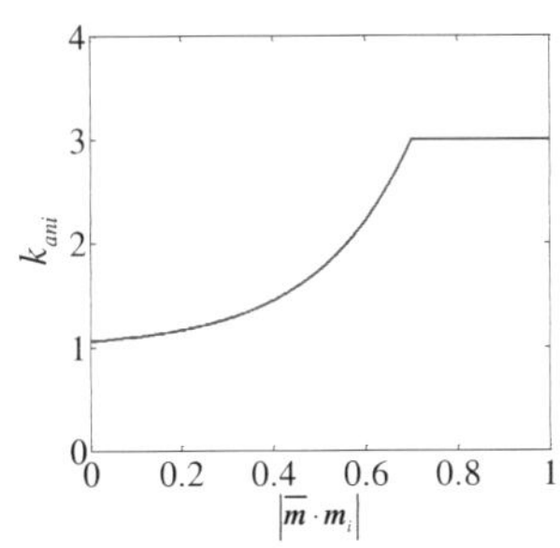

Figure 4. Relation between k_{ani} and the absolute value of an interior product of a normalized magnetization vector $\boldsymbol{m}_i$ and an average of the normalized magnetization vectors $\overline{\boldsymbol{m}}$.

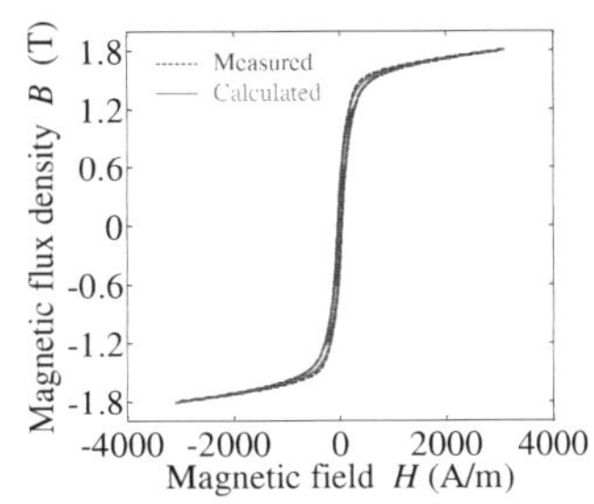

Figure 5. The dc hysteresis of the non-oriented Si steel calculated by the proposed method.

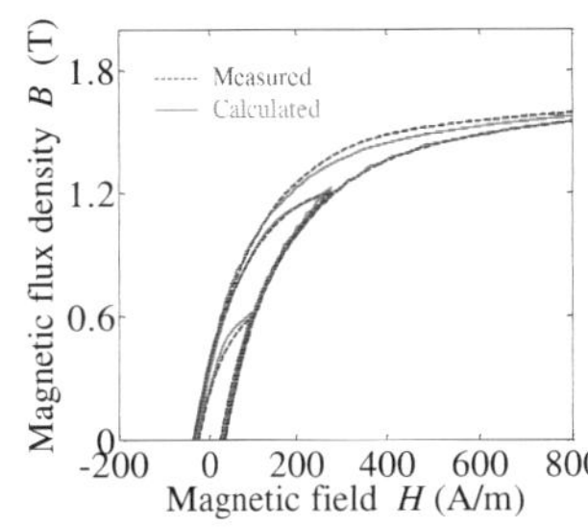

Figure 6. Enlarged view of figure 5.

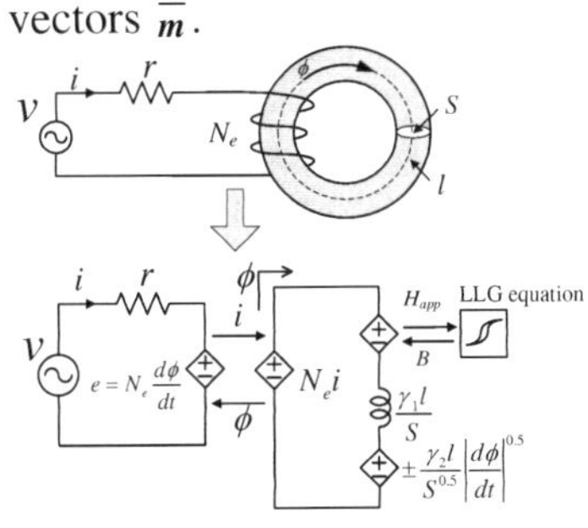

Figure 7. Magnetic circuit model combined with LLG equation.

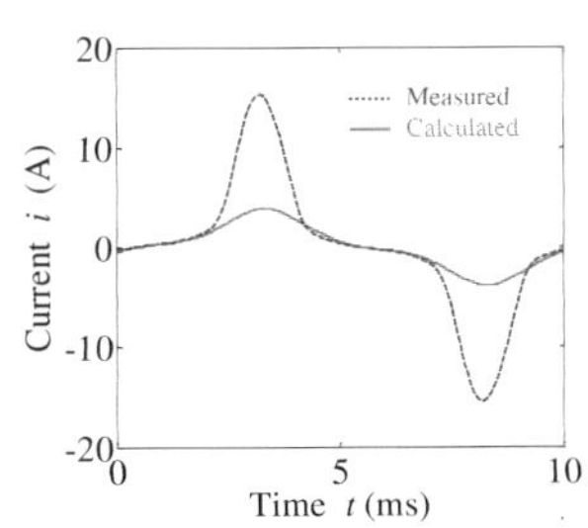

Figure 8. Current waveform calculated by the previous model.

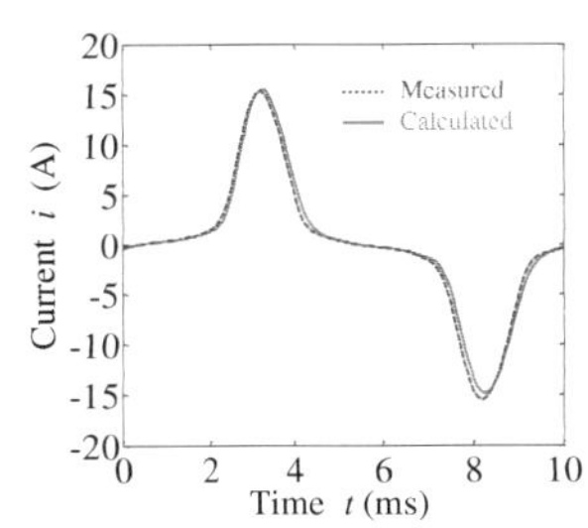

Figure 9. Current waveform calculated by the proposed model.

References

[1] Oshima H, Uehara Y, Shimizu K, Inagaki K, Furuya A, Fujisaki J, Suzuki M, Kawano K, Mifune T, Matsuo T, Watanabe K, and Igarashi H 2014 *J. Jpn. Soc. Powder Metallurgy* **61** S238-S241

[2] Tanaka H, Nakamura K, and Ichinokura O 2015 *J. Magn. Soc. Jpn* **39** 65-70

Effect of long-term thermal aging on magnetic hysteresis for low-alloy pressure vessel steel

Satoru Kobayashi[1], **Ferenc Gillemot**[2], **Ákos Horváth**[2], **Marta Horváth**[2], **Almásy László** [2]

[1] Faculty of Engineering, Iwate University, Morioka, Japan
[2] KFKI Atomic Energy Research Institute, H-1525 Budapest, Hungary

E-mail: koba@iwate-u.ac.jp

Abstract. Long-term thermal aging experiments have been performed at 290 and 550°C up to ∼ 10000 h for low-alloy pressure vessel steels with low and high Ni contents, in order to elucidate effects of thermal embrittlement on a structure-sensitive property of minor *B*-*H* loops. While a minor-loop property obtained from a power-law scaling of minor-loop parameters exhibits a small but steady increase after aging at 290°C, that exhibits a decreasing trend after aging at 550°C. The observations suggest that the evolution of nanometer precipitates plays a crucial role for a magnetic property change at 290°C, while relaxation of lattice strain in a matrix associated with precipitation may dominate at 550°C.

A development of nondestructive evaluation techniques for irradiation hardening in low-alloy steels for reactor pressure vessel has been an important issue for long-term operation of nuclear power plants. Under neutron irradiation, the formation of nanoscale defects such as Cu-rich precipitates is accelerated due to enhanced diffusion of atoms, resulting in an increase of yield stress and hardness in the low fluence regime [1]. In addition, an additional increase of mechanical property changes was found to occur at high neutron fluence above 10^{20} n/cm^2, owing to the formation of "late-blooming" Mn-Ni-Si-rich clusters. Since pressure vessel steels are generally subjected to high temperatures of 290°C during an operation of nuclear reactors, thermal embrittlement also affects the hardening. Magnetic hysteresis, which is sensitive to lattice defects, is one of possible techniques for nondestructive evaluation of neutron-irradiated materials [2]. However, combined effects of irradiation and thermal embrittlement make it difficult to understand the mechanism of magnetic property changes during neutron irradiation. In this study, we have investigated effects of long-term thermal aging on magnetic hysteresis for VVER440 and VVER1000-type low-alloy nuclear reactor pressure vessel steels, in order to get an insight of thermal embrittlement in neutron-irradiated steels.

The chemical compositions for the two steels are listed in Table 1. The compositions are similar to each other, but, VVER1000 contains a higher amount of Ni. Samples with longitudinal (LX) and transverse (TX) orientations with respect to forged direction, were prepared. The samples were aged at two different temperatures; at high temperature of 550°C up to 10080 hours to accelerate thermal effects and at an operatation temperature of 290°C up to 14000 hours. For 550°C, we prepared 1−6 samples for each aging condition, whereas for 290°C a cycle of aging and measurements was repeated for 2 samples up to 14000 h. A set of minor *B*-*H* loops with different field amplitudes up to 10 kA/m were measured with a field sweep rate of

Content from this work may be used under the terms of the Creative Commons Attribution 3.0 licence. Any further distribution of this work must maintain attribution to the author(s) and the title of the work, journal citation and DOI.
Published under licence by IOP Publishing Ltd

Table 1. Chemical compositions of measured alloys (wt.%)

Alloy	Cu	Ni	Mn	Cr	Mo	Fe
VVER440	0.09	0.07	0.54	2.70	0.68	Bal.
VVER1000	0.07	1.26	0.46	2.20	0.50	Bal.

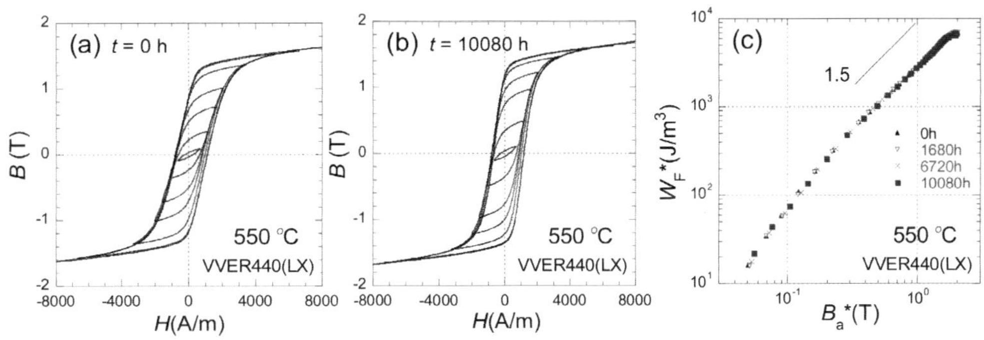

Figure 1. A set of minor B-H loops before and after aging at 550°C. The data for VVER440 (LX) at aging time of (a) 0 h and (b) 10080 h are given as an example. (c) Relationship between maximum flux density and hysteresis loss, plotted in a double logarithmic scale, taken after aging at 550°C. The data for VVER440 (LX) are given as an example.

9 kA/m/s, for plate specimens with dimensions of 4×16×0.5 mm^3, using a flux meter designed for a plate specimen [3]. Representative data for minor B-H loops taken before and after aging at 550°C for VVER440 (LX) are shown in Figs. 1(a) and 1(b), respectively. For each minor loop, we determined B-H parameters, i.e. maximum flux density, B_a^*, and hysteresis loss, W_F^*, and examined a relation between the two parameters as shown in Fig. 1(c). Using a power-law scaling relation, given by

$$W_F^* = W_F^0 \left(\frac{B_a^*}{B_s}\right)^{n_F}, \tag{1}$$

a minor-loop coefficient, W_F^0, being a sensitive indicator of defect density, was obtained. Here, B_s is normalization constant and was set to 1.8 T. An exponent n_F is generally $\sim$ 1.5 and was universally 1.44±0.03 in this study irrespective of the type of steels, sample orientation, and aging time. In this study, the coefficient, W_F^0, was used to investigate microstructural changes during thermal aging. We also performed microhardness measurements and found that a change of hardness due to thermal aging is within an experimental accuracy for both steels and is insensitive to thermal aging.

Figures 2(a)-2(d) summarize a change of W_F^0, ΔW_F^0, as a function of aging time at 550 and 290°C for VVER440 and VVER1000 steels. For aging at 550°C, ΔW_F^0 for VVER440 seems to be constant or slightly decrease with aging time, whereas for VVER1000 a decrease in ΔW_F^0 attains $\sim$ 10% after aging up to 10080 h. On the other hand, for aging at 290°C, ΔW_F^0 slightly increases with aging time and its increase attains 1−2% at the maximum aging time of 14000 h for both VVER440 and VVER1000. For both steels, ΔW_F^0 is very weakly dependent on sample orientation.

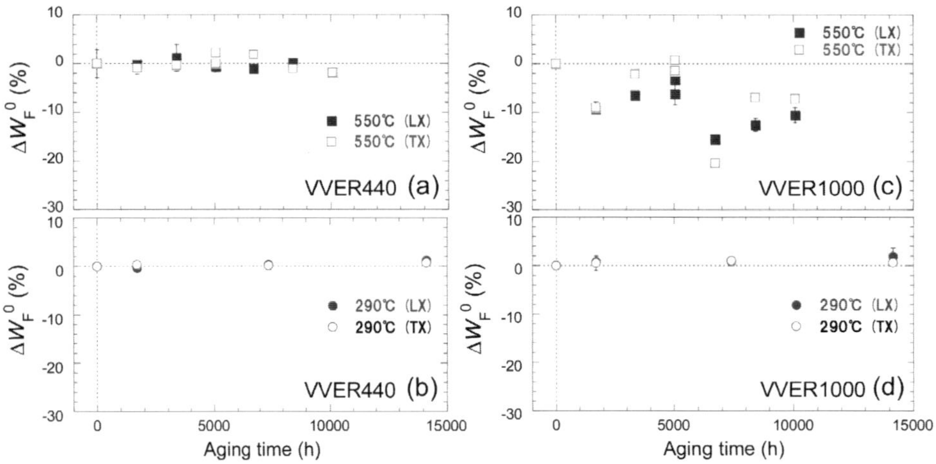

Figure 2. A change of a minor-loop coefficient W_F^0 after aging at 550 or 290°C for VVER440 [(a), (b)] and VVER1000 [(c), (d)]. Error bars are primarily due to the sample dependence.

Under irradiation, various types of lattice defects such as precipitates of a few nanometer size, carbides of a few tens nanometer size, dislocation loops, are formed. In particular, the volume fraction of precipitates after irradiation was found to be much higher for VVER1000 with high-Ni content [1]. Although such formation is generally accelerated under irradiation due to irradiation-induced diffusion, nanoscale defects would be more or less formed also during thermal aging and their density would become high at higher aging temperatures. Generally, lattice defects disturb domain wall motion and steels with high defect density have high value of W_F^0. The observed small but gradual increase of ΔW_F^0 for 290°C would reflect the formation of such nanoscale defects. Note that recent microstructural study for VVER1000 after aging at 290°C up to 10000 h showed that there is no significant evolution of number density of carbides with size greater than >10 nm [4], implying a contribution of much smaller precipitates to the magnetic property changes at 290°C.

On the other hand, at higher temperature of 550°C, ΔW_F^0 exhibits a decreasing trend with aging time. There would be two major mechanism for a decrease in W_F^0; recovery due to rearrangement of dislocations and relaxation of lattice strain in a matrix due to dissolved atoms (Ni, Mn, Cu, etc.) associated with precipitation. Considering that a decrease of ΔW_F^0 is large for high-Ni VVER1000 and Ni promotes precipitation [1], the combination of the two mechanism or the latter one dominate a magnetic property change during thermal aging at 550°C.

Acknowledgments

This work was supported by Grant-in-Aid for Scientific Research (B) (Grant No. 25289346) from JSPS, Japan.

References

[1] J. Bohmert, H.W. Viehrig, A. Ulbricht, J. Nucl. Mater. **297** (2001) 251.
[2] S. Kobayashi, F. Gillemot, A. Horvth, R. Szkely, M. Horvath, Phil. Mag. **92** (2012) 3813.
[3] S. Kobayashi, T. Yamamoto, D. Klingensmith, G. R. Odette, H. Kikuchi, Y. Kamada, J. Nucl. Mater. **422** (2012) 158.
[4] B. Gurovich, E. Kuleshova, Ya. Shtrombakh, S. Fedotova, D. Maltsev, A. Frolov, O. Zabusov, D. Erak, D. Zhurko, J. Nucl. Mater. **456** (2015) 23.

IOP Conf. Series: Journal of Physics: Conf. Series **903** (2017) 012050 doi:10.1088/1742-6596/903/1/012050

Tailoring the ultra-soft magnetic properties of sputtered FineMET thin films for high-frequency power applications

Ansar Masood*[1], Paul McCloskey[1], Cian Ó Mathúna[1, 2], and Santosh Kulkarni[1]

[1]Tyndall National Institute, Lee Maltings, Dyke parade, Cork, Ireland
[2] School of Electrical and Electronics Engineering, University College Cork, Ireland.
*ansar.masood@tyndall.ie

Abstract. This paper presents sputtered FineMET thin films with high saturation magnetic flux density and ultra-soft magnetic properties by maintaining an in-plane magnetization reversal process. The magnetic properties of FineMET thin films were investigated as a function of film thickness (40-517 nm) and annealing temperature (200-400 °C). The hysteresis loop study reveals that the films with low thickness (40 –316 nm) were perpendicularly magnetized in the as-deposited state. The ultra-soft magnetic properties (H_c=3 A/m) of the films were tailored by a thermal annealing process. A high relative permeability (μ_r) of 1763 of these films with a good loss performance up to 1 GHz, makes them attractive for high-frequency power applications.

1. Introduction

Magnetic cores with versatility in soft magnetic properties are highly anticipated in modern energy storage and transferring devices for high-frequency applications [1, 2]. The next generation of miniaturized power electronic devices requires cores with high magnetic flux density, high resistivity, and ultra-low coercivity along with ease of manufacturability [1, 2]. Amorphous and nanocrystalline materials exhibit excellent soft magnetic properties and high electrical resistivity (ρ) as compared to their crystalline counterparts, which make them a potential candidate [3, 4]. These materials are well investigated in bulk form, however, the study of these alloys in integrated thin films form that retain their exceptional properties is a recent area of research [4]. Reports suggest the soft magnetic properties of these materials in thin films are significantly poorer as compared to their bulk form [4, 5]. One of the important factors responsible for such a deterioration in magnetic properties is the generation of internal stress/strain during the fabrication process [5]. In the present work, we report the tailored soft magnetic properties of sputtered FineMET thin films with ultra-low coercive losses typically seen in their bulk form. The high magnetic flux density (1.8 T) and large ρ (145 μΩcm) of FineMET alloy are beneficial for power electronic device miniaturization.

2. Experimental Methods

Amorphous films of FineMET ($Fe_{73.5}Cu_1Nb_3Si_{13.5}B_9$, atomic %) alloy with a thickness of 40-517 nm, were deposited using DC-magnetron sputtering from a single alloy target. The films were deposited on the Si wafers. Prior to the deposition of films, an adhesive layer of 20 nm of Ti was deposited. The DC-power to the target was fixed at 100 W. The structure of the films was investigated using X-ray diffraction (XRD). The films were annealed for 60 min in an inert atmosphere. In-plane hysteresis loops were studied using a SHB (MESA 200 HF) B-H loop tracer. The high-frequency relative permeability (μ_r) was measured by Ryowa permeameter (PMM 9G).

Content from this work may be used under the terms of the Creative Commons Attribution 3.0 licence. Any further distribution of this work must maintain attribution to the author(s) and the title of the work, journal citation and DOI.
Published under licence by IOP Publishing Ltd

3. Results and Discussion

The in-plane magnetic hysteresis loops of different films (40-517 nm) were performed at room temperature (Fig. 1). The saturation magnetic flux density of the 517 nm film was estimated to be 1.8 T for a 100 mm wafer. Figure 1 illustrates that the shape of the hysteresis loops, coercivity (H_c) and anisotropic field (H_k) vary as a function of film thickness. For thinner films (40-316 nm), the shape of the hysteresis loop is not a square type. Instead, the B-H loops show the characteristics of a "transcritical loop" in which a lower magnetization slope is exhibited prior to saturation and signifies the presence of a perpendicular anisotropy component (PMA) [5]. When the thickness of the films is increased from 40 nm to 100 nm, the H_c and H_k increased to the maximum values of 360 A/m and 1.7 kA/m, respectively. The shape of the hysteresis loops suggests that PMA becomes dominant within this range of film thickness. However, as the film thickness is increased to >100 nm, H_c and H_k start to reduce to the minimum of H_c=64 A/m, H_k=128 A/m for 517 nm thick films. The Fig. 1 (a) illustrate that the shape of the hysteresis loops is entirely transformed into a square type for thicker films. From this analysis, it can be concluded that the state of magnetization is transformed from a perpendicular to an in-plane configuration with an increase of film thickness from 40-517 nm.

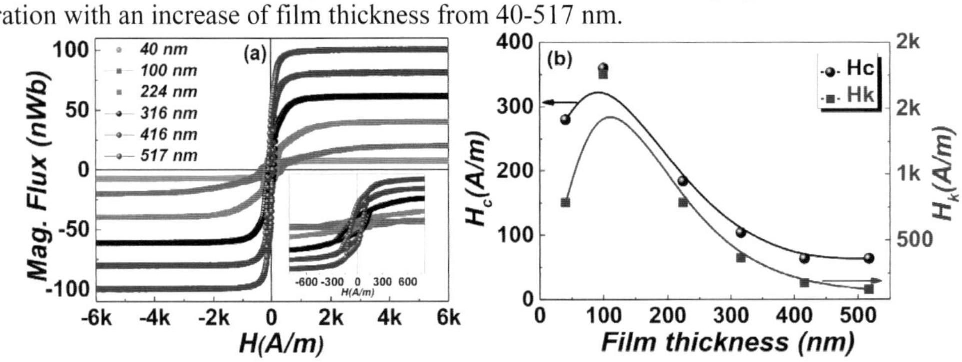

Fig. 1: (a) B-H loops of different films in the as-deposited state. (b) The coercivity (H_c) and anisotropic field (H_k) of as-deposited films as a function of thickness.

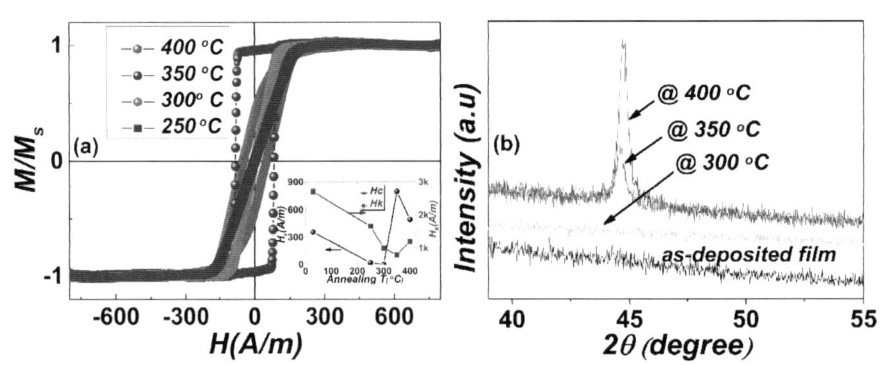

Fig. 2. (a) In-plane hysteresis loops of 100 nm film after annealing. Inset: H_c and H_k of thin films as a function of annealing temperature. (b) X-ray diffraction (XRD) spectrum of annealed films.

The effect of thermal treatment (250-400 °C) on the magnetic properties of 100 nm film is presented in Fig. 2 (a). It illustrates that the state of magnetization is affected by annealing and the transcritical loops, as a consequence, are transformed into a square type with reduced values of H_c and H_k. The hysteresis loops were soft magnetic fully square behavior after a thermal anneal at 250 °C. The BH- loop study shows that annealing at 300 °C further reduced the level of stress in thin films and as a consequence, H_c was reduced to 3 A/m. This investigation clearly shows that heat treatment radically modified the overall structure and gave rise to in-plane magnetization when annealed in the range of

250-300 °C. However, thermal anneal at higher temperatures (350-400 °C) transformed the amorphous phase into a crystalline structure as confirmed by XRD analysis presented in Fig. 2 (b). The H_c values of films with thermal anneal at higher temperatures increased dramatically and can be explained by the emergence of magnetocrystalline anisotropy of the crystalline phase embedded in the amorphous matrix.

The relative permeability (μ_r) along the hard axis direction of 100 nm thin films annealed at 300 °C was significantly higher (μ_r=1763) with a good loss performance over a wide range of frequency (f) (Fig. 3). However, the μ_r decreased with increasing film thickness. The decrease in μ_r of thicker films is due to the increase in shape dependent demagnetizing field along the in-plane direction due to increased film thickness [6].In addition, rapid decrease in μ_r as a function of f was observed for thicker films. This could be attributed to the very low value of H_k obtained for thicker films which allows the magnetization reversal process to proceed by a combination of domain wall motion and domain rotation rather than purely domain rotation alone.

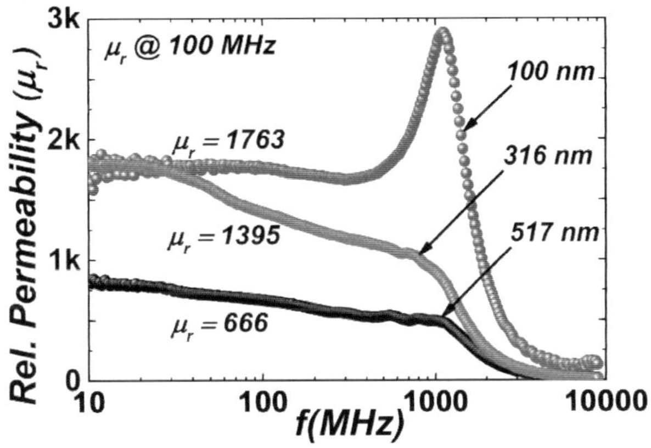

Fig. 3. Permeability as a function of frequency of different films annealed at 300 °C for an hour.

To conclude, thinner films (40-316 nm) of FineMET alloy exhibited a strong perpendicular anisotropy component with large H_c and hysteresis loops with a characteristic transcritical shape. Excellent soft magnetic properties (H_c=3 A/m) were achieved using a thermal annealing at a peak temperature of 300 °C. The high-frequency μ_r of the annealed samples decreased with increasing film thickness and attributed to the demagnetizing field which becomes dominant with film thickness.

Acknowledgements: This work was financed by CCAN, EI, and SFI Ireland. The authors would like to acknowledge Analog Devices and TEL Magnetic Solutions for supporting the program.

References:

1. Mathuna, C.O., et al., *Review of Integrated Magnetics for Power Supply on Chip (PwrSoC).* Ieee Transactions on Power Electronics, 2012. **27**(11): p. 4799-4816.
2. Kulkarni, S., et al., *Low loss thin film magnetics for high frequency power supplies.* 2014 16th European Conference on Power Electronics and Applications (Epe'14-Ecce Europe), 2014.
3. Masood, A., et al., *Effect of Ni-substitution on glass forming ability, mechanical, and magnetic properties of FeBNbY bulk metallic glasses.* Journal of Applied Physics, 2013. **113**(1).
4. Masood, A., et al., *A New Class of Materials for Magneto-Optical Applications: Transparent Amorphous Thin Films of Fe-B-Nb and Fe-B-Nb-Y Metallic Glassy Alloys.* Ieee Transactions on Magnetics, 2014. **50**(4).
5. Sharma, P., H. Kimura, and A. Inoue, *Observation of unusual magnetic behavior: Spin reorientation transition in thick Co-Fe-Ta-B glassy films.* Journal of Applied Physics, 2006. **100**(8).
6. Gardner, D.S., et al., *Integrated on-chip inductors using magnetic material (invited).* Journal of Applied Physics, 2008. **103**(7).

Influence of operating frequency on design of power electronic transformers

V C Valchev[1], A Van den Bossche[2] and T P Todorova[1]

[1] Department of Electronics, Technical University of Varna, Bulgaria
[2] Electrical Engineering Laboratory, Ghent University, Belgium

E-mail: Alex.VandenBossche@ugent.be

Abstract. The influence of operating frequency on parameters of power electronic transformers is investigated. An improved fast design algorithm is used to calculate competitive transformer designs based on ferrite cores and round wires under the same input parameters, while varying the operating frequency in a wide range (25 kHz to 400 kHz). Two sets of designs are compared: for a 400W and a1200W full bridge converter. Specific optimization approaches are proposed to decrease eddy current losses in wires by interleaving windings.

1. Introduction

The paper investigates the influence of operating frequency on parameters of power electronic transformers with ferrite cores and windings realized using round wires.

The study includes power electronic transformers in 25 kHz - 400 kHz range based on ferrites (N87) for two different output power values: 400W and 1200W and winding arrangements optimized by interleaving primary and secondary windings.

2. Design Algorithm

Calculation of losses in ferrites [1], [2] and eddy current losses in wires [3] including skin and proximity effects is decisive for the accuracy of the design.

The design algorithm applied in the paper consists of 15 steps [4]. The design starts with calculating a proper core size based on heat transfer capability under natural convection [5]:

$$a_{\rm ch}=\left(\frac{S_{\rm tot}}{A}\right)^{1/\gamma}, \quad S_{\rm tot}=\sum\nolimits_{\rm all\ windings} V_{\rm rms}\, I_{\rm rms} \tag{1}$$

where $a_{\rm ch}$ is the largest dimension of the component, used as a scaling parameter;
A is a coefficient, for ferrites, $A = (5–25) \times 10^6$ if $a_{\rm ch}$ is in (m), [4];
γ is an exponent, characterizing the material and shape of the core, $\gamma = 3$;
$S_{\rm tot}$ is the total volt-amp rating of the component.

Then, the allowable power losses of the component are obtained. The losses are equally assigned to the core and the windings. When having their values, the peak induction is found using manufacturer's graphs. Having $B_{\rm ac,peak}$ value, the necessary turns for the primary and secondary windings are calculated, followed by choosing the appropriate conductors' diameters and winding arrangements. Real power losses are found. Finally, according to the values of coefficients $k_{\rm eddy}$ and $k_{\rm cu}$, if necessary a next available (smaller or bigger) core size is chosen and the algorithm is proceeded with that core. The coefficient $k_{\rm eddy}$ gives the ratio between eddy current losses and ohmic losses and $k_{\rm cu}$ is copper filling factor.

Content from this work may be used under the terms of the Creative Commons Attribution 3.0 licence. Any further distribution of this work must maintain attribution to the author(s) and the title of the work, journal citation and DOI.
Published under licence by IOP Publishing Ltd

3. Winding Loss Calculations

Winding losses are calculated using an approach [4], based on a global loss factor k_c, which represents the ratio between the eddy current losses and the losses in the ohmic resistance:

$$P_{\mathrm{eddy}}=\left(R_0\, I_{\mathrm{ac}}^2\right) k_c \left(m_E, f_{\mathrm{op}}, d_p, \eta, \lambda\right) \tag{2}$$

where m_E is an equivalent layer, f_{op} - operating frequency, d_p - wire diameter, η – relative filling between the conductors and λ – relative filling between layers.

4. Design Results Comparison

Ten designs of power transformers are calculated based on EE ferrite cores (N87) using the 15-step algorithm [4]. Input design parameters: $V_{\mathrm{prim,rms}}$=300 V, $V_{\mathrm{sec,rms}}$=100 V, $I_{\mathrm{sec.rms}}$=12 A;4 A, f_{op}=25 kHz-400 kHz. The comparative study focuses only on using round wires. *Remarks:*

- Possible improvements applied are interleaving windings and using a few wires in parallel.
- The wire diameters are chosen so that the whole winding width is filled for both primary and secondary windings, the results are shown in Table 1.

Table 1. Design results for transformers with input design parameters: $V_{\mathrm{prim,rms}}$=300 V, $V_{\mathrm{sec,rms}}$=100 V, $I_{\mathrm{sec.rms}}$=12 A; 4 A, f_{op}=25 kHz-400 kHz, interleaving is used to decrease the losses in windings.

f_{op} (kHz)	Core (mm)	Volume (mm^3)	Component weight, (g)	$B_{ac,peak}$ (T)	Total losses (W)	Copper filling factor	Number of turns P/S	Winding arrangement
				Design 1200W				
25	65/32/27	78 650	530.200	0.212	7.76	0.260	27/9	PSP
50	55/28/21	43 900	298.305	0.167	6.862	0.284	26/9	PSP
100	56/24/19	36 400	243.782	0.123	6.355	0.288	19/7	PSP
200	56/24/19	36 400	251.179	0.078	5.870	0.323	15/5	PSP
				Design 400W				
25	42/21/15	17 300	133.585	0.263	3.521	0.31	66/22	PSPPSP
50	40/16/12	11 500	80.486	0.198	2.965	0.273	53/18	PSPPSP
100	36/18/11	9 720	85.437	0.151	2.287	0.426	45/15	PSPPSP
200	32/16/9	6 140	55.057	0.107	2.062	0.484	44/15	PSPPSP
300	32/16/9	6 140	57.476	0.079	2.03	0.531	39/13	PSPPSP
400	32/16/9	6 140	49.103	0.060	2.147	0.369	39/13	PSPPSP

4.1. Specifics of 1200W transformers design

For all the 1200W designs, the winding' realizations follow the same concept — two primary winding's layers connected in parallel and a secondary winding's layer sandwiched between them (PSP arrangement), Fig.1,a,b. The turns of the primary winding in each layer consist of a single round wire, whereas the secondary winding's turns are composed of two round parallel-connected wire.

Note that those designs are still examples. The 1200W designs could have lower losses (or a slightly smaller core) if a PSPPSP arrangement would be used, see 400W designs.

4.2. Specifics of 400W transformers design

The winding arrangement is PSP doubled, i.e. four parallel-connected primary winding's layers and two parallel-connected secondary winding's layers. The secondary layers are located between the 1st and the 2nd, and the 3rd and the 4th primary layers, respectively (PSPPSP arrangement), Fig.1,c,d.

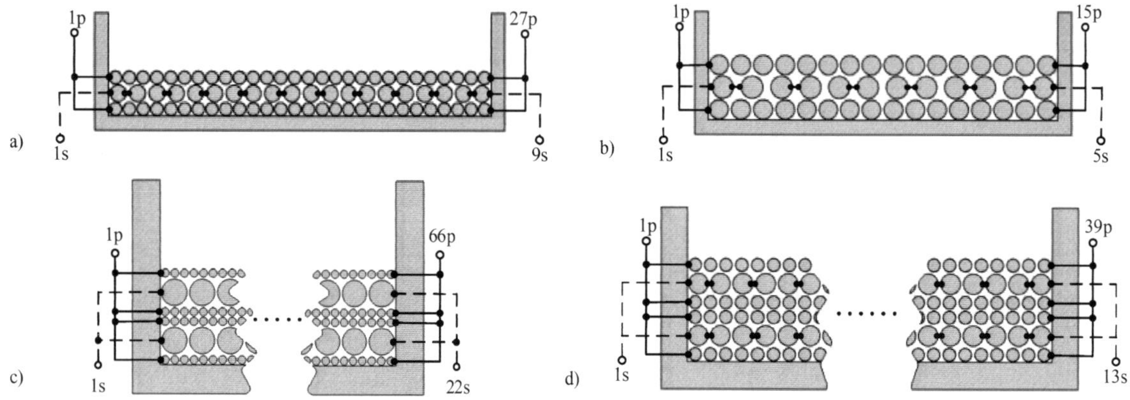

Fig.1. Winding arrangements, PSP: a) 1200W, 25kHz; b)1200W, 200kHz;
PSPPSP: c) 400W, 25kHz; d) 400W, 400kHz.

Table 2 contains the wire diameters and shows the influence of the eddy currents when increasing the frequency, presented by the coefficient k_c [4]:

$$k_c = P_{eddy} / \left(R_0\, I_{ac}^2\right) \tag{3}$$

Table 2. Wire diameters obtained and influence of eddy currents with increasing of frequency

		Design 1200W				Design 400W					
f_{op}, (kHz)		25	50	100	200	25	50	100	200	300	400
Diameter of wires,(mm)	P	1.25	1.12	1.12	1.4	0.315	0.280	0.400	0.355	0.400	0.400
	S	2x1.6	2x1.4	2x1.4	2x1.6	1.000	0.800	1.120	1.120	1.250	2x0.63
k_c	P	0.88	1.514	2.384	4.873	0.005	0.013	0.205	0.385	0.400	1.296
	S	0.192	0.404	0.952	1.897	0.041	0.059	0.595	1.331	1.250	0.817

5. Conclusion

The analysis of the discussed cases leads to the following design considerations:

- Increasing frequency influences strongly eddy currents losses. For example, in 1200W, 50kHz and 1200W,100kHz designs the same wires are used, but k_c is increased from 1.514 to 2.384 (primary winding) and from 0.404 to 0.9592 (secondary winding).
- At high powers the transformer design become more complicated as the currents are higher and the conductors' diameters must have larger effective cross-sectional area.
- A way to reduce the eddy current losses is using thinner wires connected in parallel.
- In cases with interleaving, the best approach is all of the turns to fit in a single winding width.
- Interleaving windings provides significant reduction of eddy current and total losses.

References

[1] Gulbis A, Malinovska D and Stepins D, "Influence of losses on the output voltage of ferrite transformer in case of strong magnetic field in the core", 2016 IOP Conf. Series: Materials Science and Engineering 108 (2016) 012023

[2] Han Y, Cheung G, Li A, Sullivan C R and Perreault D J, 'Evaluation of Magnetic Materials for Very High Frequency Power Applications', IEEE Transactions on Power Electronics, Vol. 27, No. 1, pp. 425-435, Jan. 2012

[3] Dixon L H, "Eddy current losses in transformer windings and circuit wiring", Texas Instruments, Incorporated [SLUP 197], 2000.

[4] Van den Bossche A and Valchev V C, "Inductors and transformers for power electronics", 2005, CRC, Boca Raton, USA.

[5] Valchev V C and Van den Bossche A, "Accurate natural convection model for magnetic components", Journal of Microelectronic Reliability, Elsevier, UK, 2003, Vol.3, pp. 797-802.

Micromagnetic simulation of nanomagnets with geometry-tuned domain wall nucleation

W. Kaiser, M. Kiechle, G. Žiemys, D. Schmitt-Landsiedel and S. Breitkreutz-v. Gamm

Institute for Technical Electronics, Technical University of Munich, 80333 Munich, Germany

E-mail: waldemar.kaiser@tum.de

Abstract. In this work we show the feasibility of tuning the domain wall nucleation by changing the geometry of nanomagnets using shadowing effects. Therefore, micromagnetic simulations based on a three-dimensional model of the nanomagnetic structures have been performed. Atomic force microscopy and anisotropy measurements on fabricated Co/Pt-nanomagnets are used to calibrate the model. It is shown that an increasing gradient due to shadowing effects decreases the switching field of the nanomagnets. By varying the nanomagnet geometry, the switching field can be further decreased and the spot of nucleation can be controlled making these technique feasible for the use in nanomagnetic logic circuitry.

Index terms— Nanomagnetic Logic, Domain wall nucleation, Micromagnetic simulation, Shadowing effects, Artificial nucleation center

1. Introduction

Perpendicular Nanomagnetic Logic (pNML) is a promising beyond-CMOS emerging technology exploiting three-dimensional magnetic field coupling between nanostructured Co/Pt-multilayers [1, 2]. The switching process is based on domain wall (DW) nucleation in the weakest spot followed by DW propagation through the entire magnet [3]. The switching energy is provided by an alternating perpendicular magnetic clocking field [4]. Controlled DW nucleation is ensured by creating an artificial nucleation center (ANC) with local focused ion beam (FIB) irradiation [3, 5], which however causes large fabrication variations, requires an additional process step and is difficult to implement on wafer processes [6, 7]. Kimling shows in [8] the local reduction of the perpendicular magnetic anisotropy (PMA) due to shadowing effects during sputter deposition of the multilayers, generating an anisotropy gradient at the edges of the magnetic structures. Hence, the nucleation field can be tuned by changing the geometry of the magnetic wire ends [8, 9]. The origin of the effects occurring due to the shadowing process is determined by a simulation model in [8]. We improve this model by including the gradient, measured by atomic foce microscopy (AFM), to all sides of the nanomagnets, as naturally observed during fabrication. Multilayer films with scaled thicknesses are measured to obtain the scaling of the anisotropy at the gradient. A three-dimensional model of the nanomagnets is designed using the experimental data. Simulations including different gradients and geometries of the nanomagnets are performed using OOMMF [10] to obtain the switching field and the spot of domain wall nucleation.

Content from this work may be used under the terms of the Creative Commons Attribution 3.0 licence. Any further distribution of this work must maintain attribution to the author(s) and the title of the work, journal citation and DOI.
Published under licence by IOP Publishing Ltd

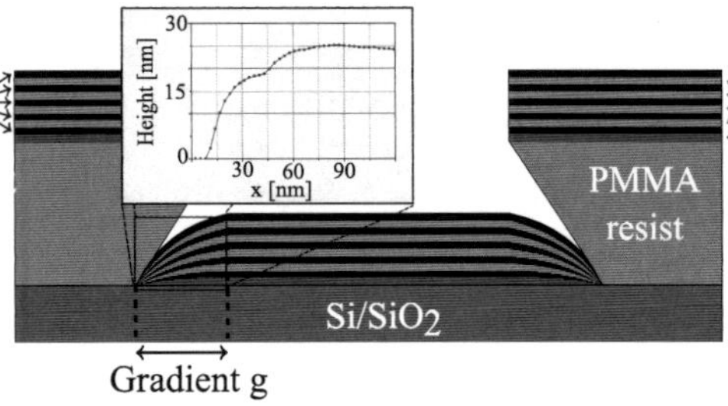

Figure 1. Illustration of the shadowing process and the gradient measured by AFM.

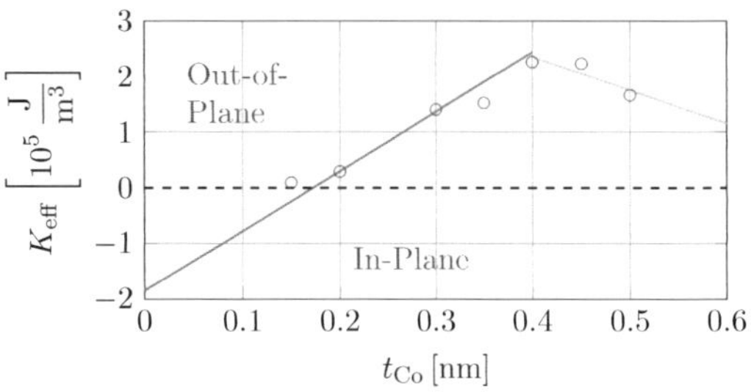

Figure 2. Effective anisotropy of scaled Co/Pt-multilayers.

2. Experimental measurements of the Co/Pt-multilayers

Co/Pt-multilayers and magnetic structures have been fabricated and characterized to feed the micromagnetic model with realistic geometry and magnetic properties resulting from shadowing effects. The different magnetic characteristics of this geometry need to be considered in the switching behavior of the nanomagnets. In Fig. 1 the cross-sectional view of the shadowing effect illustrates the lowered height of the individual layers at the edges due to the lift-off process [8]. To measure the magnetic anisotropy of the changed geometry a series of Co/Pt-multilayer stacks with a composition of $Si/(SiO_2)/Ta_{3nm}/Pt_{x\cdot 5.0nm}/4\times[Co_{x\cdot 0.5nm}+Pt_{x\cdot 2.0nm}]/Pt_{1nm}$ has been fabricated under variation of the scaling factor $0 < x < 1$. Realistic dimensions of process-related lift-off edges of fabricated nanomagnets are investigated by AFM as highlighted in Fig. 1. Anisotropy measurements based on extraordinary Hall voltage measurement [11] show a significantly lowered anisotropy for samples with an individual cobalt layer thickness t_{Co} below 0.4 nm, as depicted in Fig. 2. The decreasing anisotropy for thinner layers origins from a lower density of the magnetic material. Since the layer thickness is in the range of the diameter of a single Co atom, the deposited magnetic material generates atomic islands instead of complete layers [12]. The anisotropy drop above $t_{Co} = 0.4$ nm has its nature in the volume contribution of the effective anisotropy dominating in thicker layers [12].

3. Micromagnetic Modeling and Simulation

We investigate the DW nucleation behavior of tip nanomagnets with a tip opening angle Θ of 20° to 150°, shown in Fig. 3a. As fabrication results of tip nanomagnets showed an uncontrolled rounding of the tip, it is more beneficial to investigate nanomagnets with flat ends, as shown in Fig. 3b. Hereby, the width w is varied between 10 nm and 100 nm at constant length. We perform non-thermal simulations using OOMMF [10] to extract the induced shadowing effects. Magnetic field pulses of 10 ns length and a rise time of 2 ns are applied. In our 3D-model, a gradient is added to all sides of the nanomagnets, as observed by AFM measurements, which is depicted in Fig. 4. The net saturation magnetization is kept constant at $M_{S,net} = 7.23\times10^5$ A/m for each cell. The effective anisotropy is scaled according to Fig. 2. Fig. 5 shows the simulation results for the nucleation field B_{nuc} of the magnets in dependency of Θ. The nucleation field is strongly decreasing for increasing gradients g, as the anisotropy energy slope from the edge to the center of the magnet is reduced. At a constant gradient, B_{nuc} decreases for lower angles, passes a minimum and increases slightly for low angles. For small tip opening angles the amount of magnetic material with in-plane easy axis increases, thus the anisotropy step increases and DW nucleation requires more energy. The DW nucleates for all angles at the desired ANC. A similar behavior is obtained for the simulated nanomagnets with flat tips. For decreasing widths the nucleation field is lowered. The geometries of the flat magnets with DW nucleation at undesired spots are marked by the shaded region in Fig. 5b. For geometries in the unshaded region, we observe a DW nucleation at the desired ANC, like shown in the inset of Fig. 5b. Thus,

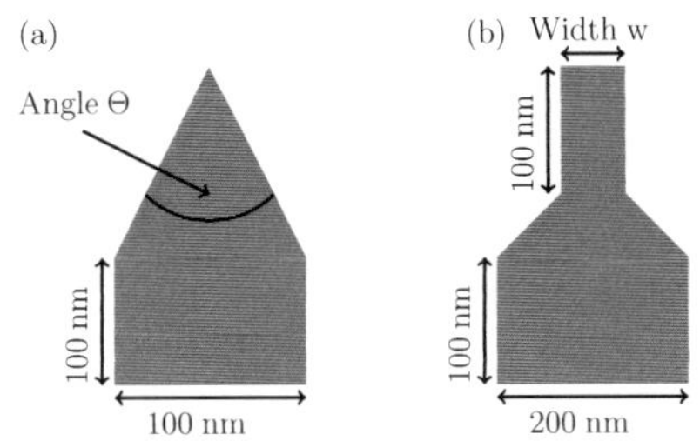

Figure 3. Top view of simulated magnets: (a) Tip magnets, (b) Flat end magnets.

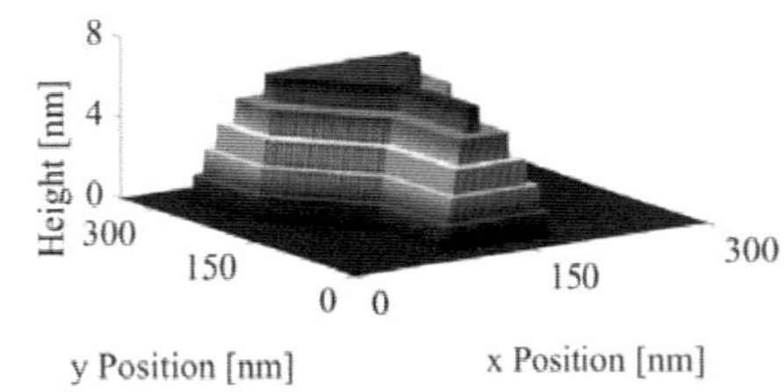

Figure 4. Model of a flat end magnet (gradient $g = 40\,\text{nm}$, width $w = 90\,\text{nm}$).

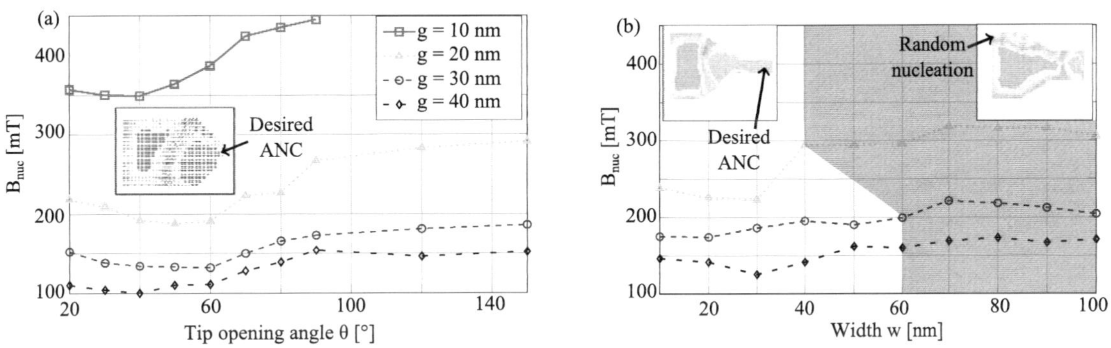

Figure 5. DW nucleation field of (a) tip magnets and (b) flat end magnets. The shaded area marks the geometries undesired DW nucleation occurs.

for application in pNML circuitry nanomagnets geometry and gradient need to be controlled to ensure signal flow directionality.

4. Conclusion

In this work, a study on the DW nucleation of geometry-tuned nanomagnets has been performed using 3D simulations based on fabrication results. It is shown, that including a gradient at a tip using shadowing effects lowers the anisotropy and creates an ANC working as required in pNML devices. This shows the potential of substituting the ANC creation by the use of shadowing during the sputtering process, instead of an additional FIB irradiation step.

Acknowledgments

The authors thank M. Becherer (TUM) and G. Csaba (ND) for their fruitful support.

References

[1] Nikonov D E and Young I A 2015 *IEEE J. Explor. Solid-State Computat. Devices Circuits* **1** 3-11
[2] Becherer M *et al.* *EUROSOI-ULIS, 2015* 121-124
[3] Breitkreutz S *et al.* 2012 *J. Appl. Phys.* **111** 07A715
[4] Becherer M *et al.* 2013 *Proceedings of the ESSDERC* 276-279
[5] Breitkreutz S *et al.* 2011 *Proceedings of the ESSDERC* 323-326
[6] Eichwald I *et al.* 2012 *IEEE Transactions on Magnetics* **11** 4332-4335
[7] Engelen J B C *et al.* 2009 *Nanotechnology* **21** 035703
[8] Kimling J *et al.* 2013 *J. Appl. Phys.* **113** 163902
[9] Mansell R *et al.* 2015 *Appl. Phys. Lett.* **107** 092405
[10] *The Object Oriented MicroMagnetic Framework (OOMMF)* http://math.nist.giv/oommf
[11] Moon K W *et al.* 2009 *Review of Scientific Instruments* **80** 113904
[12] Johnson M T *et al.* 1996 *Reports on Progress in Physics* **59** 1409

Angular dependence of the domain wall depinning field in sensors with segmented corners.

D. Heinze[1], B. Borie[1,2], H. Grimm[2], J. Wahrhusen[2] and M. Kläui[1]
[1]Institut für Physik, Johannes Gutenberg-Universität Mainz, Staudingerweg 7, Mainz D-55128, Germany
[2]Sensitec GmbH, Hechtsheimer Str. 2, Mainz D-55131, Germany

E-mail: klaeui@uni-mainz.de

Abstract. Rotating domain wall based sensors that have recently been developed are based on a segmented looping geometry. To determine the crucial pinning of domain walls in this particular geometry, we investigate the depinning under different angles of an applied magnetic field and obtain the angular dependence of the depinning field of the domain walls. Due to the geometry, the depinning field not only exhibits an 180°-periodicity but a more complex dependence on the angle. The depinning field depends on two different angles associated with the initial state and the segmented geometry of the corner. We find that depending on the angle of the applied field two different switching processes occur and we can reproduce the angular dependence using a simple model calculation.

1. Introduction

Magnetic domain walls in soft magnetic material nanostructures [1, 2] are of interest to the sensor industry for their ability to be stable at room temperature and since they can be easily moved by magnetic fields [3, 4]. Currently, only one sensor based on domain walls is produced and sold [5]. One of the difficulties that need to be overcome to use domain walls in more devices is the complex dynamics of the domain wall behaviour in structures with different geometries [6, 7, 8, 9, 10, 11]. The sensor devices are characterized by a magnetic operating window set by the nucleation field (upper bound) and the depinning or propagation field (lower bound) in which the sensors function. The propagation and depinning of domain walls are affected by various internal and external effects. The edge roughness [12, 13, 14, 15], the crystallite size [16], variations of the topography are the main affecting factors. Also, the Walker Breakdown [17, 18] changes the spin structure of the domain wall rendering the determination of the pinning challenging. In the case of straight wires, kinks and rings, the motion and depinning of domain walls was intensively investigated [19, 20, 21]. However, polygons and segmented shapes can bring different results and different domain wall spin structures [22] as compared with smooth geometrical variations, and these geometries are used in the sensor devices. In earlier studies, it was found that the depinning field exhibits a pronounced angular dependence [20, 21, 23]. In [20], it was described that the switching field is lowest when the field is tangential to the circumference of the structure at the domain wall position. The measurement was performed in that case on circular rings without sharp variations in the path of the domain wall. However, the influence of sharp corners in segmented loops will impact the pinning and depinning processes and needs to be determined to ascertain the most reliable structures for domain wall displacements.

Content from this work may be used under the terms of the Creative Commons Attribution 3.0 licence. Any further distribution of this work must maintain attribution to the author(s) and the title of the work, journal citation and DOI.
Published under licence by IOP Publishing Ltd

In this paper, we report the Kerr-microscopy measurement of domain wall depinning from corners in a segmented looping structure under different angles of an applied magnetic field. We show that the depinning field is a function of the initial position and the segmented geometry and explain it based on an analysis of the domain wall motion for different angles.

The samples were produced on a wafer of Si/AlO_x and deposited in a magnetron sputtering tool. The sputtered magnetic stack consists of 1 nm of $Co_{90}Fe_{10}$, 29 nm of $Ni_{80}Fe_{20}$ and a capping layer of 4 nm of Ta. The shape of the device was then obtained through a photolithography step and an ion milling using Ar ions. A width of 350 nm is obtained with an edge roughness of typically 10 nm. A sketch of the geometry is shown in Figure 1.

2. Measurement of the angular dependent depinning field

The measurements were performed on magnetic domain wall based rotation sensors, which have a square-shaped geometry with segmented corners. They consist of 16 loops in total starting with a nucleation pad connected to the outer wire (Figure 1). The technique used to image is a magneto-optical Kerr effect microscope in the longitudinal mode [24]. The orientation of the sensor under the microscope is kept fixed for the experiment as shown in Figure 1. A magnetic field is applied, and the angle between the field and the imaged stripes is defined as 0° points in the x-direction and at 90° is the -y direction.

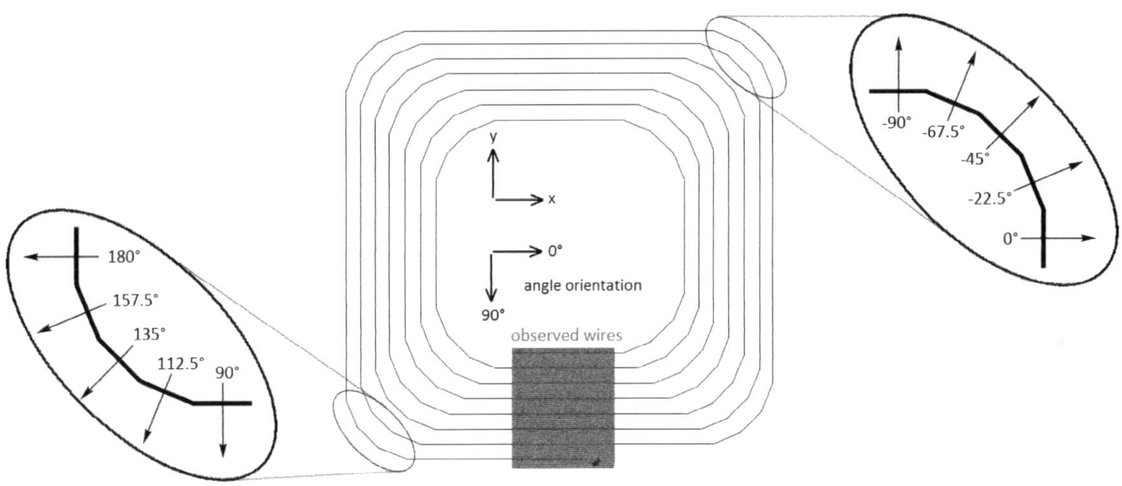

Figure 1: Sketch of a sensor with zoomed in bottom left and top right corners. The angle orientation and the definition of the x- and y-axis to simplify explanations is also given. A Kerr microscopy imaged shows the observed wires with the magnetic contrast visible.

During the depinning experiment, domain walls are nucleated in the structure and positioned in the segmented corners. This latter is done by the application of a strong magnetic field (80 mT) along with an angle of 135° (Figure 1). When the magnetic field is reduced to zero, the magnetization relaxes to a state with domain walls in the top right and bottom left corner that is shown in Figure 1. Depending on the slight misalignment of the field with the structure the domain walls are going to stabilize at one of the two kinks of the 135° and -45°.

After the initialization, the angle of the applied field with the imaged stripes was set to a constant value between 90° and -90° and the field was increased until a contrast change could be observed in the observed wires. The angle was then varied by 2° and the operation was repeated. For each angle this procedure was repeated ten times, so at each angle, there was an average of 80 values due to the eight wires observed by the Kerr microscopy imaging (see Figure 1). For our purposes, the mean of the depinning field was taken from the observed eight wires for each angle. The result of the measurement of the field at which the wires reverse as a function of the applied field direction can be seen in Figure 2.

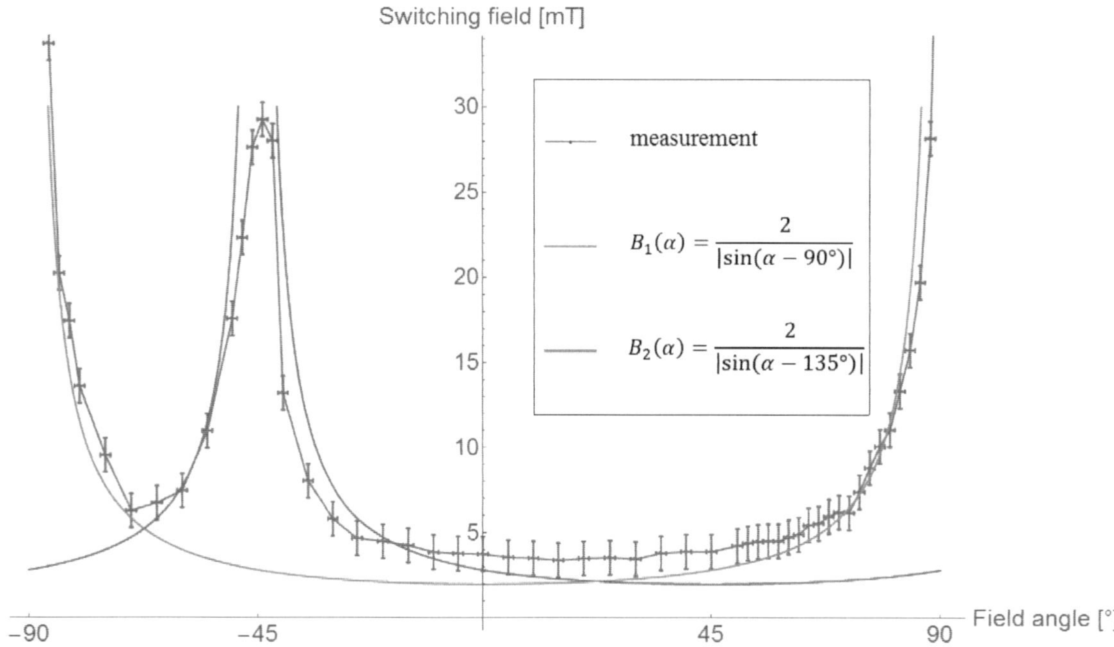

Figure 2: Measurement and fit of the angular dependency of the switching field.

To simplify the analysis, only the abrupt reversal in the observed wires was detected, and the field values for this event are plotted. First, we see that the switching field diverges at ±90°, this can be explained by the Zeeman energy and is due to the probed field of view of the sensor.

$$E_{\text{Zeeman}} = -\mu_0 \int_V \vec{M} \cdot \vec{H}_{\text{ext}} \, \mathrm{d}V \tag{1}$$

In Equation 1, $\vec{M}$ is the magnetization and $\vec{H}_{\text{ext}}$ is the applied external magnetic field. For the magnetization of the wires being aligned along the wire axis due to the shape anisotropy, the applied field is orthogonal to the magnetization. Therefore, neither depinning nor nucleation of the domain walls can be imaged in the observed wires for these field directions. To explain the angular dependence, we divide the graph into different sections and in particular the angle range between 90° and -90° is divided into three different parts based on the movement of the different domain walls and the resulting switching processes occurring for the different field angles.

(I) In the first range between 90° and 0°, the -y-component is in the same direction as the vertical parts of the loops. The domain walls located in the corners will move horizontally as schematically shown in Figure 3a where the displacement direction of the walls is shown by green arrows.

(II) In the range between 0° and -45°, the x-component of the applied magnetic field is stronger than the y-component and both are in the direction opposite to the current magnetization configuration at the domain wall position. Therefore, the domain walls at the bottom left corner move first to the right and the ones in the top right to the left thus changing the magnetization direction in the observed wires (Figure 3a). The motion is thus anti-clockwise. In a second step, the domain wall in the top wire moves down, and the one in the bottom wire moves up. This part is not schematically shown as it is not imaged because the magnetization in the observed stripes (Figure 1) is not affected.

(III) Finally, between -45° and -90°, the domain walls also move in a two-step process. First, they move in the vertical direction because the y-component of the field is stronger than the x-component (the first step in Figure 3b). Then, if the x-component of the applied field is sufficiently high, the domain walls can also depin from the new position in the corners in a second step (also shown in Figure 3b) and move horizontally and therefore a contrast change in the observed wires can be detected. The sense of rotation of the wall movement for this angle range is clockwise.

(IV) An interesting feature appears at -45°. At this angle, the applied magnetic field is precisely antiparallel to the initial position of the domain walls. Therefore, at -45°, the Zeeman force pushing the domain wall horizontally is equal to the one pushing it vertically thus the total acting torques lead to the domain wall being in a metastable equilibrium position. But the measured field values for the switching do not completely diverge. There are two possible explanations for this behaviour: The nucleation of new domain walls or the depinning due to certain asymmetries in the geometry and domain wall structure. The detected field value in our system is 29 mT, which is also the field needed to nucleate new domains starting from a spin structure without domain walls so that we identify nucleation as the process.

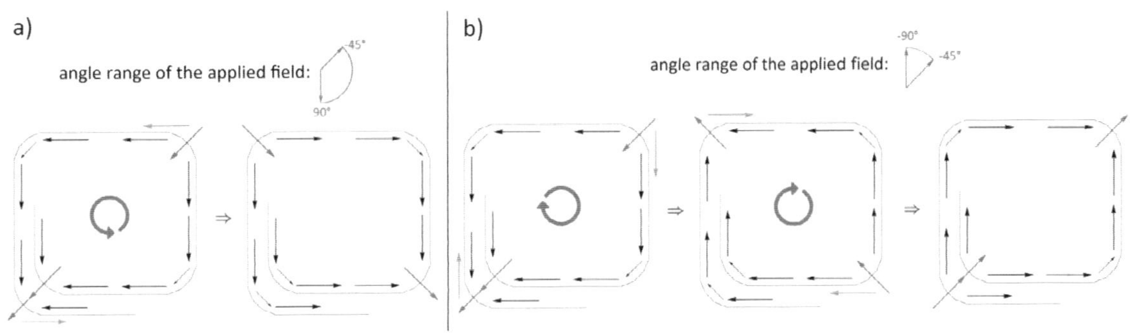

Figure 3: a) Sketch of the domain wall propagation in the range of 90° to -45°. b) Sketch of the two-step domain wall propagation process in the range of -45° to -90°. The domain walls are indicated by red arrows and their movement directions by green arrows. The black arrows indicate the magnetization directions in the looped wires. The opposite displacement direction for a) and b) is indicated as well in thick red.

Furthermore, we see that the local minimum in the switching field at -67.5° is not as low as at -22.5°. This asymmetry can be explained by the fact that the depinning field from the -45° angle goes down while the field is rotated to -90°. However, as the field approaches -90°, the detection of the depinning of domain walls in the observed wires becomes increasingly difficult.

The depinning field depends on the component acting in the direction of the displacement (sine dependence) as shown in [20]. As a result of the geometry, the switching field depends here on two angles: Since the sensors are square-shaped, the angular dependence has an 180°-periodicity, and due to the position of the observed wires, there is a divergence at ±90° (Equation 2). Furthermore due to the positioning of the domain walls in the initial state (IV) there is the divergence at 135° and -45°, which results in another switching field, depending on this angle (Equation 3):

$$B_1(\alpha) = \left| \frac{2\,\mathrm{mT}}{\sin(\alpha - 90^\circ)} \right| \qquad (2) \qquad\qquad B_2(\alpha) = \left| \frac{2\,\mathrm{mT}}{\sin(\alpha - 135^\circ)} \right| \qquad (3)$$

We see that the depinning field is always the maximum of the two fields, and this reproduces the experiment well (Figure 2). The latter also shows that, in our geometry, the depinning is different from previous studies with simpler geometries [20]. There, the minimal depinning field was with the applied field tangential to the wire, where the domain wall is located. In our case, that would be at an angle of 45°. This deviation is a result of the different shape of the sensor. In the round wire (ring), all directions are equal. In contrast, in the square-shaped sensor and the segmented corner, the pinning is governed by the directions of the segments (see Figure 1), which results in the lowest depinning field at 22.5°.

3. Conclusion and prospects

In summary, the angular dependence of the depinning and propagation of magnetic domain walls is highly geometry related. In particular, the observed wires (Figure 1) reverse by the domain walls arriving for different angular ranges from various directions. Between 90° and -45° the domain walls propagate directly through the observed wires, due to the initial domain wall position ((I),(II)). In the range from -45° to -90° there is a two-step process necessary to change the magnetization in the observed wires (III), which leads to a local switching minima on the MOKE area at -67.5°. With our square-shaped sensors and the initializing field at 135°, at the two angles ±90° and -45° no reversal by depinning and propagation of domain walls is observed. In contrast to ±90°, at -45° the nucleation of domain walls in the wires governs the switching of the magnetization. The superposition of the depinning fields for these angles yields a minimum at 22.5°, which is not tangential to the wires, where the domain walls are initialized, showing a clear difference to the behaviour of circular rings, which can be ascribed to the segmented corners.

Further investigations have to reveal if the peak at -45° is only related to the initialization. This effect can be achieved by initializing the sensor for example at 112.5° instead of 135° (Figure 1). If the peak occurs at -67.5°, the theory of the correlation between initial state and the peak will be confirmed for this sensor geometry. The understanding of the depinning mechanism in segmented loops will improve the magnetic operating window. Indeed, an engineering of the number of segments is likely to lead to a lower depinning field and a larger operating window.

The authors kindly thank the company Sensitec GmbH for providing the samples, in particular, J. Paul for fruitful discussions. The work and results reported in this publication were obtained with research funding from the European Community under the Seventh Framework Programme - The people Programme, Multi-ITN "WALL" Contract Number Grant agreement no.: 608031, the ERC (MultiRev ERC-2014-PoC (665672)) and the DFG (SFB TRR 173 Spin+X).

References

[1] R D McMichael and M J Donahue. Head to head domain wall structures in thin magnetic strips. *IEEE Transactions on Magnetics*, 33(5):4167–4169, Sep 1997.

[2] M Kläui. Head-to-head domain walls in magnetic nanostructures. *J. Phys.: Condens. Matter*, **20**(31):313001, 2008.

[3] Marco Diegel, Roland Mattheis, and Ernst Halder. 360° domain wall investigation for sensor applications. *IEEE Trans. Magn.*, **40**:2655, 2004.

[4] Marco Diegel, Sacha Glathe, Roland Mattheis, Manfred Scherzinger, and Ernst Halder. A new four bit magnetic domain wall based multiturn counter. *IEEE Trans. Magn.*, **45**:3792, 2009.

[5] Novotechnik : Rsm-2800 multiturn. `http://www.novotechnik.de/produkte/winkelsensoren/`. Accessed: 2016-07-17.

[6] A Allwood, G Xiong, C C Faulkner, D Atkinson, D Petit, and R P Cowburn. Magnetic domain-wall logic. *Science*, **309**:1688, 2005.

[7] T. Ono, H. Miyajima, K. Shigeto, K. Mibu, N. Hosoito, and T. Shinjo. Propagation of a magnetic domain wall in a submicrometer magnetic wire. *Science*, **284**:468, 1999.

[8] D Atkinson, D A Allwood, G Xiong, M D Cooke, C C Faulkner, and R P Cowburn. Magnetic domain-wall dynamics in a submicrometre ferromagnetic structure. *Nat. Mater.*, **2**:85, 2003.
[9] R P Cowburn, A D Allwood, G Xiong, and M D Cooke. Domain wall injection and propagation in planar Permalloy nanowires. *J. Appl. Phys.*, **91**:6949, 2002.
[10] André Bisig, Martin Stärk, Mohamad-Assaad Mawass, Christoforos Moutafis, Jan Rhensius, Jakoba Heidler, Felix Büttner, Matthias Noske, Markus Weigand, Stefan Eisebitt, et al. Correlation between spin structure oscillations and domain wall velocities. *Nat. Commun.*, **4**:2328, 2013.
[11] Aakash Pushp, Timothy Phung, Charles Rettner, Brian P. Hughes, See-Hun Yang, Luc Thomas, and Stuart S. P. Parkin. Domain wall trajectory determined by its fractional topological edge defects. *Nat. Phys.*, **9**:1745, 2013.
[12] F Cayssol, J L Menéndez, D Ravelosona, C Chappert, J P Jamet, J Ferré, and H Bernas. Enhancing domain wall motion in magnetic wires by ion irradiation. *Appl. Phys. Lett.*, **86**:022503, 2005.
[13] M T Bryan, D Atkinson, and R P Cowburn. Experimental study of the influence of edge roughness on magnetization switching in Permalloy nanostructures. *Appl. Phys. Lett.*, **85**:3510, 2004.
[14] E Martinez, L Lopez-Diaz, L Torres, C Tristan, and O Alejos. Thermal effects in domain wall motion: Micromagnetic simulations and analytical model. *Phys. Rev. B*, **75**(17):174409, 2007.
[15] T W Chiang, L J Chang, C Yu, S Y Huang, D C Chen, Y D Yao, and S F Lee. Demonstration of edge roughness effect on the magnetization reversal of spin valve submicron wires. *Appl. Phys. Lett.*, **97**(2):022109, 2010.
[16] Michele Voto, Luis Lopez-Diaz, and Luis Torres. Effects of grain size and disorder on domain wall propagation in CoFeB thin films. *J. Phys. D: Appl. Phys.*, **49**(18):185001, 2016.
[17] Norman L Schryer and Laurence R Walker. The motion of 180° domain walls in uniform dc magnetic fields. *J. Appl. Phys.*, **45**(12):5406, 1974.
[18] Sang-Koog Kim, Jun-Young Lee, Youn-Seok Choi, Konstantin Yu. Guslienko, and Ki-Suk Lee. Underlying mechanism of domain-wall motions in soft magnetic thin-film nanostripes beyond the velocity-breakdown regime. *Appl. Phys. Lett.*, **93**:052503, 2008.
[19] S Glathe and R Mattheis. Magnetic domain wall pinning by kinks in magnetic nanostripes. *Phys. Rev. B*, **85**:024405, 2012.
[20] Daniel Bedau, Mathias Kläui, Ulrich Rüdiger, Carlos AF Vaz, J Anthony C Bland, Giancarlo Faini, Laurent Vila, and Wolfgang Wernsdorfer. Angular dependence of the depinning field for head-to-head domain walls at constrictions. *J. Appl. Phys.*, **101**(9):09F509, 2007.
[21] Héctor Corte-León, Anthony Beguivin, Patryk Krzysteczko, Hans Werner Schumacher, Alessandra Manzin, Russell P Cowburn, Vladimir Antonov, and Olga Kazakova. Influence of geometry on domain wall dynamics in Permalloy nanodevices. *IEEE Trans. Magn.*, **51**(1):1–4, 2015.
[22] C W Sandweg, N Wiese, D McGrouther, S J Hermsdoerfer, H Schultheiss, B Leven, S McVitie, B Hillebrands, and J N Chapman. Direct observation of domain wall structures in curved permalloy wires containing an antinotch. *J. Appl. Phys.*, **103**(9):093906, 2008.
[23] S Glathe, I Berkov, T Mikolajick, and R Mattheis. Experimental study of domain wall motion in long nanostrips under the influence of a transverse field. *Appl. Phys. Lett.*, **93**:162505, 2008.
[24] Rudolf Schäfer. Investigation of domains and dynamics of domain walls by the magneto-optical Kerr-effect. In H Kronmüller and S S P Parkin, editors, *Handbook of Magnetism and Advanced Magnetic Materials*. John Wiley & Sons, Ltd, 2007.

Magnetization of a Single Carbonaceous Grain Obtained by Field-Induced Acceleration

Keiji Hisayoshi and Chiaki Uyeda

Institute of Earth and Space Science, Graduate School of Science, Osaka University, Toyonaka, Osaka, Japan

hisayoshi@ess.sci.osaka-u.ac.jp

Abstract. Diamagnetic susceptibility χ_{DIA} of single carbonaceous grains were detected by observing their translations induced by field-gradient force in an area of microgravity. Using the above method, χ_{DIA} of a small carbonaceous particle is obtained with no interfering signal of the sample holder; it is unnecessary to know the mass of sample. The χ_{DIA} values of various materials obtained by the above method agreed fairly well with their published values. By comparing the obtained χ_{DIA} value with a list of published values, the material of an unidentified organic grain can be determined without consuming the sample. The principle of magnetic transition is applicable to investigate the magnetic properties of nano-size carbonaceous materials.

1. Introduction

The Magnetization of a weak magnetic material becomes difficult to measure with the reduction of sample size because of the following two reasons. First of all, it is difficult to know the precise mass of a sample which has reduced sizes. Secondly, the interference of background signals emitted from the sample holder is not negligible when sample size is at sub-mm level. The above 2 difficulties was successively avoided by observing an acceleration of sample caused by a field gradient force produced by a permanent magnet [1], [2]; here the sample translated through a diffused area within a short duration of microgravity condition (~ 0.5 s). Diamagnetic susceptibility χ_{DIA} was obtained from the above translation without knowing the value of sample mass, because variation of field-induced potential between two sample positions was completely converted to kinetic energy. The mass independent property of the translation was observed because the motion was caused by a volume force. Up to now, the above-mentioned measurement of χ_{DIA} has been performed only on inorganic materials.

The difference of magnetic potential has been conventionally used in a "Faraday method" to measure magnetization. Magnetic susceptibility of small samples were recently detected by observing their translation caused by a large field-gradient at a level of ~1T/cm [3], [4]. These experiments were performed in terrestrial gravity, and sample motion was affected by a viscous drag of the medium. It is not widely recognized that dynamic motion of ordinary diamagnetic solid is caused by a small potential of permanent magnet. Various dynamical effects that were reported on weak magnetic materials required a strong magnetic field at a level of 10 Tesla [5].

Content from this work may be used under the terms of the Creative Commons Attribution 3.0 licence. Any further distribution of this work must maintain attribution to the author(s) and the title of the work, journal citation and DOI.
Published under licence by IOP Publishing Ltd

In the present report, efficiency to detect χ_{DIA} by the field-induced acceleration is examined for several carbonaceous materials. Based on the obtained results, various advantages are considered in clarifying the magnetic properties of single organic materials having reduced sizes.

2. Procedure

The field gradient used in the present study was produced by a compact magnetic circuit composed of a pair of NeFeB plates [2]. The direction of magnetic line of force produced between the N pole to S pole of the circuit was defined as a [+z]-axis of a co-ordinate, and a direction which had the largest gradient of monotone decreasing field was defined as a [+x]-axis. The field centre of the circuit was defined as the origin of the co-ordinate (B = 0.65 T at field centre). Sub-mm size crystals of 4 carbonaceous materials, namely graphite, diamond, coronene and naphthacene were prepared for observing the field-induced translation. The weight of the samples are listed in Table I along with their published χ_{DIA} values [6]. The crystals were placed on a sample stage that was located in the area of monotone decreasing field along the x-axis. In order to realize the diffused condition of the translating sample, the above was enclosed in a glass tube, and inner pressure of the tube was reduced to 30Pa. A short μg condition (duration ~ 0.5 s) was produced in a compact drop capsule, which was operated in a shaft of 1.5 m in length. Shortly after μg was achieved, the sample stage was rapidly removed from its initial position in the [-z]-direction. The rapid motion of stage was necessary to minimize the initial momentum of the released sample, which was essential for precise measurement of the sample translation. The reduction of medium pressure was effective as well to diminish the turbulence cause by the fast motion of the sample stage. The translation of the samples were observed from the [-y]-direction by a high-speed camera (CASIO EX-F1), which could observe the sample motions with a spatial resolution of 0.004 cm and a time resolution of 0.033 fps.

3. Results and Discussion

Time dependent images of a coronene crystal are shown in Fig.1. It is seen that the sample is accelerated in the direction of decreasing field. Similar accelerations were observed for other 3 materials. As described before, a field-induced translation was expected to follow a conservation rule considered for a magnetic potential and a kinetic energy [2]. Specifically, the energy conservation of a particle between two arbitrary positions, namely x_1 and x_2 located along the x-axis, is described as,

$$½m\chi_{DIA}B_1^{\,2} + ½mv_1^{\,2} = ½m\chi_{DIA}B_2^{\,2} + ½m_2v_2^{\,2}. \qquad (1)$$

Here field intensity at positions x_1 and x_2 are defined as B_1 and B_2, respectively; v_1 and v_2 denote the sample velocities at x_1 and x_2, respectively. In order to examine the linear relationship between $v_2^2 - v_1^2$ and $B_2^2 - B_1^2$ that is deduced from the above equation [2], about 10 sets of $v(x)$ and $B(x)$ values at different sample positions were obtained from the time-dependent images as described in Fig.1; the field intensity along the x-axis was measured before performing the translation experiments. Using the above-mentioned data sets of [$v(x)$, $B(x)$], the relationships between $v_2^2 - v_1^2$ and $B_2^2 - B_1^2$ were examined, and the linear correlations as expected in eq.(1) were reproduced in the 4 materials. According to eq.(1), the gradient of the above linear correlation is equivalent to χ_{DIA} of the translating particle. As seen in Table 1 and Fig.2, the experimental χ_{DIA} values obtained from the observed linear correlations agreed fairly well with the published values [6]. Errors of the experimental χ_{DIA} values are estimated from the standard deviation of the linear regression that was observed between $v_2^2 - v_1^2$ and $B_2^{\,2} - B_1^2$.

The above consistencies between experimental and published values indicate that the method of measuring χ_{DIA} based on field-induced translation is approved for an organic material. The deviations of the measured χ_{DIA} are attributed to an attachment of some small paramagnetic particle on the sample. Currently, the spatial resolution of the proposed method in observing the translating sample is about 20μm, which is determined by the resolution of the high-speed camera; this resolution gives the present sensitivity limits in sample size. Note that, in principle, χ_{DIA} measurement based on eq.(1) is possible if the field-induced translation of the sample is observable; here it is unnecessary to know the m value. Hence, in a future study, χ_{DIA} of a μ m-size sample can be measured by installing an optical microscope in the system. The sensitivity of χ_{DIA} is improved by reducing the field gradient and by increasing the μ g duration, which is necessary to collect sufficient number of [$v(x)$, $B(x)$] data.

As mentioned before, the field-induced translation was previously performed for inorganic materials. According to a data book on diamagnetism, their χ_{DIA} values are usually smaller than $5\text{x}10^{-7}$ emu/g, whereas most of the χ_{DIA} values reported for organic materials range between 4.5 to $50\text{x}10^{-7}$ emu/g [6]. Therefore the field-induced translation may serve as a simple method to distinguish an organic grain from an inorganic grain, when they are included in an ensemble of heterogeneous particles. Furthermore, by comparing the χ_{DIA} value obtained by translation with the list of published values, the material of an unidentified organic grain can be determined without consuming the grain. In studying the property of an individual particle, it is desirable to identify the material of the grain by a non- destructive and simple method.

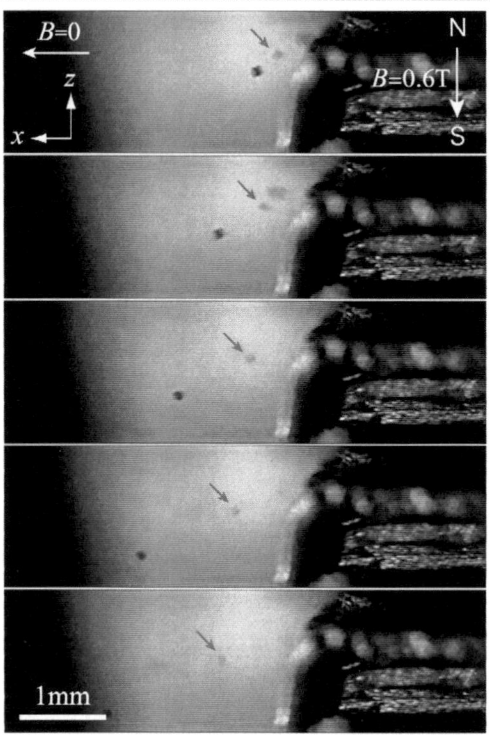

Figure 1. Visual images of translational motion of coronene particle. Field decreases monotonously [x]-direction. The images are arranged in the sequence of time from top to bottom. Time interval between the images are 0.01sec.

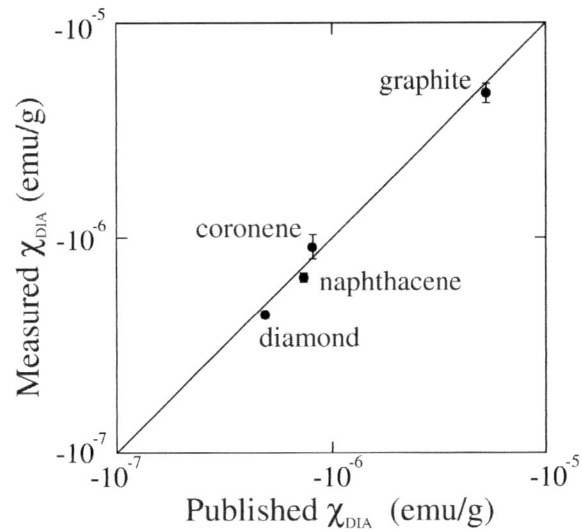

Figure 2. Relationship between measured and published diamagnetic susceptibilities.

Table 1. Numerical parameters of the carbonaceous material measured in the present study. The experimental χ_{DIA} values were obtained by inserting the measured values is eq.(1). The errors are derived from the standard deviation of the regression.

Material		mass [g]	measured χ_{DIA} [x10^{-7}emu/g]	published χ_{DIA} [x10^{-7}emu/g]
Graphite	C	4.7x10^{-6}	-47±5	-52
Diamond	C	0.07	-4.4±0.1	-4.9
Coronene	$C_{24}H_{12}$	5.8x10^{-6}	-9.1±1.2	-8.1
Naphthacene	$C_{18}H_{12}$	2.3x10^{-5}	-6.6±0.2	-7.4

The degree of unsaturation can be estimated from the χ_{DIA} value of an organic grain, when its chemical composition is unknown. This is because an intrinsic χ_{DIA} value of an organic molecule is nearly equivalent to a simple sum of diamagnetic susceptibility assigned to individual chemical bonds such as the C-H, C-C, C=C as well as the C≡C bonds [6].

Precise χ_{DIA} data of single organic crystals may provide quantitative information on the distortion of crystal structure in the nano-size region [2]. This is because diamagnetic magnetization is directly related to the 3-dementional configuration of electron orbital in the crystal; the $\Delta\chi_{DIA}$ values of organic crystals were explained in terms of a molecular- orbital method [7]. The degree of the above distortion is considered to increase with reduction of particle size. Necessity clarifying the structure of a single nano-sized organic particle is increasing in both industrial and fundamental researches.

References

[1] Hisayoshi K, Kanou S and Uyeda C 2009 *J.Phys.:Conf.Ser.* **156** 012021
[2] Uyeda C, Hisayoshi K and Kanou S 2010 *J.Phys.Soc.Jpn.* **79** 064709

[3] Suwa M and Watarai H 2002 *Anal. Chem.* **74** 5027
[4] Takahashi K, Mogi I, Awaji S and Watanabe K 2009 *J. Phys. Conf. Ser.* **156** 0120203
[5] Maret G and Dransfield K 1985 *Topics Appl. Phys.* **57** 144
[6] Guputa R 1983 *Landort Bornstein New Series II* 445
[7] Pauling L 1936 *J. Chem. Phys.* **4** 672

IOP Conf. Series: Journal of Physics: Conf. Series **903** (2017) 012055 doi:10.1088/1742-6596/903/1/012055

Probing domain walls in cylindrical magnetic nanowires with electron holography

M Staňo [1], S Jamet [1], J C Toussaint [1], S Bochmann [2], J Bachmann [2], A Masseboeuf [3], C Gatel [3] and O Fruchart [1,4]

[1] Univ. Grenoble Alpes - CNRS, Inst NEEL, F-38000 Grenoble, France
[2] Univ. Erlangen, D-91054 Erlangen, Germany
[3] CNRS, CEMES, F-31400 Toulouse, France
[4] Univ. Grenoble Alpes - CNRS - CEA, SPINTEC, F-38000 Grenoble, France

E-mail: michal.stano@atlas.cz

Abstract. We probe magnetic domain walls in cylindrical soft magnetic nanowires using electron holography. We detail the modelling of expected contrast for both transverse and Bloch point domain walls and provide comparison with experimental observations performed on NiCo nanowires, involving also both magnetic and electrostatic contribution to the electron holography map. This allows the fast determination of the domain wall type without the need for uneasy and time-consuming experimental removal of the electrostatic contribution. Finally, we describe and implement a new efficient algorithm for calculating the magnetic contrast.

1. Introduction

In this study we were interested in imaging magnetic domain walls (DWs) in rather soft magnetic cylindrical nanowires (NWs) by means of electron holography [1]. Aside from fundamental research interest, IBM proposed a concept of three-dimensional solid-state magnetic race-track memory based on shifting domain walls in magnetic NWs [2]. In our work, the focus was on creating and localizing a DW and identifying its type with the help of numerical simulations.

Biziere et al. [3] applied electron holography to Ni NWs and identified both a Transverse Wall (TW) and a TW with significant curling for NWs with diameters 55 nm and 85 nm, respectively. For larger diameters another DW may also be found, with lower energy than TWs. It is the so called Bloch Point Wall (BPW), characterized by a purely orthoradial curling. We have recently provided the first experimental observation of the BPW [4] by polarized synchrotron X-rays.

Here we considered $Ni_{60}Co_{40}$ NWs with diameter 100-150 nm, so as to favour BPWs. Nevertheless, we observed only TWs, following nucleation with a transverse field. We show how it is necessary to consider the electrostatic potential, as well as the azimuthal degree of freedom of the transverse wall, to formally reproduce and thus identify the experimental contrast.

2. Electron holography

Electron holography enables us to determine quantitatively both amplitude and phase of the electron wave that passes through a thin sample, thanks to interference and phase analysis on fringes frequency. The phase shift of the electron wave with respect to vacuum is affected by both electrostatic and magnetic potentials. Note that only in-plane (perpendicular to the electron

Content from this work may be used under the terms of the Creative Commons Attribution 3.0 licence. Any further distribution of this work must maintain attribution to the author(s) and the title of the work, journal citation and DOI.
Published under licence by IOP Publishing Ltd

beam) magnetic induction components contribute to the magnetic phase shift. Therefore the sample needs to be tilted to get information about the remaining induction component.

3. Experimental procedures

Electron holography imaging was conducted on the I^2TEM microscope (Lorentz mode; using a transfer lens corrector as imaging lenses). A double biprism was used to enlarge field of view and remove Fresnel fringes. We considered $Ni_{60}Co_{40}$ nanowires with diameter 100-150 nm, length at least 20 µm, terminated at either end with 200nm-diameter parts to deter the DW annihilation. The NWs were dispersed on a Cu grid with a lacey carbon thin film. The objective lenses of the microscope were used to apply field of 1 T transverse to the wire axis to nucleate a DW and then switched off to allow imaging at remanence. In some cases the electrostatic contribution to the phase was experimentally removed via subtracting images with a flipped sample.

4. Simulation

We calculated expected phase maps based on two micromagnetic configurations with a domain wall - TW or BPW. This can be done for different wire tilts and DW orientation (azimuth for the TW) for both magnetic and electrostatic contributions to the electron phase shift. First we compute the relaxed 3D micromagnetic configuration of the DW in the wire using FeeLLGood [5], a home-built code based on the temporal integration of the Landau-Lifshitz-Gilbert equation in a finite-element scheme, i.e. using tetrahedra cells. Only exchange and dipolar interactions were taken into account with the following parameters: saturation polarization $\mu_0 M_s = 1\,T$ and exchange stiffness $A = 10^{-11}\,J/m$. The tetrahedron size was 5 nm or smaller. We consider a wire with the length sufficiently longer than the DW width, however much shorter than in the experiment. Magnetic charges are removed at either end, thus mimicking an infinitely long wire.

The relaxed micromagnetic configuration is then used as an input for the calculation of the electron holography phase maps. Our technique combines both computational efficiency and precision which is uncommon for a numerical method, mainly due to the transformation of a 3D problem to a 2D one. Instead of performing the calculation in the whole (x,y,z) space like in [3], the magnetic phase shift is proportional to the z-component of the magnetic vector potential produced by a 2D equivalent magnetic system whose in-plane magnetization components are given by the path integral $\int dz\ (M_y, -M_x)$, where z denotes the beam direction. The latter expression corresponds to the integrated in-plane magnetization of the 3D original system rotated by 90° around the beam axis. We assume the electrostatic part to be proportional to the sample thickness and to a volume-averaged Mean Inner Potential of 22 V with the constant of proportionality $c_{300\,keV} = 6.5262 \cdot 10^6\ rad/(V \cdot m)$ [1].

5. Results and discussion

The electron phase shift maps for a wire with a domain wall (reconstructed from the electron hologram) and their numerical modelling are summarized in Fig. 1. The electrostatic part which may make the DW identification more difficult can be either removed experimentally or included in the simulations. The first option paves the way towards quantitative matching of the DW pattern but it is time consuming and not straightforward. On the other hand, even with a rather simple model for the electrostatic part we can reproduce the experiment numerically without this nuisance. The experimental magnetic phase shift is slightly smaller than in the simulations; we attribute this to a reduced saturation magnetization due to surface oxidation. Quantitative agreement for both magnetic and electrostatic contributions is possible yet difficult due to the model simplicity and electrostatic contributions arising from an inhomogeneous supporting carbon film, defects and crust of impurities on the wire [Fig. 1 (a)].

In all our experiments we identified the TW with various orientation (azimuth). Fig. 2 shows that depending on this azimuth, the same DW produces different magnetic phase maps.

IOP Conf. Series: Journal of Physics: Conf. Series **903** (2017) 012055 doi:10.1088/1742-6596/903/1/012055

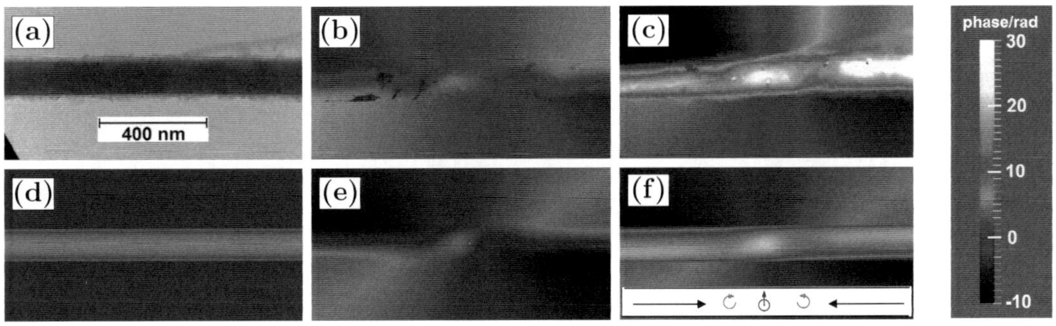

Figure 1. Electron holography. (a) hologram (wire+interference fringes); experimental phase maps: (b) magnetic only and (c) electrostatic+magnetic. Simulated phase: (d) electrostatic only, (e) magnetic only and (f) both contributions with a scheme of the magnetization pattern.

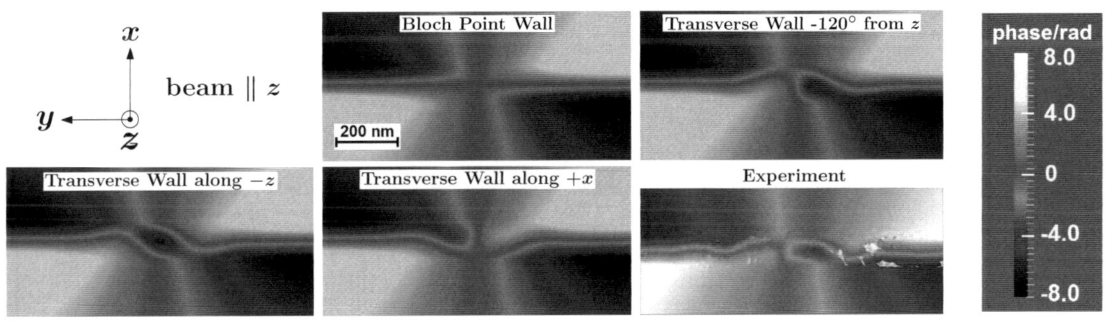

Figure 2. Magnetic phase-shift maps for both DWs and different azimuth of the TW. Here transverse wall with azimuth of -120 ° (vs z) was identified.

Note that different magnetic configurations can produce similar phase maps - e.g. BPW and ideal TW without curling pointing along the beam (exist only for small diameters [4]). We can distinguish BWP and TW along the beam thanks to an additional curling that has opposite sense at either side of the TW [3]. For the BPW the curling sense is the same on both sides.

6. Conclusion

Electron holography is a great tool for fast determination of a domain wall presence and together with simulations enables identification of the DW and determination of its structure. Our new efficient algorithm for simulations of the electron holography phase maps shows very good agreement with the experiment. Both magnetic and electrostatic contributions to the phase shift can be modelled for various magnetic nanostructures as demonstrated here on nanowires.

Acknowledgements

The experiments were conducted via the French research federation METSA and received funding from the European Union Seventh Framework Programme (FP7/2007-2013) under grant agreement n° 309589 (M3d) and from the Laboratoire d'excellence LANEF in Grenoble (ANR-10-LABX-51-01).

References

[1] Lichte H and Lehmann M 2008 *Rep. Prog. Phys.* **71** 016102
[2] Parkin S S, Hayashi M and Thomas L 2008 *Science* **320** 190–194
[3] Biziere N, Gatel C, Lassalle-Balier R, Clochard M C, Wegrowe J E and Snoeck E 2013 *Nano Lett.* **13** 2053–2057
[4] Da Col S *et al.* 2014 *Phys. Rev. B* **89**(18) 180405
[5] http://feellgood.neel.cnrs.fr

IOP Conf. Series: Journal of Physics: Conf. Series **903** (2017) 012056 doi:10.1088/1742-6596/903/1/012056

New calculations of Gilbert damping in ferromagnetic transition metals

A. Umerski
Department of Mathematics and Statistics, Open University, Milton Keynes, MK7 6AA, UK
E-mail: Andrey.Umerski@open.ac.uk

D. M. Edwards
Department of Mathematics, Imperial College London, SW7 2BZ, UK

Abstract. Calculations of the Gilbert damping parameter alpha for bulk Ni are presented as a function of the spin-dependent Lorentzian broadening of the one-electron states. The Kambersky formula for alpha is used, correctly restricted to second order in the spin-orbit parameter so that intra band terms are suppressed, and the spin-dependent broadening is deduced from the recent calculations of electron lifetimes due to electron-phonon scattering by Carva et al. The calculated value of alpha for Ni at room temperature is in reasonable agreement with experiment.

1. Introduction

New calculations are presented of the Gilbert damping parameter α in bulk ferromagnetic transition metals which arises from spin-orbit coupling (SOC). We follow previous work [1, 2] by using the Kambersky formula [3] in which the effect of impurity and phonon scattering is simulated by a Lorentzian broadening Γ of the one-electron states. However it has recently been shown [4] that the Kambersky formula is only valid to second order in the spin-orbit parameter ξ, so that the one-electron states in the formula must be calculated in the absence of SOC. Failure to apply this restriction, as in most existing calculations [1, 2], results in α diverging in the limit of a pure metal at $T = 0$ which corresponds to $\Gamma \to 0$. The striking difference at low Γ between calculations of α with and without SOC in the band structure is shown for Fe and fcc Co in [2]. Qualitative comparison between calculations of this type and the observed temperature dependence of α [5] may be made with the reasonable assumption that Γ increases with T owing to phonon scattering. The observed rapid increase of α as T decreases below 100K in Ni and hcp Co is often quoted [3, 1] in support of calculations using the Kambersky formula with SOC included in the band structure. However the observed temperature independence of α in Fe is in agreement with the calculation with SOC not included [2].

Here we report on new results for bulk Ni, which correctly exclude SOC in the band structure and use special techniques to calculate α down to the very low values of Γ appropriate to a pure metal at low temperature. It is found that α increases as Γ decreases to low values which is consistent with the observed temperature dependence. Additionally, in our calculations we include a spin-dependent broadening of electron states, obtained from first principles calculations of electron lifetimes due to electron-phonon scattering [6].

Content from this work may be used under the terms of the Creative Commons Attribution 3.0 licence. Any further distribution of this work must maintain attribution to the author(s) and the title of the work, journal citation and DOI.
Published under licence by IOP Publishing Ltd

2. Calculation of the Gilbert damping constant

The calculation of the Gilbert damping constant proceeds from the Kambersky formula

$$\alpha = \frac{\pi}{\mu_s} \sum_{n,m} \frac{1}{\Omega_{BZ}} \int d\mathbf{k}\, |A_{nm}(\mathbf{k})|^2\, L(\varepsilon_F - \varepsilon_n(\mathbf{k}), \Gamma_n)\, L(\varepsilon_F - \varepsilon_m(\mathbf{k}), \Gamma_m), \tag{1}$$

which is the same as that used in [2], for bulk materials, when the temperature $T = 0$. Here μ_s denotes the spin magnetic moment per atom (in units of the Bohr magneton μ_B), ε_F is the Fermi energy, and the integral is over the three dimensional Brillouin zone of volume Ω_{BZ}. The sum is over all energy eigenstates n, with energy $\varepsilon_n(\mathbf{k})$ (which includes both majority and minority state electrons). The spin-orbit torque matrix element A_{nm}, is given by

$$A_{nm}(\mathbf{k}) = \xi < \mathbf{k}, n|[S^-, H_{\rm SO}]|\mathbf{k}, m >,$$

where $S^- = S_x - iS_y = \frac{1}{2}(\sigma_x - i\sigma_y)$ is the spin lowering operator and $\xi H_{\rm SO}$ is the one electron spin-orbit interaction with coupling constant ξ. The Lorentzian functions are defined as $L(\varepsilon, \Gamma_n) = (\Gamma_n/2\pi)/(\varepsilon^2 + \Gamma_n^2/4)$ where Γ_n is the spin-dependent electron-scattering rate.

It is important to realise that, as shown in [4], this expression for α is only valid to second order in ξ and therefore, the eigenstates $|\mathbf{k}, n>$ must be calculated in the absence of SOC. In previous publications [3, 1] this point has been overlooked leading to non-zero diagonal (intra-band) terms A_{nn}. Such terms give rise to the result that α diverges as $\Gamma \to 0$, which corresponds to the unphysical prediction that Gilbert damping becomes infinite at $T = 0$ in a pure metal.

In this communication we calculate equation (1) within the tight-binding model, using s, p and d atomic orbitals, obtained by fitting *ab-initio* energy bands of bulk metals [7]. Even within this relatively straightforward approach, the calculation of α is computationally extremely demanding for small Γ. This is because the Lorentzians become very narrow and hence the contribution to the integral over $\mathbf{k}$ becomes localised to a small region around the Fermi-surface in the three dimensional Brillouin zone. As a result, we typically require 800^3 $\mathbf{k}$-points for a tolerable convergence of the integral for $\Gamma \geq 0.002$eV.

However the three-dimensional integral, over (k_x, k_y, k_z), can be converted to a two-dimensional one, over (k_x, k_y), by use of a novel trick. For each (k_x, k_y), the positions of the Fermi-surface branches can be determined exactly by finding the eigenvalues of the matrix

$$\mathbf{X} = \begin{pmatrix} \mathbf{0} & \mathbf{t}^{-1} \\ -\mathbf{t}^{\dagger} & (\varepsilon - \mathbf{u})\mathbf{t}^{-1} \end{pmatrix}$$

where $\mathbf{u}$ is the tight-binding on-site potential and $\mathbf{t}$ is the hopping matrix [8]. As a result, the peaks of the Lorentzians can be determined exactly and analytic approximations to the integral over k_z can be used. This results in a significant reduction in the number of k-points required to achieve convergence for small Γ. For example, to obtain convergence for $\Gamma \geq 0.002$eV, we now only require 800^2 $\mathbf{k}$-points.

3. Discussion of results

In figure 1, we see the effect of wrongly including the intra-band transitions in the calculation of α for bulk Ni (black curve). In this figure, we take a spin-independent Lorentzian broadening $\Gamma = \Gamma_\uparrow = \Gamma_\downarrow$, and as $\Gamma \to 0$, the value of α is seen to dramatically increase (without limit) below $\Gamma = 0.05$eV. On the other hand, the value of α correctly calculated to second order in ξ (red curve) remains relatively flat but displays a slight minimum around $\Gamma \approx 0.02$eV.

In figure 2 we plot our new results for α over an enlarged scale, both with spin-independent (red curve) and with spin-dependent (black curve) Γ included. The spin dependent values of Γ at the Fermi surface are obtained from the spin resolved electron lifetimes $\tau_\sigma = \{\tau_\uparrow, \tau_\downarrow\}$, given

IOP Conf. Series: Journal of Physics: Conf. Series **903** (2017) 012056 doi:10.1088/1742-6596/903/1/012056

in references [6] and [9], by using the relation $\Gamma_\sigma = \hbar/2\tau_\sigma$. We observe that even without the intra-band terms α begins to increase for $\Gamma \leq 0.02$eV.

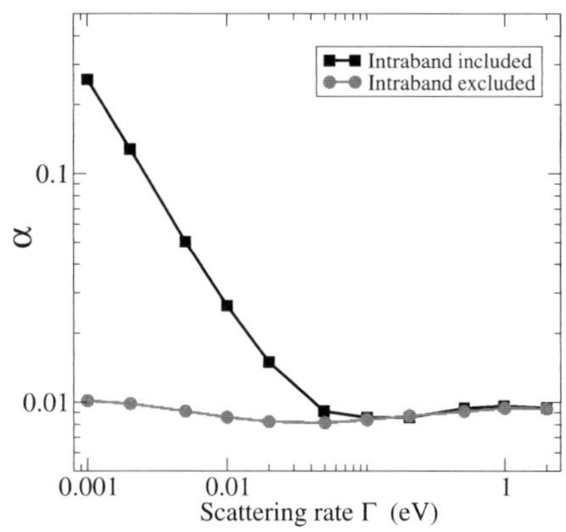

Figure 1. α against scattering rate for bulk Ni, with and without intra-band terms included. Here $\Gamma = \Gamma_\uparrow = \Gamma_\downarrow$.

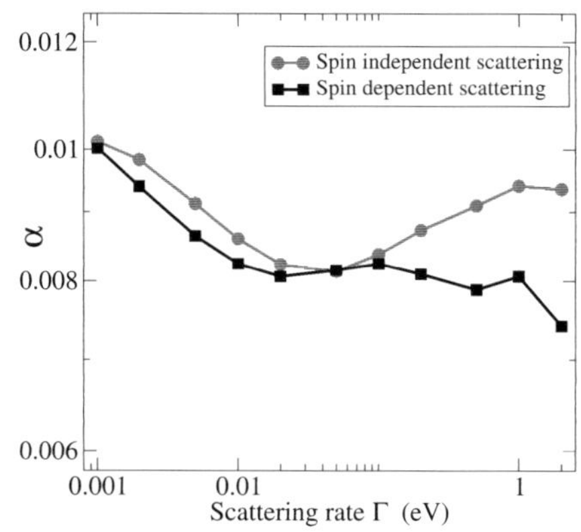

Figure 2. α against scattering rate for bulk Ni, with and spin-dependent and spin-independent scattering. Here the horizontal axis denotes $\Gamma = \Gamma_\uparrow = 0.53\Gamma_\downarrow$.

We note that, using the results of references [6] and [9] for bulk Ni, room temperature corresponds to a value of $\Gamma_\uparrow = 0.0029$eV. Hence our value of α at room temperature is predicted to be $\alpha_{RT} \approx 0.01$ (see figure 2). Experimental values for α can be deduced from the ferromagnetic relaxation frequencies λ measured by Bhagat and Lubitz [5]. They find that for pure Ni at room temperature $\lambda \approx 2.3 \times 10^8$ sec^{-1}, which corresponds to a measured value of $\alpha_{expt} \approx 0.025$, in reasonable agreement with our predicted result.

4. Conclusion

We have presented calculations of the Gilbert damping parameter α for bulk Ni as a function of the Lorentzian broadening Γ of the one-electron states. These calculations are based on Kambersky's formula, correctly evaluated to second order in the spin-orbit coupling parameter ξ, using both spin dependent and spin-independent values of Γ. In both cases we find that α increases below $\Gamma \approx 0.02$eV. We intend to calculate α at lower values of Γ than conventionally possible, using a novel technique, in order to determine its predicted behaviour as $T \to 0$.

References

[1] K. Gilmore, Y.U. Idzerda and M.D. Stiles, Phys. Rev. Lett. **99** 027204 (2007)
[2] E. Barati, M. Cinal, D. M. Edwards and A. Umerski, Phys. Rev. B **90** 014420 (2014)
[3] V. Kambersky, Czech. J. Phys B **26** 1366 (1976)
[4] D. M. Edwards, J. Phys. Condens. Matter **28** 086004 (2016)
[5] S. M. Bhagat and P. Lubitz, Phys. Rev. B **10** 179 (1974)
[6] K. Carva, M. Battiato, D. Legut and P.M. Oppeneer, Phys. Rev. B, **87**, 184425 (2013).
[7] D. A. Papaconstantopoulus, *Handbook of the Band Structure of Elemental Solids*, Plenum, New York, 1986.
[8] A. Umerski, Phys. Rev. B **55** 5266 (1997)
[9] We are grateful to K. Carva for supplying a corrected value of the spin-lifetime for majority Ni.

Liquid-liquid phase separation and cluster formation at deposition of metals under inhomogeneous magnetic field

O Yu Gorobets[1], Yu I Gorobets[1,2], V P Rospotniuk[1], V I Grebinaha[1], A A Kyba[1]

[1] National Technical University of Ukraine "KPI", Kyiv, Ukraine
[2]Institute of Magnetism NAS of Ukraine, Kyiv, Ukraine

E-mail: rospotnyk@ukr.net

Abstract. The formation and dynamic of expansion and deformation of the liquid-liquid interface of an electrolyte at deposition of metals at the surface of the magnetized steel ball is considered in this paper. The electrochemical processes were investigated in an external magnetic field directed at an arbitrary angle to the force of gravity. These processes are accompanied by the formation of effectively paramagnetic clusters of electrochemical products – magnions. Tyndall effect was used for detection of the presence of magnions near the magnetized steel electrode in a solution. The shape of the interface separating the regions with different concentration of magnions, i.e. different magnetic susceptibilities, was described theoretically based on the equation of hydrostatic equilibrium which takes into account magnetic, hydrostatic and osmotic pressures.

1. Introduction

The effect of inhomogeneous magnetic fields (MF) on electrochemical processes is the subject of current research in the field of magnetoelectrolysis [1-8]. It was discovered experimentally [1-4] that the course of these processes depends on the gradient of MF and its distribution on the electrode surface. In particular, the formation of regions with higher concentration of electrochemical reaction products near the magnetic poles of a magnetized steel spherical sample was studied in [5-7] at the etching [5] and the deposition of metals on its surface [6, 7] under the influence of constant MF. The approach of the impact of gradient magnetic force not on single paramagnetic ions, but on the clusters of electrochemical reaction products (magnions) was proposed in [5-7]. Magnions can represent paramagnetic or effectively paramagnetic bubbles of gases, nano- and/or microparticles with their ionic environment [5-8]. Also the concept of effective magnetic susceptibility of magnions $\chi = \chi_m - \chi_s$ as the difference between magnetic susceptibilities of a magnion χ_m and a solution χ_s was introduced in this model [5-7].

The aim of this work is to study the process of the formation of a heterogeneous state of an electrolyte which is observed at the metal deposition on the surface of a spherical magnetized ferromagnetic electrode in an external MF (EMF) directed at an arbitrary angle to the direction of the force of gravity (FG) without passing of an external electric current through the solution. The theoretical model of the paper is based on the equation of hydrostatic equilibrium consisting of magnetic, osmotic and gravitational pressures and the pressure of the centrifugal force which causes the electrolyte rotation near the electrode surface with the velocity about 1 cm/s [7]. This equation allows calculating the form of the interface in liquid near the electrode surface that separates the regions with higher concentration of magnions and the bulk solution.

Content from this work may be used under the terms of the Creative Commons Attribution 3.0 licence. Any further distribution of this work must maintain attribution to the author(s) and the title of the work, journal citation and DOI.
Published under licence by IOP Publishing Ltd

IOP Conf. Series: Journal of Physics: Conf. Series **903** (2017) 012057 doi:10.1088/1742-6596/903/1/012057

2. Experimental results

Experimental observation of electrolyte phase separation at the deposition of copper ions on the surface of a steel spherical sample was carried out at angles of 90, 80, 70, 60, 50 and 40 degrees between the directions EMF to FG. The Tyndall effect was used to visualize areas with higher magnions concentrations similarly to [6]. The set of photos of liquid-liquid separation at different stages are shown in Fig. 1. Regions with higher magnion concentration are forming near the magnetic poles of the steel spherical electrode where the strength of MF (as the superposition of an uniform EMF and dipole MF of a magnetized steel spherical electrode) is the largest. In addition, the anisotropic sedimentation of copper ions near the magnetic equator of the steel spherical electrode (where the gradient MF is minimal) is observed. As it was shown in [6] the shape of interface remains almost unchanged for about 1000 seconds after its relatively fast formation. This allows considering such state of the system as the quasi-stationary one.

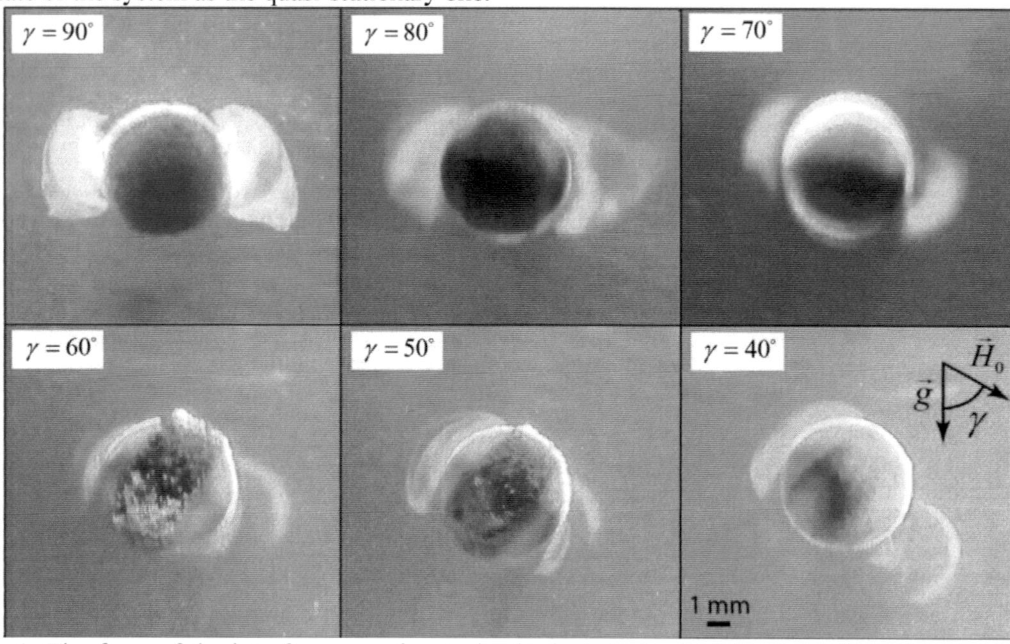

Fig. 1. The form of the interface near the steel spherical electrode with the radius of 3 mm in the $CuSO_4$ solution in an EMF H_0 = 1.8 kOe directed at different angles γ to the direction of FG.

3. Theoretical model

The hydrostatic equilibrium in an electrolyte depends on the balance of pressures at the interface between regions with different concentrations of electrochemical reaction products:

$$\Delta P_m + \Delta P_o + \Delta P_g + \Delta P_\omega = 0, \quad (1)$$

where $\Delta P_o = nk_BT$ is the difference of osmotic pressures between two medium, n is magnion concentration, k_B is the Boltzmann constant, T is the absolute temperature; $\Delta P_m = -\chi \vec{H}^2/2$ is the magnetic pressure, $\vec{H} = \vec{H}_0 + \vec{H}_m$ is the MF strength which is the vector sum of EMF $\vec{H}_0$ and MF induces by a magnetized electrode $\vec{H}_m$; $\Delta P_g = \Delta\rho g r\cos(\theta - \gamma)$ is the pressure of FG, $\Delta\rho$ is the difference of densities of separated electrolyte regions, θ is the angle between $\vec{H}_0$ and radius-vector $\vec{r}$, γ is the angle between $\vec{H}_0$ and the acceleration of free fall $\vec{g}$; $\Delta P_\omega = \int \Delta\rho\vec{\omega}^2\vec{r}\sin\theta d\vec{r}$ is the pressure of centrifugal force, $\vec{\omega}$ is the angular velocity of solution rotation.

It is easy to see that the contribution of rotational effects on the shape of the interface is negligibly small. Particularly, the ratio $\Delta P_\omega/\Delta P_g \cong v^2/gR_0$ has the order of magnitude $10^{-2} \div 10^{-3}$. It was found in the paper [6] that terms ΔP_o and ΔP_g have the same order of magnitude. This allows neglecting the last terms in the equation (1). The effect of the centrifugal force will be significant for smaller electrode sizes and higher electrolyte rotation velocities. Thus, the equation (1) in the case of a spherical ferromagntic electrode can be written as follows:

$$\cos 2\gamma \cos^2(\theta-\gamma)+\sin 2\gamma \sin(\theta-\gamma)\cos(\theta-\gamma)+B(x)\cos(\theta-\gamma)+A(x)=0,$$

where $A(x)=\left(\frac{8\pi}{3x^3}\left[\xi\left(3\sin^2\gamma-1\right)+\frac{2\pi}{3x^3}\left(3\sin^2\gamma+1\right)\right]+\xi^2-C_o\right)\Big/\frac{8\pi}{x^3}\left(\xi+\frac{2\pi}{3x^3}\right)$, $B(x)=C_g x\Big/\frac{8\pi}{x^3}\left(\xi+\frac{2\pi}{3x^3}\right)$,

$x=r/R_0$, R_0 is ball radius, $\xi=H_0/M_0$, $C_g=\frac{2\rho_0 g R_0}{\chi M_0^2}$, $C_o=\frac{2nk_BT}{\chi M_0^2}$.

The comparison of theoretical modeling with experimental data is shown in Fig. 2.

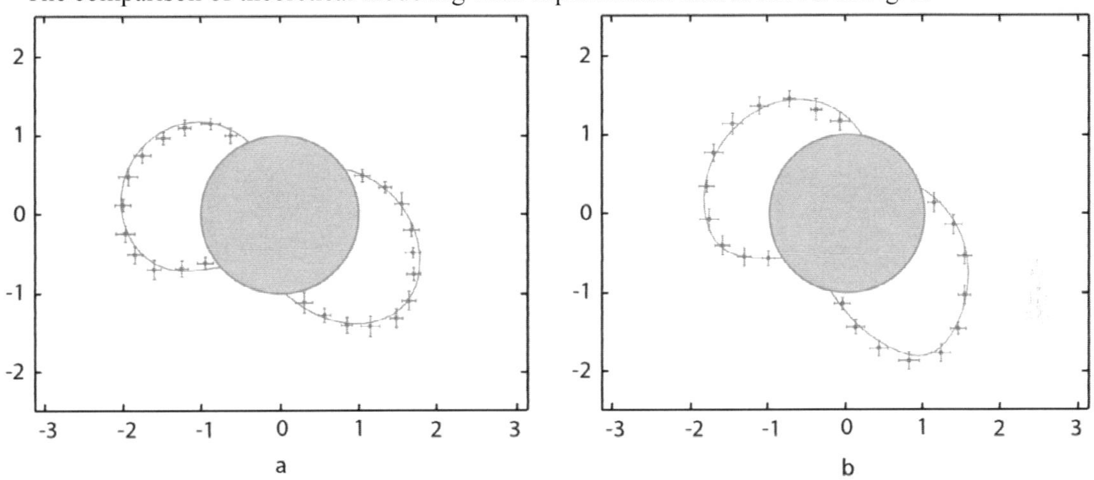

Fig. 2. The comparison of experimental data (points with errors) with theoretical modeling (solid line). The shape of the interface forming at copper deposition when EMF is directed at 70° (a) and 50° (b) to FG. a - $C_g=6.1$, $C_o=30$, b – $C_g=7.4$, $C_o=31$. Distances are given in radius of spherical ferromagnetic sample.

4. Conclusions

The formation of quasi-stationary heterogeneous state of an electrolyte and liquid-liquid phase separation were experimentally detected at the metal deposition on the surface of a magnetized ferromagnetic spherical electrode in an inhomogeneous constant MF. This state of the system exists around 10^3 sec and is characterized by the appearance of regions in the electrolyte with different magnetic susceptibilities. The change of the interface shape was investigated depending on the relative angle between the directions of EMF and FG. Metal deposition in gradient MF is accompanied by the formation of nano- and/or microclusters containing 10^4-10^5 paramagnetic ions – magnions.

This project has received funding from the European Union's Horizon 2020 research and innovation programme under the Marie Skłodowska-Curie grant agreement No 644348 (MagIC).

References

[1] Fahidy T Z 1973 *Electrochim. Acta.* **18 (8)** 607
[2] Yang X, *et. al.* 2012 *J. Phys. Chem. Lett.* **3 (23)** 3559
[3] Dunne P, Mazza L and Coey J M D 2011 *Phys. Rev. Lett.* **107** 024501
[4] Dunne P and Coey J M D 2012 *Phys. Rev. B* **85** 224411
[5] Gorobets O Yu, Gorobets Yu I, Bondar I A and Legenkiy Yu A 2013 *JMMM* **330** 76
[6] Gorobets O Yu, *et. al.* 2015 *J. Solid State Electrochem.* **19 (10)** 3001

[7] Gorobets O Yu, Gorobets Yu I and Rospotnuik V P 2014 *Condens. Matter Phys.* **17(4)** 43401
[8] Tyrrell J W G and Attard P 2001 *Phys. Rev. Lett.* **87** 176104

IOP Conf. Series: Journal of Physics: Conf. Series **903** (2017) 012058 doi:10.1088/1742-6596/903/1/012058

Terahertz excited plasmon-magnon interaction and magnetoplasmon-enhanced energy transfer within the framework of generalized spin Hamiltonian

Vyacheslav S.Gritzaenko[1], **Dmitry I.Bazhanov**[2,3], **Oleg V.Farberovich**[1,4,5]

[1]Voronezh State University, Voronezh 394000, Russia
[2]Faculty of Physics, Moscow State University, GSP-1, Lenin Hills, 119991 Moscow, Russia
[3]Institution of Russian Academy of Sciences, Dorodnicyn Computing Centre of RAS, Vavilov st. 40, 119333 Moscow, Russia
[4]School of Physics and Astronomy, Beverly and Raymond Sackler Faculty of Exact Sciences, Tel Aviv University, Tel Aviv 69978, Israel
[5]International Center "Smart Materials", Southern Federal University, Zorge 5, 344090, Rostov-on-Don,Russia

E-mail: slava.gricaenko@yandex.ru

Abstract.
Research into ultrafast terahertz technology attracts a significant interest nowdays in the fields of plasmonics and magnonics, since a recent progress in both fields unveils new fundamental physics and opens up new opportunities to engineering high-speed spintronic devices. However insufficient exchange of knowledge between these research fields still remains. The goal of this paper is to bridge such a gap by presenting a new theoretical approach to study magnetoplasmon-enhaced energy transfer between supported magnetic nanoparticles in the framework of generalized spin Hamiltonian (GSH) under ultrafast THz-wave excitation.

1. Introduction

Ultrafast spin dynamics induced by transport of spin polarized carriers has attracted considerable interest over last decade [1, 2]. It is motivated by the fundamental interest in magnetic excitations and applications like spintronics and data storage[3, 4]. To achieve an understanding of the underlying elementary processes at microscopic scale, that typically occur on femtosecond time scales, a time dependent approach has been developed that probes the spin dynamics induced by plasmon-magnon excitations[5]. Surface magnetoplasmons (SMP) in the THz range are technologically important because they provide a possibility to develop subwavelength-size devices and offer an opportunity to investigate nanoscale systems at THz frequencies time-domain spectroscopy. In conjunction with pump-probe measurements it offers a powerful tool to examine energy transfer and ultrafast spin dynamics in nanosystems. The THz spectral range provides a high degree of sensitivity for research of a spin structure, which at very strong exchange complex generally dominate the charge response[6]. Further, with an ultrafast optical excitation, this approach permits one to probe non-equilibrium systems with picosecond to sub-picosecond time resolution. These unique capabilities of the optical-pump/THz-probe

Content from this work may be used under the terms of the Creative Commons Attribution 3.0 licence. Any further distribution of this work must maintain attribution to the author(s) and the title of the work, journal citation and DOI.
Published under licence by IOP Publishing Ltd

spectroscopy give the possibilities for the investigation of energy transfer phenomena that have remained inaccessible with conventional techniques.

Magnons can be interpreted as collective excitations associated to the electron spin structure and plasmons can be directly linked to high-energy oscillations of the electron density in the cluster. In this context, a magnon is a spin wave with zero phase, and it can be understood as a quantum of a spin wave. Both quasiparticles (magnon and plasmon) and their collective excitations are of great importance for better understanding of the electronic origin of magnetic phenomena and, therefore, the study of their properties is a key prerequisite regarding the applications in charge-based electronics and spintronics[7].

In this work we describe the process of Förster transfer (FT) between nanoparticles in the presence of a metal nanocrystal subsystem [8]. We focus here on an optical processes, which take place on a Ni/Cu(100) surface in the vicinity of $Gd_3N@C_{80}$ endohedral metallofullerenes nanoparticles (EMFNPs). Since, it was shown, that such nanoparticles can strongly modify the optical process [9]. EMFNPs on a Ni/Cu(001) surface strongly confine carriers and do not permit efficient tunnel coupling. However, instead of direct tunnel coupling, EMFNPs permit FT of magnetic generated magnons. The FT mechanism comes from the inter-EMFNP Coulomb interaction and does not require the tunnel coupling between EMFNPs. The enhancement of Förster transfer can occur due to the effect of plasmon-assisted amplification of electric fields inside the nanoscale assembly [10]. Here, we show how the theory of Förster resonant energy transfer (FRET) may be generalized through inclusion the spin structure of a d-,f-shell nanoparticles. In particular, here we demonstrate that the photoluminescence spectra should be carefully analyzed since the plasmon-enhanced Förster effect can appear together with strong magnon energy dissipation.

2. Theoretical approach

To study magnetoplasmon-enhanced energy transfer between interacted EMFNPs on a $Ni/Cu(001)$ surface, we used numerical scheme constructed from two parts. The first part is represented by the re-expanded model [11] used to describes the electrostatic interaction between two charged (polarized) metallofullerene Gd_3N nanoparticle spheres surrounded by a homogeneous dielectric medium ϵ_d. Let us to consider thsese nanoparticle spheres charged by q_B and q_A with relative dielectric functions $\epsilon_{EMFNP_B}(\omega)$ and $\epsilon_{EMFNP_A}(\omega)$, and radii R_B and R_A, respectively. The center-center distance between spheres is labeled by R and the spheres are surrounded by a medium with dielectric constant ϵ_d, as shown in Fig. 1. Since we study the energy transfer between a donor (d) Ni ion and an acceptor (a) Cu ion on copper substrate in the vicinity of two EMFNPs, the donor and acceptor ions will be represented by point electric dipoles $\mathbf{p}_d$ and $\mathbf{p}_a$.

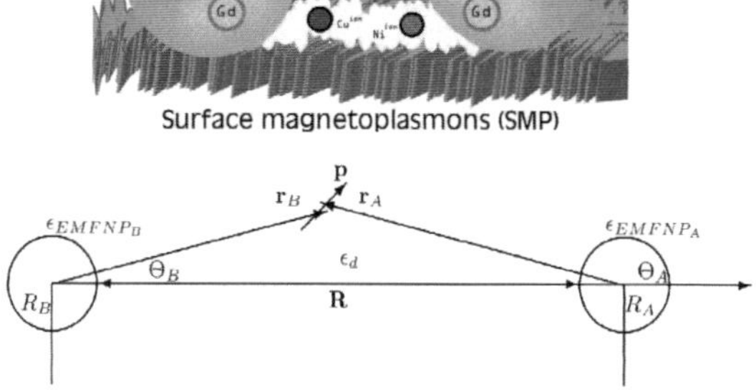

Figure 1. (Color online) Sketch of the two spherical coordinate systems (A and B) used in the re-expanded method to describe the mutual electrostatic interaction between two metallofullerens nanospheres.

When two charged or polarized spheres are located close together, an infinite number of multipoles are induced in the spheres. These multipoles lead to the difficulties in the precise evaluation of the enhancement factor between these two spheres [12]. Therefore we used the re-expansion of a given two Legendre expansion for the electrostatic field in a Taylor series around

IOP Conf. Series: Journal of Physics: Conf. Series **903** (2017) 012058 doi:10.1088/1742-6596/903/1/012058

the two centers of the nanoparticle spheres [11]. If a dipole **p** is located at the position P, it will induce polarization charges on both spheres. The potential at the position of the dipole charge is then given by:

$$\Phi_{out} = \sum_{l=0}^{\infty}\left[c_l^B\left(\frac{R_B}{r_B}\right)^{l+1}P_l(cos\Theta_B) + c_l^A\left(\frac{R_A}{r_A}\right)^{l+1}P_l(cos\Theta_A)\right] + \\ +\frac{1}{2}\sum_{l=0}^{\infty}\left[\delta c_l^B\left(\frac{R_B}{r_B}\right)^{l+1}P_l(cos\Theta_B) + \delta c_l^A\left(\frac{R_A}{r_A}\right)^{l+1}P_l(cos\Theta_A)\right], \tag{1}$$

where P_l denote associated Legendre functions and c_l^A/B are coefficients in decompositions. The first sum is the contribution from the sphere–sphere interaction, which can be determined by solving next equations:

$$(\mathbf{H}_{A/B} - \mathbf{I})\mathbf{c}^{A/B} = \mathbf{w}^{A/B}, \tag{2}$$

where $\mathbf{I}$ is the unit matrix; $\mathbf{c}^A$ and $\mathbf{c}^B$ are two column vectors with elements c_l^A and c_l^B, respectively. The matrix and vector elements are:

$$h_{lm}^{A/B} = f(\epsilon_{EMFNP_{A/B}}, l)\left(\frac{R_{A/B}}{R}\right)^{l+m+1}\left(\frac{R_{B/A}}{R}\right)\sum_{j=1}^{j_{max}} f(\epsilon_{EMFNP_{B/A}}, j)\frac{(j+l)!(j+m)!}{j!l!j!m!}\left(\frac{R_{B/A}}{R}\right)^{2j}, \tag{3}$$

$$w_l^{A/B} = f(\epsilon_{EMFNP_{A/B}}, l)\frac{1}{4\pi\epsilon_d R}\left(\frac{R_{A/B}}{R}\right)^l\left[q_{B/A} - q_{A/B}\left(\frac{R_{B/A}}{R}\right)\sum_{j=1}^{j_{max}} f(\epsilon_{EMFNP_{B/A}}, j)\frac{(j+l)!}{j!l!}\left(\frac{R_{B/A}}{R}\right)^{2j}\right], \tag{4}$$

$$f(\epsilon, j) = \frac{j(\epsilon/\epsilon_d - 1)}{j(\epsilon/\epsilon_d + 1) + 1}. \tag{5}$$

In the same manner we obtained two matrix equations:

$$(\delta\mathbf{H}_{A/B} - \mathbf{I})\delta\mathbf{c}^{A/B} = \delta\mathbf{w}^{A/B} \tag{6}$$

where

$$\delta\mathbf{H}_{A/B} = \mathbf{H}_{A/B}$$

$$\delta w_l^B = f(\epsilon_{EMFNP_B}, l)\frac{Q(r)}{4\pi\epsilon_d r}\left[\frac{1}{r}\left(\frac{R_B}{r}\right)^l - \\ -\frac{R_A}{R^2 - Rr}\left(\frac{R_A}{R}\right)^l\sum_{j=1}^{j_{max}} f(\epsilon_{EMFNP_A}, j)\frac{(j+l)!}{j!l!}\left(\frac{R_A^2}{R^2 - Rr}\right)^j\right] \tag{7}$$

$$\delta w_l^A = f(\epsilon_{EMFNP_A}, l)\frac{Q(r)}{4\pi\epsilon_d(R - r)}\left[\frac{1}{R - r}\left(-\frac{R_A}{R - r}\right)^l - \\ -\frac{R_B}{Rr}\left(-\frac{R_A}{R}\right)^l\sum_{j=1}^{j_{max}} f(\epsilon_{EMFNP_B}, j)\frac{(j+l)!}{j!l!}\left(\frac{R_B^2}{Rr}\right)^j\right] \tag{8}$$

IOP Conf. Series: Journal of Physics: Conf. Series **903** (2017) 012058 doi:10.1088/1742-6596/903/1/012058

The second part of our numerical scheme shows how FRET can be generalized with including the spin structure of a d-,f-shell nanoparticles. The influence of the electric and magnetic component of the electromagnetic (EM) field can be described in the framework of GSH:

$$\widehat{H} = \widehat{H}^{(o)}_{ion} + \widehat{H}^{(o)}_{spin} + \widehat{H}'(t) \tag{9}$$

where

$$\widehat{H}^{(o)}_{ion} = -\nabla^2 + V^{ion}_{eff}, \widehat{H}^{(o)}_{spin} = \widehat{H}_{ex} + \widehat{H}_{an} + \widehat{H}_{ZEE}, \widehat{H}'(t) = \mathbf{E}_o \mathbf{r} e^{-i\omega t} - \gamma \widehat{\mathbf{S}} \mathbf{H}^{LLG}_{eff} e^{-i\omega t}. \tag{10}$$

Here V^{ion}_{eff} represents the effective single-electron potential of ion associated with $Q(r) = -rV^{ion}_{eff}(r)/2$. Hamiltonians $\widehat{H}_{ex}$, $\widehat{H}_{an}$, and $\widehat{H}_{ZEE}$ denote the Heisenberg isotropic exchange interaction, the anisotropic exchange interaction due to the axial single-ion anisotropy, and the interaction between the spin system and the external magnetic field, repectively. The last term of equation describes consequently the electric and magnetic contributions of EM field, where where γ is the gyromagnetic ratio coming from the relation between spin $\widehat{\mathbf{S}}$ and the effective magnetic field $\mathbf{H}^{LLG}_{eff}$. The field $\mathbf{H}^{LLG}_{eff}$ is a result of Landau-Lifshitz-Gilbert (LLG) equation solution [13, 14]:

$$\frac{\partial \langle \widehat{\mathbf{S}} \rangle}{\partial t} = \frac{1}{1+\lambda^2} \langle \widehat{\mathbf{S}} \rangle \times \mathbf{H}^{LLG}_{eff} - \frac{\lambda}{1+\lambda^2} \langle \widehat{\mathbf{S}} \rangle \times (\langle \widehat{\mathbf{S}} \rangle \times \mathbf{H}^{LLG}_{eff}), \tag{11}$$

where λ is damping constant. The eigenstates $\mid SM \rangle$ of GSH are given by the linear combinations of the basis states $\mid S^{(\mu)} M^{(\mu)} \rangle$, where $M = -S, .., S$, $\mu = 1, 2, ..., N-1$, and the coefficients $c_{\mu\nu}$ can be evaluated once the spin Hamiltonian of the system has been diagonalized. With the algebra of the spin operators we can obtain the expectation values for $\widehat{H}^{(o)}_{spin}$:

$$\langle \hat{S}_x \rangle = \frac{1}{2} \sum_{\mu=1}^{N} c^2_{\mu\mu} A_\mu, \langle \hat{S}_y \rangle = -\frac{i}{2} \sum_{\mu=1}^{N} c^2_{\mu\mu} B_\mu, \langle \hat{S}_z \rangle = \sum_{\mu=1}^{N} c^2_{\mu\mu} M^{(\mu)}, \tag{12}$$

where

$$A_\mu = \sqrt{S^{(\mu)}(S^{(\mu)}+1) - M^{(\mu)}(M^{(\mu)}+1)} + \sqrt{S^{(\mu)}(S^{(\mu)}+1) - M^{(\mu)}(M^{(\mu)}-1)}$$

and

$$B_\mu = \sqrt{S^{(\mu)}(S^{(\mu)}+1) - M^{(\mu)}(M^{(\mu)}+1)} - \sqrt{S^{(\mu)}(S^{(\mu)}+1) - M^{(\mu)}(M^{(\mu)}-1)}.$$

Now we turn to the discussion about the photoresponse of EMFNPs to presence of an external EM field in terms of the frequency-dependent polarizability $\alpha(\omega)$. For this purpose we analyze at first the dispersion properties of the surface magnetoplasmons (SMP) in the EMFNPs structure under an applied external magnetic field. The dielectric response in the THz frequency region originates from the intraband motion of conduction electron and spins. Therefore we can use Drude model to describe the THz electron response and the Kramers-Heisenberg dispersion formula for $\alpha^{spin}(\omega)$. In this case, a dielectric function $\epsilon(\omega)$ for magnetoplasmon system is:

$$\epsilon_{(x,y,z)}(\omega) = \tilde{\epsilon}^{Drude}_{(x,y,z)}(\omega) + \alpha^{spin}_{(x,y,z)}(\omega) \tag{13}$$

In the framework of Drude model, the components of the dielectric tensor in presence of magnetic field are the following:

$$\tilde{\epsilon}_{xx}^{Drude} = \epsilon_b \left[1 - \frac{\omega_p^2(\omega^2 + i\omega\Gamma)}{(\omega^2 + i\omega\Gamma)^2 - \omega^2\omega_c^2} \right], \tilde{\epsilon}_{xy}^{Drude} = \frac{i\omega_p^2 \epsilon_b \omega\omega_c}{(\omega^2 + i\omega\Gamma)^2 - \omega^2\omega_c^2}, \tilde{\epsilon}_{zz}^{Drude} = \epsilon_b \left[1 - \frac{\omega_p^2}{\omega^2 + i\omega\Gamma} \right], \tag{14}$$

where ϵ_b is the background dielectric constant, ω_p is the effective plasma frequency, which lies in the terahertz regime, ω_c is the cyclotron frequency, and Γ is the damping constant.

The Kramers-Heisenberg dispersion formula can be applied to calculate $\alpha^{spin}(\omega)$[10]. So far as, the Hamiltonian $\widehat{H}_{spin}^{(o)}$ is induced by transitions between different spin substates of the magnetic molecule, let us denote:

$$M_{if}^{(x)} = \sum_{\nu=1}^{N} c_{i\nu} c_{\nu f} A_\nu; M_{if}^{(y)} = \sum_{\nu=1}^{N} c_{i\nu} c_{\nu f} B_\nu; M_{if}^{(z)} = \sum_{\nu=1}^{N} c_{i\nu} c_{\nu f} M^\nu$$

where i is initial spin energy state $\mid SM\rangle_i$ and f is final spin energy $\mid SM\rangle_f$ of GSH in the spin structure. Then

$$\alpha_{xx}^{spin}(\omega) = 2\gamma^2 \sum_{i=1}^{N_i} \sum_{f=1}^{N_f} \mid H_{eff}^{LLG(x)} M_{if}^{(x)} \mid^2 \frac{\omega_f - \omega_i}{(\omega_f - \omega_i)^2 - (\omega - i\Gamma)^2}, \tag{15}$$

$$\alpha_{xz}^{spin}(\omega) = 2\gamma^2 \sum_{i=1}^{N_i} \sum_{f=1}^{N_f} H_{eff}^{LLG(x)} H_{eff}^{LLG(z)} M_{if}^{(x)} M_{if}^{(z)} \frac{\omega_f - \omega_i}{(\omega_f - \omega_i)^2 - (\omega - i\Gamma)^2}. \tag{16}$$

Thus, for the dielectric response in the THz frequency region we obtain:

$$\epsilon_{xx} = \tilde{\epsilon}_{xx}^{Drude}(\omega) + (4\pi\alpha_{xx}^{spin}(\omega) + 1), \epsilon_{xz} = \tilde{\epsilon}_{xz}^{Drude}(\omega) + (4\pi\alpha_{xz}^{spin}(\omega) + 1) \tag{17}$$

In this EMFNP structure the Gd_3N cluster is gyroelectric anisotropic with permittivity tensor:

$$\overleftrightarrow{\epsilon}_{EMFNP} = \begin{pmatrix} \epsilon_{xx} & 0 & \epsilon_{xz} \\ 0 & \epsilon_{yy} & 0 \\ -\epsilon_{xz} & 0 & \epsilon_{xx} \end{pmatrix} \tag{18}$$

Further, we describe energy transfer between Ni (a donor point electric dipole $\mathbf{p}_d$ at locations $\mathbf{r_d}$) and Cu (an acceptor point electric dipole $\mathbf{p}_a$ at locations $\mathbf{r_a}$) ions in the vicinity of EMFNPs on a $Ni/Cu(001)$ surface. For this purpose, we propose here a theory of the plasmon-magnon enhancement of the many body phenomena resulting for the SMP-dressed Coulomb interaction. The interaction energy between the Ni and Cu ions (dipoles $\mathbf{p}_a$ and $\mathbf{p}_d$, respectively) can be obtained from the expression:

$$U = -\frac{1}{2}(\mathbf{p}_a \cdot \mathbf{E}_a + \mathbf{p}_d \cdot \mathbf{E}_d), \tag{19}$$

where $\mathbf{E}_a$ is the electric field at the position of the accepter ion and $\mathbf{E}_d$ is the corresponding field at the donor ion. These fields are obtained using

$$\mathbf{E} = -\nabla\Phi_{out} \tag{20}$$

and omitting the terms Φ_a, when computing $\mathbf{E}_a$, and Φ_d, when computing $\mathbf{E}_d$. Thus we may write:

$$\Delta U(\omega) = -\frac{1}{2}[\mathbf{p}_a \cdot \mathbf{E}_d(r_a, \omega) + \mathbf{p}_d \cdot \mathbf{E}_a(r_d, \omega)], \tag{21}$$

where $\mathbf{E}_j(r_i, \omega)$ is the field of the jth particle at the point r_i.

Finally, we examine our numerical scheme in direct calculations of FRET between interacted EMFNPs including SMP, as illustrated schematically in Fig. 1. For selected interparticle distance and radii we observed an energy transfer between considered particles mediated predominantly via SMP (see Fig 2). We revealed a significant enhancement of magnetoplasmon-enhanced energy transfer if the particles increase thier size (see Fig. 3). Thus, the obtained results clearly show that even a small variation in distance can effect an energy transfer between magnetic nanoparticles, that could be important to develop the ultrafast spintronic devices.

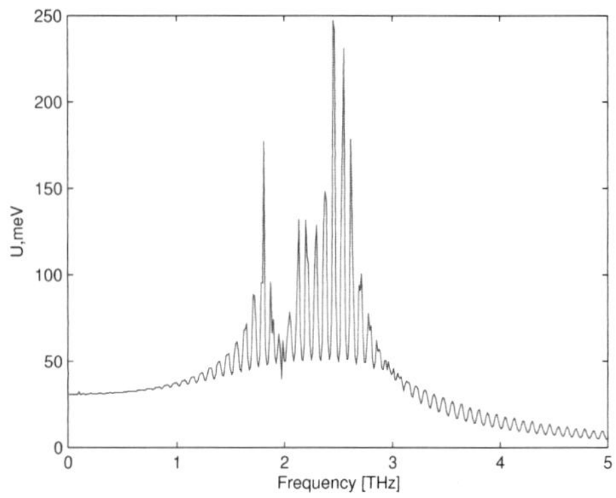

Figure 2. (Color online) The energy transfer between $EMFNP1$ ($R_B = 0.22nm$) and $EMFNP2$ ($R_A = 0.22nm$) on a $Ni/Cu(001)$ surface

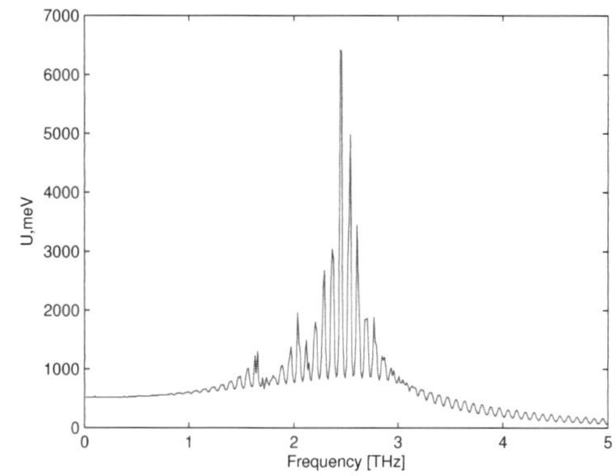

Figure 3. (Color online) The energy transfer between $EMFNP1$($R_B = 0.24nm$) and $EMFNP2$($R_A = 0.24nm$) on a $Ni/Cu(001)$ surface

References

[1] T. Kampfrath, M. Battiato, P. Maldonado, G. Eilers, J. Nötzold1, S. Mährlein, V. Zbarsky, F. Freimuth, Y. Mokrousov, S. Blügel, M.Wolf, I. Radu, P. M. Oppeneer and M. Münzenberg, *Nat. Nanotechnol.* **8**, 256 (2013); T. J. Huisman, R. V. Mikhaylovskiy, J. D. Costa, F. Freimuth, E. Paz, J. Ventura, P. P. Freitas, S. Blügel, Y. Mokrousov, Th. Rasing and A. V. Kimel, *Nat. Nanotechnol.* **11**, 455 (2016).
[2] Z. Jin, A. Tkach, F. Casper, V. Spetter, H. Grimm, A. Thomas, T. Kampfrath, M. Bonn, M. Kläui and D. Turchinovich, *Nature Phys.* **11**, 761 (2015).
[3] N. J. Tao, *Nature Nanotech.* **1**, 173 (2006).
[4] L. Bogani and W. Wernsdorfer, *Nat. Mater.* **7**, 179 (2008).
[5] M. Tonouchi, *Nat. Photonics* **1**, 97 (2007).
[6] E. Beaurepaire, G. M. Turner, S. M. Harrel, M. C. Beard, J. Y. Bigot, and C. A. Schmuttenmaernmaer, *Appl. Phys. Lett.* **84**, 3465 (2004).
[7] I. Crassee, M. Orlita, M. Potemski, A. L. Walter, M. Ostler,Th. Seyller, I. Gaponenko, J. Chen,and A. B. Kuzmenko, *Nano Lett.* **12**, 2470 (2012).
[8] Alexander O. Govorov, Jaebeom Lee, and Nicholas A. Kotov, *Phys. Rev. B* **76**, 125308 (2007).
[9] Christian F. Hermanns, Matthias Bernien, Alex Kruger, Christian Schmidt, Soren T. Waerroth, Gelavizh Ahmadi, Benjamin W. Heinrich, Martin Schneider, Piet W. Brouwer, Katharina J. Franke, Eugen Weschke, and Wolfgang Kuch, *Phys. Rev. Lett.* **111**, 167203 (2013).
[10] J. I. Gersten, A. Nitzan, *Chem. Phys. Lett.* **104**, 31 (1984).
[11] Yoji Nakajima, Takashi Sato, *J. Elect.* **45**, 213 (1999).
[12] X.M.Hua, J.I.Gersten,A.Nitzan, *J. Chem. Phys.* **83**, 3650 (1985).
[13] V. L. Mazalova, O. V. Farberovich, *J. Mag. Mag. Mater.* **405**, 169 (2016).
[14] Oleg V.Farberovich, Victoria L.Mazalova, Alexander V.Soldatov, *J. Mag. Mag. Mater.* **394**, 422 (2015).

Magnetic response of surface plasmon polaritons in silver / $Co_2FeSi_{0.6}Al_{0.4}$ Heusler alloy / silver films

Yoshito Ashizawa, Yuya Fukumoto, and Katsuji Nakagawa

College of Science and Technology, Nihon University, 7-24-1 Narashinodai, Funabashi, Chiba, 274-8501 Japan

ashizawa.yoshito@nihon-u.ac.jp

Abstract. Reflectivity change caused by an excitation condition of surface plasmon polaritons (SPPs) by an external magnetic field is important for a magnetic sensing technique to use the SPPs. The magneto-plasmonic effect utilizing the $Co_2FeSi_{0.6}Al_{0.4}$ Heusler alloy, which is one of candidates for the high spin-polarized material even at room temperature for spintronic devices, as a magnetic responsible layer was investigated. The SPPs were excited in the $Co_2FeSi_{0.6}Al_{0.4}$ Heusler alloy by using a sandwiched structure between two silver layers. The maximum value of absolute magnetic activity $|\Delta R|^{max}$ of 0.14 were obtained.

1. Introduction

It is important for realizing a magnetic sensor using magnetic response of surface plasmon polaritons (SPPs) that a large reflectivity change caused by an excitation condition of SPPs is observed by an external magnetic field. The magnetic response of SPP, which is called magneto-plasmonic effect, has been reported using films combined a noble metal with a ferromagnetic material [1-6]. Although half-metallic ferromagnetic materials, such as Heusler alloys, are paid attention to spintronic devices because of their high spin polarization [7, 8], their magneto-plasmonic effect has not been reported yet. Therefore, the magneto-plasmonic effect using a $Co_2FeSi_{0.6}Al_{0.4}$ Heusler alloy, which is one of candidates for the high spin-polarized material even at room temperature [8,9], as a magnetic responsible layer was investigated for development of a spintronic and magneto-plasmonic device.

2. Experimental procedure

Film structure for magneto-plasmonic effect was a glass substrate / Ag / $Co_2FeSi_{0.6}Al_{0.4}$ / Ag. Film thicknesses of Ag and $Co_2FeSi_{0.6}Al_{0.4}$ layers were varied from 3 to 30 nm and from 1 to 4 nm, respectively. The specimens were fabricated by rf magnetron sputtering at ambience temperature on a glass substrate. Base pressure and process Ar gas pressure of sputtering were less than 1×10^{-4} Pa and 1 Pa, respectively. Deposition rates of Ag and $Co_2FeSi_{0.6}Al_{0.4}$ layers were 1 and 2 nm/s, respectively. The specimens were annealed at 500 °C in 1×10^{-3} Pa of nitrogen during 0.5 hours after air exposure.

The SPPs were excited by the attenuated total reflection (ATR) method in the Kretschmann-Raether configuration [10] using light with wavelength of 700 nm. External magnetic fields of 4 and 0 kOe were applied normal to the film plane using a permanent magnet. This is Voigt configuration in Cyclotron Resonance Inactive (CRI) mode. The reflectivity in the configuration, which strongly depends on the boundary condition of the SPP excitation, is magnetically changed by a change of diagonal component, which is parallel or perpendicular to the magnetization direction, in dielectric tensor of a magnetic

Content from this work may be used under the terms of the Creative Commons Attribution 3.0 licence. Any further distribution of this work must maintain attribution to the author(s) and the title of the work, journal citation and DOI.
Published under licence by IOP Publishing Ltd

substance[4]. The magneto-plasmonic properties was evaluated on the basis of the normalized reflectivity change using the following formula:[1,11]

$$\Delta R = \{R(H) - R(0)\} / R(0), \quad (1)$$

where $R(H)$ and $R(0)$ correspond to reflectivity with the magnetic field H of 4 and 0 kOe, respectively. Crystal structure of the films was analyzed by X-ray diffraction (XRD) with Cu K_α line. The magnetic properties of the films were measured using a vibrating sample magnetometer (VSM) applied normal to the film plane.

3. Results and Discussion

A reflectivity R, for a $Co_2FeSi_{0.6}Al_{0.4}$ Heusler alloy single layer film as deposited with 35 nm-thick fabricated on a glass substrate as a function of incident angle θ, is shown in Fig. 1. Wavelength λ, is 700 nm. The reflectivity is monotonously decreased at the angle higher than the critical angle of 42.4°. The reflectivity dip, which corresponds to the excitation of SPPs, cannot be observed in the profile. This fact indicates that a large amount of light is absorbed in the $Co_2FeSi_{0.6}Al_{0.4}$ Heusler alloy film.

Therefore, a thinner $Co_2FeSi_{0.6}Al_{0.4}$ Heusler alloy layer was sandwiched between two silver layers, which can easily excite SPPs. The thickness of the $Co_2FeSi_{0.6}Al_{0.4}$ layer was varied from 3 to 7 nm. The reflectivities R, for a glass sub. / Ag (25) / $Co_2FeSi_{0.6}Al_{0.4}$ (d) / Ag (5 nm) multilayer films as a function of incident angle θ, are shown in Fig. 2. All reflectivity curves show a steep dip at around 43.5° by using the Ag layers as the SPP excitation layer. The difference between the maximum and minimum reflectivities, R^{max}-R^{min}, and the maximum absolute inclination of the reflectivity curve, $|dR/d\theta|^{max}$, for the $Co_2FeSi_{0.6}Al_{0.4}$ with 5 nm-thick sandwiched film are 58.8 % and 56.8 %/°, respectively.

The magnetic activities ΔR, which is calculated using the formula (1) by the reflectivity profiles measured at λ = 500 and 700 nm, for the glass sub. / Ag (25) / $Co_2FeSi_{0.6}Al_{0.4}$ (5) / Ag (5 nm) film are shown in Fig. 3. The maximum values of absolute magnetic activity $|\Delta R|^{max}$ measured at 500 and 700 nm show 0.11 and 0.14, respectively. The maximum values of absolute magnetic activity $|\Delta R|^{max}$ of 0.14 is not so high compared with previous reports [5, 6]. One of the methods to increase the $|\Delta R|^{max}$ is to adjust the each layer thickness because it can improve plasmonic properties of the film.

To confirm the magnetic and structural properties of the $Co_2FeSi_{0.6}Al_{0.4}$ films, magnetic hysteresis loop and XRD were measured. A magnetic hysteresis loop of the glass sub. / Ag (25) / $Co_2FeSi_{0.6}Al_{0.4}$ (5) / Ag (5 nm) multilayer film measured by a VSM with a magnetic field normal to the film plane is shown in Fig. 4. The magnetization of the $Co_2FeSi_{0.6}Al_{0.4}$ layer is almost saturated at 4 kOe, which is as same magnitude as the applied field at ATR measurement. The XRD profiles for the $Co_2FeSi_{0.6}Al_{0.4}$ films with 200 nm-thick fabricated on a glass substrate to examine the crystal orientation and degree of

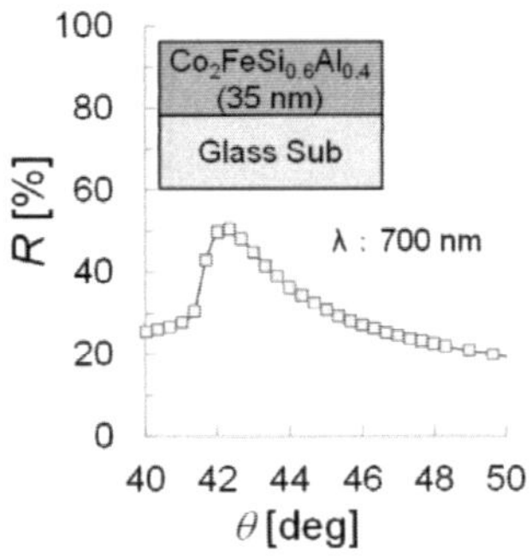

Fig. 1 Reflectivity R for a $Co_2FeSi_{0.6}Al_{0.4}$ single layer with 35 nm-thick fabricated on a glass substrate as a function of incident angle θ. Wavelength λ is 700 nm.

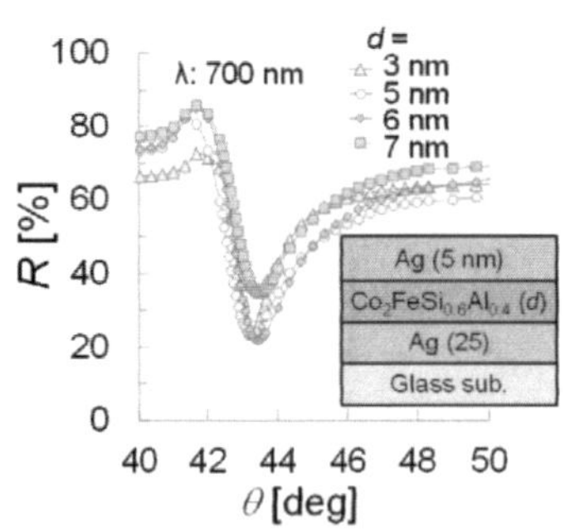

Fig. 2 Reflectivity R for a glass substrate / Ag (25) / $Co_2FeSi_{0.6}Al_{0.4}$ (d) / Ag (5 nm) multilayer films as a function of incident angle θ. Wavelength λ is 700 nm

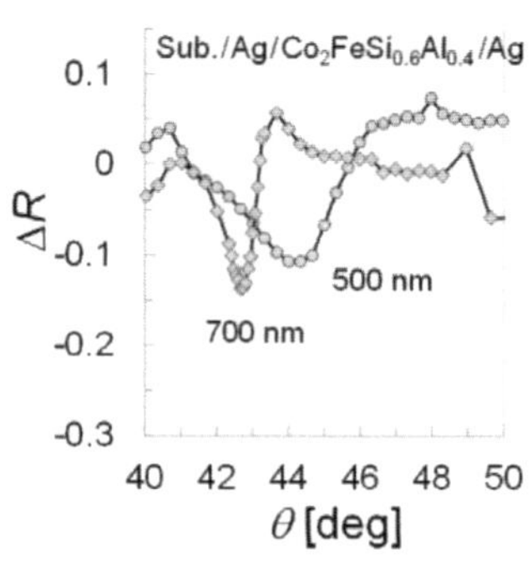

Fig. 3 Normalized reflectivity change ΔR for a glass sub. / Ag (25) / $Co_2FeSi_{0.6}Al_{0.4}$ (5) / Ag (5 nm) single layer as a function of incident angle θ. Wavelengths λ are 500 and 700 nm.

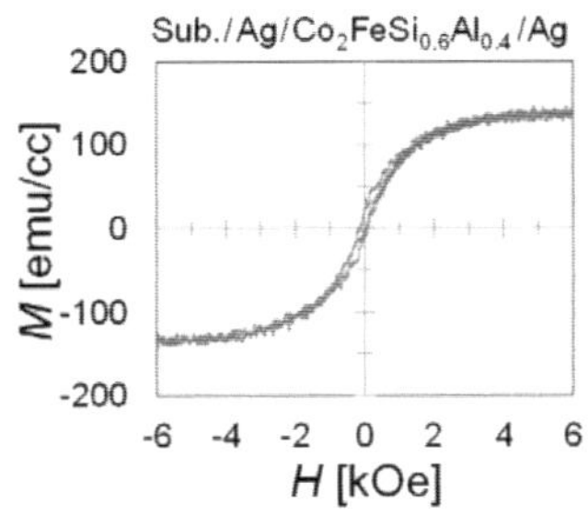

Fig. 4 Hysteresis loop for a glass substrate / Ag (25) / $Co_2FeSi_{0.6}Al_{0.4}$ (5) / Ag (5 nm) film. External magnetic field H was applied normal to the film plane.

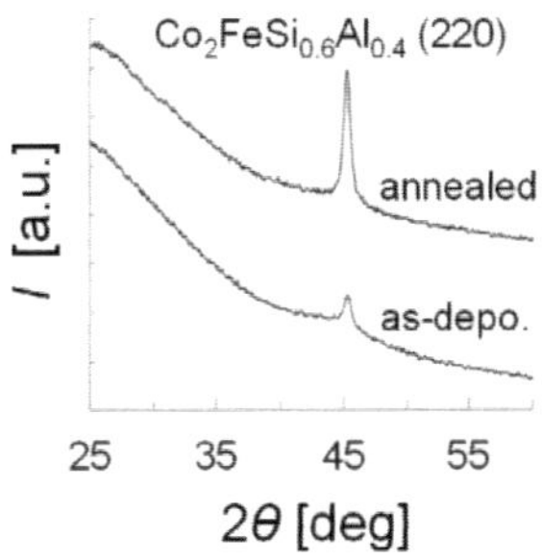

Fig. 5 XRD profiles for a $Co_2FeSi_{0.6}Al_{0.4}$ films fabricated on a glass substrate. Samples are as deposited and annealed at 500 °C in 30 min.

order are shown in Fig. 5. One is annealed at 500 °C and the other is as-deposited. Diffracted lines can be observed at around $2\theta = 44.3°$ which corresponds to the diffraction angle from the $Co_2FeSi_{0.6}Al_{0.4}$ (220) plane in both profiles. The diffracted line intensity from the (220) plane of the annealed film is stronger than that of the as-deposited film. There are, however, no diffracted lines originated from the ordered phase with $B2$ and $L2_1$ structures even in the annealed film. Both $Co_2FeSi_{0.6}Al_{0.4}$ films are oriented (220) plane with the $A2$ disordered structure. Magneto-plasmonic effect of the ordered $B2$ and $L2_1$ structures of $Co_2FeSi_{0.6}Al_{0.4}$ should be also examined for development of a spintronic and magneto-plasmonic device.

4. Conclusion

Plasmonic and magneto-plasmonic properties for a $Co_2FeSi_{0.6}Al_{0.4}$ Heusler alloy film were investigated for a magnetic sensing technique to use the SPPs. The SPPs were excited in the $Co_2FeSi_{0.6}Al_{0.4}$ Heusler alloy by using a sandwiched structure between two silver layers. The maximum value of absolute magnetic activity $|\Delta R|^{max}$ of 0.14 were obtained.

References

[1] González-Díaz J B, García-Martín A, Armelles G, García-Martín J M, Clavero C, Cebollada A, Lukaszew R A, Skuza J R, Kumah D P, and Clarke R 2007 *Phys. Rev. B* **76** 153402
[2] Temnov V V, Armelles G, Woggon U, Guzatov D, Cebollada A, Garcia-Martin A, Garcia-Martin J-M, Thomay T, Leitenstorfer A, and Bratschitsch R, 2010 *Nat. Photonics* **4**, 107
[3] Clavero C, Yang K, Skuza J R, and Lukaszew R A, 2010 *Opt. Lett.* **35**, 1557
[4] Ashizawa Y, Shinohara S, Nawata T, and Nakagawa K 2011 Abstr. Magnetics and Optics Research Int. Symp. P 103
[5] Tachikawa T, Ashizawa Y, and Nakagawa K 2014 *J. Magn. Soc. Jpn.* **38** 135
[6] Narushima K, Ashizawa Y, Brachwitz K, Hochmuth H, Lorenz M, Grundmann M, and Nakagawa K, 2016 *Jpn. J. Appl. Phys.* **55** 07MC05
[7] Sakuraba Y, Hattori M, Oogane M, Ando Y, Kato H, Sakuma A, Miyazaki T and Kubota H 2006 *Appl. Phys. Lett.* **88** 192508
[8] Tezuka N, Ikeda N, Sugimoto S, and Inomata K 2006 *Appl. Phys. Lett.* **89** 252508
[9] Tezuka N, Ikeda N, Miyazaki A, Sugimoto S, Kikuchi M, and Inomata K 2006 *Appl. Phys. Lett.* **89**, 112514
[10] Kretschmann E and Raether H 1968 *Z. Naturforsch. A* **23**, 2135
[11] Sepulveda B, Carrascosa L G, Regatos D, Otte M A, Farina D, and Lechuga L M 2009 *Proc. SPIE* **7397**, 73970Y

Acknowledgement

This work was partially supported by MEXT-Supported Program for the Strategic Research Foundation at Private Universities, 2013-2017.

IOP Conf. Series: Journal of Physics: Conf. Series **903** (2017) 012060 doi:10.1088/1742-6596/903/1/012060

Development of in situ magneto-ellipsometry for studying correlation between the optical and magneto-optical properties of ferromagnetic thin films

O A Maximova[1,2,3], N N Kosyrev[1,4], S N Varnakov[1,2], S A Lyashchenko[1,2], I ATarasov[1,2], I AYakovlev[1,2], O M Maximova[3], D V Shevtsov[1,2] and S G Ovchinnikov[1,2,3]

[1]Kirensky Institute of Physics, Krasnoyarsk, 660036, Russia
[2]Reshetnev Siberian State Aerospace University, Krasnoyarsk, 660014, Russia
[3]Siberian Federal University, Krasnoyarsk, 660041, Russia
[4]Achinsk Branch of Krasnoyarsk State Agrarian University, Achinsk, 662150, Russia

E-mail: maximo.a@mail.ru

Abstract. In this work we present the way of nanostructured films study by means of magneto-ellipsometry. The method of interpretation of *in situ* magneto-optical ellipsometry spectra from the *in situ* molecular beam epitaxy setup with an integrated magneto-ellipsometric real time synthesis control is described. The method has been successfully tested on $Fe/SiO_2/Si$ nanostructures within the model of a homogeneous semi-infinite medium. As a result, the dielectric tensor components for Fe layer were calculated using a developed approach.

1. Introduction

In recent years, the magnetic materials for data storage and spintronic devices have deserved significant attention. There is a problem of an *in situ* real time control of nanomaterials synthesis [1] and their properties investigation because the in air investigation of these structures is often impossible due to the high chemical activity of many materials used in this field. One of the best solutions of this problem is to use the optical and magneto-optical techniques. They are powerful, do not influence on the sample and have some flexibility when using *in situ*, directly in ultrahigh vacuum chamber. We suggest that magneto-ellipsometry [2] meets all these requirements. This experimental technique combines ellipsometry [3] and the magneto-optical Kerr effect measurement within one setup with an ultra-high vacuum chamber and the electromagnet for magnetization reversal of the sample [4].

In this paper, we suggest a new approach of real time control of obtaining material parameters of magnetic thin films right in the process of molecular beam epitaxy by means of the *in situ* magneto-optical ellipsometry. In the end, we demonstrate how our method works by presenting the results of Fe(layer)/SiO_2(layer)/Si(substrate) film study.

2. Magneto-ellipsometry data analysis

Here we describe the method of interpretation of the magneto-ellipsometric measurements data. We consider the case of electromagnetic wave incidence from non-magnetic dielectric medium (characterized by the refraction index N_0) onto ferromagnetic metal (the refraction index N) in the visible light range. The magnetization vector is z-axis directed. YX is a plane of incidence, YZ is a

Content from this work may be used under the terms of the Creative Commons Attribution 3.0 licence. Any further distribution of this work must maintain attribution to the author(s) and the title of the work, journal citation and DOI.
Published under licence by IOP Publishing Ltd

boundary plane The key idea of the proposed approach is reported in [5], where it was applied toward the particular case of the use of small parameters. Here we consider a general case of experimental data processing for the model of a homogeneous semi-infinite medium without any constraints.

We suggest that magneto-ellipsometry technique gives an opportunity to determine all elements of the dielectric permittivity tensor ε of the magnetized ferromagnetic metal [6]

$$[\varepsilon]=\begin{bmatrix}\varepsilon_{11} & \varepsilon_{12} & 0\\ \varepsilon_{21} & \varepsilon_{22} & 0\\ 0 & 0 & \varepsilon_{33}\end{bmatrix}=\begin{bmatrix}\varepsilon'_{11}-i\varepsilon''_{11} & -i(\varepsilon'_{12}-i\varepsilon''_{12})(Q_1-iQ_2) & 0\\ i(\varepsilon'_{12}-i\varepsilon''_{12})(Q_1-iQ_2) & \varepsilon'_{11}-i\varepsilon''_{11} & 0\\ 0 & 0 & \varepsilon'_{11}-i\varepsilon''_{11}\end{bmatrix}, \quad (1)$$

where ε' and ε'' are a real and imaginary parts of medium permittivity respectively, $Q=Q_1-iQ_2$ is a proportional to magnetization magneto-optical parameter. Diagonal tensor elements are responsible for refractive index and extinction coefficient, off-diagonal tensor elements are related to magneto-optical effects. Let us denote the ellipsometric parameters in the non-magnetic condition as ψ_0 and Δ_0. In the case of magneto-ellipsomertic characterization of the sample the surface transverse magneto-optical Kerr effect results in the ellipsometric angles corrections $\delta\psi$ and $\delta\Delta$. Thus, the ellipsometric parameters become $\psi_0+\delta\psi$, $\Delta_0+\delta\Delta$. It means that we have four measured independent real-valued quantities (ψ_0, $\delta\psi$, Δ_0, $\delta\Delta$), as a result, we can derive four real-valued quantities (ε'_{11}, ε''_{11}, ε'_{12}, ε''_{12}).

Table 1. The notations used in the developed approach.

1	2	3	4
$\alpha_1=(A^2+4B^2)^{-1}$	$C=2(NS-PT)$	$\xi_0=b^2-d^2$	$T=K_1(2W+V)+K_2(2U-2X)$
$\alpha_2=(C_2^2+4D_2^2)^{-1}$	$D=2(NT+PS)$	$\xi_1=na+n_0b$	$S=K_1(2U-2X)-K_2(2W+V)$
$A=\xi_1^2+\xi_2^2+\xi_3^2+\xi_4^2$	$\gamma_1=(ab-cd)$	$\xi_2=nc+n_0d$	$K_1=Q_1(n_0^2-k_0^2)-2n_0k_0Q_2$
$A_1=\xi_5\xi_6-\xi_7\xi_0+2\xi_8\xi_9$	$\gamma_2=nk_0+n_0k$	$\xi_3=ka+k_0b$	$K_2=Q_2(n_0^2-k_0^2)+2n_0k_0Q_1$
$A_2=\xi_6\xi_7-\xi_5\xi_0-2\xi_8\xi_9$	$\gamma_3=n_0a+nb$	$\xi_4=kc+k_0d$	$N=Re(\sin\varphi_0)a-Im(\sin\varphi_0)c$
$B=-\xi_1\xi_2-\xi_3\xi_4$	$\gamma_4=n_0c+nd$	$\xi_5=n^2+k^2$	$P=-Re(\sin\varphi_0)c-Im(\sin\varphi_0)a$
$B_1=-2\xi_5ac+2\xi_7bd+2\xi_8\gamma_1$	$\gamma_5=k_0a+kb$	$\xi_6=a^2-c^2$	$U=nk\xi_6+n_0k_0\xi_0+\gamma_1\gamma_2$
$B_2=-2\xi_7ac+2\xi_5bd-2\gamma_1\xi_8$	$\gamma_6=k_0c+kd$	$\xi_7=n_0^2+k_0^2$	$V=\xi_1^2-\xi_2^2-\xi_3^2+\xi_4^2$
$C_2=\gamma_3^2-\gamma_4^2+\gamma_5^2-\gamma_6^2$	$\cos\varphi_0=a+ic$	$\xi_8=n_0k-nk_0$	$W=-2nkac-2n_0k_0bd-\gamma_2\xi_9$
$D_2=-\gamma_3\gamma_4-\gamma_5\gamma_6$	$\cos\varphi_1=b+id$	$\xi_9=bc+ad$	$X=-\xi_1\xi_2+\xi_3\xi_4$

There are four steps of data analysis.

1 Carrying out spectral ellipsometry (ψ_0, Δ_0) and magneto-optical Kerr effect measurements ($\psi_0+\delta\psi$, $\Delta_0+\delta\Delta$).

2 Calculation of spectral dependences of refractive index (n) and extinction coefficient (k)

$$N=n-ik=N_0\sin\varphi_0(1+tg^2\varphi_0(1-(tg\psi_0\exp(i\Delta_0)))^2(1+(tg\psi_0\exp(i\Delta_0)))^{-2})^{1/2} \quad (2)$$

3 Theoretical calculation of the ellipsometric parameters ψ_0, $\delta\psi$, Δ_0, $\delta\Delta$.

Here we rewrite the basic equation of ellipsometry in the following way:

$$tg(\psi_0+\delta\psi)\exp(i(\Delta_0+\delta\Delta))=R_pR_S^{-1}=(R'_p-iR'')(R'_S-iR'')^{-1}, \quad (3)$$

where R_p and R_s are complex reflection coefficients corresponding to in-plane p-polarization and out-of-plane s-polarization respectively, real parts are marked by ', imaginary by ". According to mode conversion from the p to the s polarized channel we can write that

$$R_p=R_{pp}+R_{ps}=R'_{p0}+R'_{p1}-i(R''_{p0}+R''_{p1});\qquad R_S=R_{SS}+R_{sp}=R_{S0}=R'_{S0}-iR''_{S0} \quad (4)$$

IOP Conf. Series: Journal of Physics: Conf. Series **903** (2017) 012060 doi:10.1088/1742-6596/903/1/012060

$$R'_{p0} = \alpha_1(A_1A + 2B_1B);\quad R'_{p1} = \alpha_1(A^2C - 4B^2C + 4ABD);\quad R'_{S0} = \alpha_2(A_2C_2 + 2B_2D_2) \qquad (5)$$

$$R''_{p0} = \alpha_1(AB_1 - 2BA_1);\quad R''_{p1} = \alpha_1(A^2D - 4B^2D - 4ABC);\quad R''_{S0} = \alpha_2(B_2C_2 - 2A_2D_2) \qquad (6)$$

where we have distinguished the magnetic field contribution and marked it by subscript 1, non-magnetic summands– by subscript 0. By substituting the determined values of n and k into equations (5, 6) and using the notations presented in Table 1, where φ_0 and φ_1 are the angles of incidence and refraction respectively, we obtain all necessary expressions for ellipsometric angles calculation.

4 Fitting to the experimental ellipsometric angles by the Nelder–Mead method. As a result it yields the spectral dependences of real (Q_1) and imaginary (Q_2) parts of magneto-optical parameter Q. Thus, we have information about all elements of the dielectric permittivity tensor.

3. Results and discussion

In order to demonstrate the method of interpretation of *in situ* magneto-ellipsometry measurements data, the sample in the form of polycrystalline Fe layer on the surface of SiO_2/Si(100) was studied. Fe film was made by ultrahigh vacuum thermal evaporation with deposition on the cool substrate.

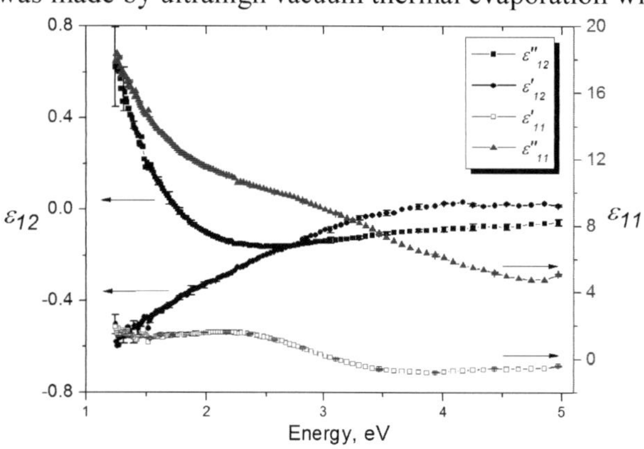

Figure 1. The calculated values of real and imaginary parts of the Fe diagonal tensor element ε_{11} and off-diagonal tensor element ε_{12} (during measurements the angle of incidence was fixed at 56°, the magnetization reversal of the sample in the ± 2 kOe field, the thicknesses of SiO_2 and Fe layers were 3.84 nm and 160.5 nm, respectively).

In conclusion, the algorithm for the interpretation of magneto-ellipsometric measurement data has been proposed for the model of a homogeneous semi-infinite medium. The values of refractive index, extinction coefficient and magneto-optical parameter Q of Fe layer in Fe/SiO_2/Si structure were obtained using the presented approach. Comparison of the Fe magneto-optical parameter Q with [7] shows a good agreement. We have completely determined all elements of the dielectric permittivity tensor (Figure 1). Thus, the opportunity of simultaneous characterization of optical and magneto-optical properties of nanostructures has been demonstrated by means of magneto-ellipsometry.

4. Acknowledgements

The work was supported partly by the Russian Foundation for Basic Research Grant No. 16-32-00209 mol_a, Grant No. 14-02-01211, Grant No. 16-42-24083; the Complex program of SB RAS № II.2P, project 0358-2015-0004; the Ministry of Education and Science of the RF (State task No. 16.663.2014K); Grant of the President of the Russian Federation (Scientific School No. 7559.2016.2).

References

[1] Kumar D, Potdar S, Reddy V R and Gupta A 2008 *J. Phys.: Conf. Ser.* **114** 012034
[2] Fördös T et al. 2011 *J. Phys.: Conf. Ser.* **303** 012041
[3] Fujiwara H 2007 Spectroscopic Ellipsometry Principles and Applications (Chichester: John Wiley & Sons) p. 392
[4] Rykhlitskii S V et al 2012 *Prib. Tekh. Eksp.* **2** 165
[5] Maksimova O A et al. 2014 *Journal Of Structural Chemistry* **55**. No. 6. 1134–41
[6] Sokolov A V 1961 *Optical Properties of Metals* (Moscow: GIFML) p.322
[7] Neuber G et al. 2005 *Rev. Sci. Instrum.* **76** 023910

Influence of Mn diffusion on IrMn thickness threshold for the onset of exchange bias in IrMn/Co bilayers

G Vinai[1], L Frangou[2], C Castan-Guerrero[1], V Bonanni[1,3], B Gobaut[4], S Auffret[2], I L Prejbeanu[2], B Dieny[2], V Baltz[2], P Torelli[1]

[1] Laboratorio TASC, IOM-CNR, S.S. 14 km 163.5, Basovizza, I-34149 Trieste, Italy
[2] SPINTEC, Univ. Grenoble Alpes / INAC-CEA / CNRS, F-38000 Grenoble, France
[3] Department of Physics, University of Milan, Via Celoria 16, I-20133 Milan, Italy
[4] Sincrotrone Trieste S.C.p.A., S.S.14 km 163.5, Area Science Park, 34149 Trieste, Italy

E-mail: vinai@iom.cnr.it

Abstract. In this work, we studied the influence of the buffer layer composition on the IrMn thickness threshold for the onset of exchange bias in IrMn/Co bilayers. By means of magnetometry, x-ray absorption and x-ray photoelectron spectroscopy, we investigated the magnetic and chemical properties of the stacks. We demonstrated a higher diffusion of Mn through the Co layer in the case of a Cu buffer layer. This is consistent with the observation of larger IrMn thickness threshold for the onset of exchange bias.

1. Introduction

Reducing the thickness of exchange biased stacks is of technological interest for applications in hard disk drives to allow larger data capacity. In this context, earlier publications demonstrated that an antiferromagnetic (AF) layer (e.g. IrMn) sandwiched between two ferromagnetic (F) layers, with out-of-plane (e.g. Pt/Co) and in-plane (e.g. Co) anisotropy respectively, exhibits lower AF thickness threshold (t_C) for the onset of exchange bias compared to simple IrMn/Co bilayers [1]. Here, further investigations were carried out in order to determine whether other mechanisms also contribute. We compared the chemical and magnetic properties of X/IrMn/Co exchange coupled thin films for different buffer layers (X). In particular, we focused our attention on Mn diffusion through the Co layer by x-ray absorption (XAS), x-ray magnetic circular dichroism (XMCD) and x-ray photoelectron spectroscopy (XPS) measurements.

2. Experimental procedure

A series of samples was deposited by dc-magnetron sputtering with an Ar pressure of 2.5 10^{-3} mbar onto thermally oxidized silicon substrates, Si/SiO_2//. The stack was $Ta_3/X/IrMn_{thickness}/Co_5/Pt_2$ (nm). Ta layer was used to decouple the Si/SiO_2 substrate from the following layers. The buffer layers X were Cu_2, Pt_2, $Pt_{7.2}$, $Ru_{7.2}$, $(Pt_{1.8}/Ru_{0.6})_{x3}$ and $(Pt_{1.8}/Co_{0.6})_{x3}$ layers. IrMn layer was deposited from an $Ir_{20}Mn_{80}$ target (at. %). The top layer of Pt was used to prevent oxidation in air. The samples were annealed after deposition at 300°C in 10^{-6} mbar pressure and field cooled down to room temperature under an

Content from this work may be used under the terms of the Creative Commons Attribution 3.0 licence. Any further distribution of this work must maintain attribution to the author(s) and the title of the work, journal citation and DOI.
Published under licence by IOP Publishing Ltd

applied in-plane magnetic field of 2 kOe. Hysteresis loops were measured on all samples at room temperature with a vibrating sample magnetometer.

XAS, XMCD and XPS measurements were performed at APE beamline at Elettra synchrotron [2]. Samples were measured at room temperature, with the beam impinging on the surface of the samples at 60°. Absorption spectra have been taken in circular polarization, clockwise (σ^+) and anti-clockwise (σ^-). The sum between the two spectra provides information on the chemical properties of the elements, whereas the difference represents the magnetic dichroic signal. Measurements were taken in total electron yield (TEY) mode, normalizing the sample current with the current coming on the mesh of the beamline. XAS and XMCD were performed at remanence after having saturated the samples along the direction of the setting field, by reversing the polarization of the circular light. In this way, no training effects on exchange bias had to be taken into account.

3. Results and discussion

Figure 1 shows the values of exchange bias (H_{ex}) and coercive field (H_C) for the whole series of samples. The variation of H_{ex} as a function of IrMn thickness follows the one expected for similar IrMn/Co stacks [3] for all buffer layers. For IrMn layers below t_C, no exchange bias is measured. For values between 2 and 4 nm, H_{ex} increases whereas H_C exhibits a maximum. The highest value of exchange bias appears between 4 and 5 nm, and then decreases for larger thicknesses. This trend is consistent with a granular model of exchange bias, which is commonly used for sputtered samples [4].

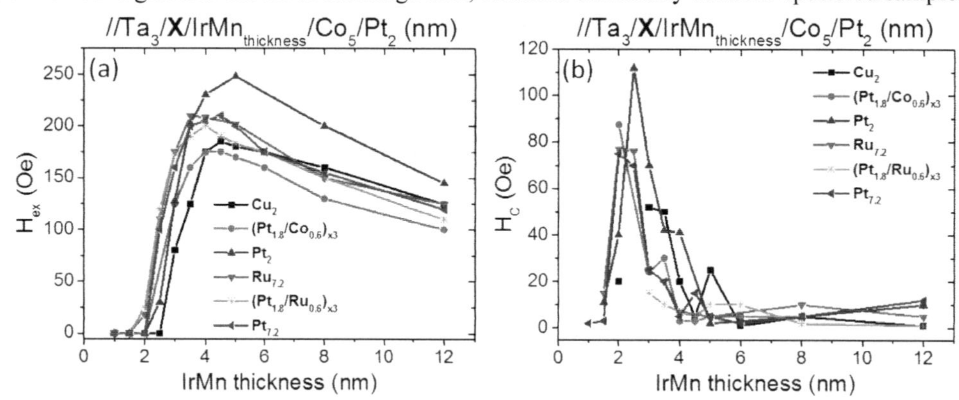

Figure 1. Exchange bias (a) and coercive (b) fields for the different stacks at room temperature.

From the full series of samples, those with an IrMn thickness of 3 nm (i.e. as the exchange bias increases with the IrMn thickness) have been selected. At this thickness, values of H_{ex} range from 80 Oe, for Cu_2 buffer layer, to twice this value, 175 Oe, for $Ru_{7.2}$ and $(Pt_{1.8}/Ru_{0.6})_{x3}$ buffer layers.

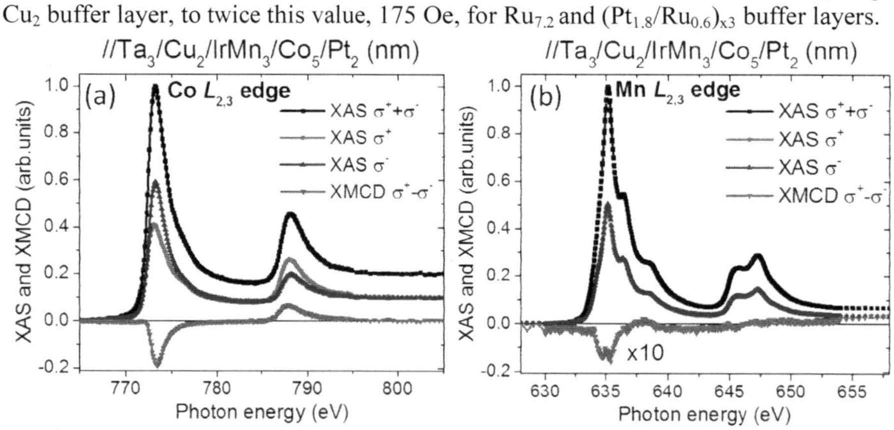

Figure 2. XAS and XMCD spectra at Co (a) and Mn (b) $L_{2,3}$ edges for $Ta_3/Pt_{7.2}/IrMn_3/Co_5/Pt_2$ sample.

Figure 2 shows the XAS and XMCD spectra at Co (a) and Mn (b) $L_{2,3}$ edges for the $Ta_3/Pt_{7.2}/IrMn_3/Co_5/Pt_2$ sample. XAS and XMCD curves have been normalized to the sum of the two circularly polarized spectra. Co absorption spectrum is characteristic of metallic Co and shows no traces of oxidation [5], which indicates the successful protection by the 2 nm Pt capping. Different buffer layers yield no changes in Co spectra shape. Mn absorption spectrum (figure 2b) also reveals no signs of oxidation, as a fully metallic Mn spectrum [6]. Considering that the penetration depth of TEY measurements is of only a few nm, the weak XAS signal measured at Mn edge, because of the covering by the 2 nm of Pt and 5 nm of Co layers, comes from the Mn atoms at the IrMn/Co interface.

From magnetic point of view, we can notice that both Co and Mn XMCD spectra have the same sign, which corresponds to a ferromagnetic coupling across the IrMn/Co interface, as reported in the literature [7]. This has been confirmed for all buffer layers. Co dichroic signal, by taking into account the angle of 60° of the sample with respect to the beam and a polarization level of 75%, is around 29%. By applying the sum rules [8], we evaluate a total magnetic moment of Co of 2.3 μ_B, which corresponds to typical bulk value. Similar values were obtained for all samples. Regarding Mn, the XMCD spectrum has been magnified by a factor 5 in order to be clearly visible. Its dichroic intensity of around 4% with comparison with the one of Co confirms that only a slight percentage of spins participate to the exchange. In this case, signal intensity and signal-over-noise ratio give error bars too large to apply sum rules. XAS and XMCD measurements at Mn edge were also performed with the samples oriented normal to the beam incidence. In this configuration, no magnetic dichroic signal was measured. This means that the uncompensated Mn spins are aligned in-plane with the Co spins, with no out-of-plane canting.

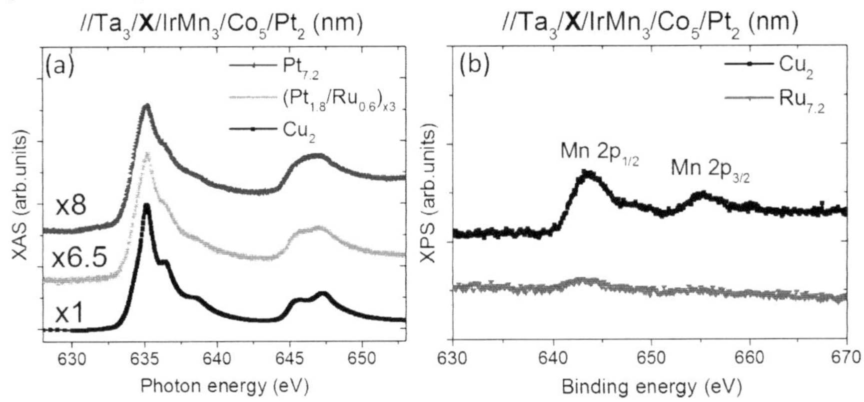

Figure 3. (a) Comparison between XAS spectra at Mn edges for different buffer layers. Curves have been normalized to Cu_2 buffered sample; (b) Mn $2p$ XPS spectra for Cu_2 and $Ru_{7.2}$ buffer layers.

Concerning the absorption spectra at Mn edge, $Ta_3/Cu_2/IrMn_3/Co_5/Pt_2$ sample differed from all other buffer layers. In figure 3(a), we show the XAS curves for three different buffer layers. Curves have been normalized to the signal intensity of Cu_2 buffered sample. In order to do so, the spectra of $Pt_{7.2}$ and $(Pt_{1.8}/Ru_{0.6})_{x3}$ buffered samples had to be multiplied by a factor 8 and 6.5 respectively. It has to be stressed that all measurements were performed in identical beamline conditions, thus the comparison between background-to-peak intensity al L_3 edges between different samples gives information on the location of Mn atoms through the thickness of the sample, since the signal intensity exponentially decreases while penetrating into the sample. Moreover, the shape of Mn absorption spectra for Cu_2 buffered sample strongly differs from other samples. It presents features which can be attributed to a partial oxidation of Mn, which appears in the $3d^5$ configuration of Mn^{2+} in octahedral symmetry [9]. One possible reason for such differences among the Mn spectra is the uncomplete covering of the Pt capping layer for the Cu_2 buffered sample. In order to verify this, nominally identical samples deposited in different periods with Cu_2 buffer layer were compared, giving identical XAS. We thus consider that such feature is intrinsic to the sample.

In order to investigate the reason of such differences, XPS measurements were performed on all samples of the series setting the beamline energy at $hv = 1000$ eV. This kind of measurement is more surface sensitive than XAS, with a penetration depth of roughly one nm. As it can be seen from figure 3(b), in case of Cu_2 buffer layer Mn $2p$ levels were measured, whereas in the case of $Ru_{7.2}$ no sign of Mn was observed. Curves have been normalized at Pt $4f$ levels.

By comparing XAS raw measurements of figure 3(a) and XPS Mn $2p$ spectra of figure 3(b), we can state that in case of Cu_2 buffer layer Mn diffused throughout the Co layer up to the surface, resulting in a higher level of oxidation compared to other samples. The interdiffusion of Mn most probably took place during the annealing process at 300°C, necessary to set the exchange bias coupling. Mn atoms which diffused up to the surface experience an oxidation process, as observed by XAS measurement. In this case, most of the absorption signal comes from the Mn atoms close to the surface, hiding the signal of metallic Mn atoms present at the IrMn/Co interface. Interdiffusion is a critical aspect from technological point of view, because of the annealing steps necessary in the devices processing. It has been shown that interdiffusion takes place preferentially along grain boundaries [10], with levels of diffusions which vary with annealing temperature.

From the XAS and XPS measurements performed, we observe that not only the layers over IrMn play a role on Mn diffusion to the surface, but also buffer layers do. Whereas Pt and Ru layers help limiting the Mn diffusion, Cu does not. Indeed, Cu tends to form large grains separated by pronounced grain boundaries favoring atomic mobility. This has consequences on exchange bias field value. The Cu_2 buffer yields the largest t_C characterizing the onset of H_{ex}, likely because of a strong reduction in the AF intergrain exchange coupling across the grain boundaries. For all other samples, wherein IrMn is unoxidized and interdiffusion is very limited, H_{ex} is maximized.

4. Acknowledgments

This work has been partly performed in the framework of the nanoscience foundry and fine analysis (NFFA-MIUR Italy Progetti Internazionali) project.

References

[1] Vinai G, Moritz J, Bandiera S, Prejbeanu I L and Dieny B 2013 *J. Phys. D.: Appl. Phys.* **46** 322001

[2] Panaccione G, Vobornik I, Fujii J, Krizmancic D, Annese E, Giovanelli L, Maccherozzi F, Salvador F, De Luisa A, Benedetti D, Gruden A, Bertoch P, Polack F, Cocco D, Sostero G, Diviacco B, Hochstrasser M, Maier U, Pescia D, Back C H, Greber T, Osterwalder J, Galaktionov M, Sancrotti M and Rossi G 2009 *Rev. Sci. Instrum.* **80** 043105

[3] Ali M, Marrows C, Al-Jawad M, Hickey B, Misra A, Nowak U and Usadel K 2003 *Phys. Rev. B* **68** 214420

[4] O'Grady K, Fernandez-Outon L E and Vallejo-Fernandez G 2009 *J. Magn. Magn. Mater.* **322** 883

[5] Regan T, Ohldag H, Stamm C, Nolting F, Lüning J, Stöhr J and White R 2001 *Phys. Rev. B* **64** 214422

[6] Torelli P, Sirotti F and Ballone P 2003 *Phys. Rev. B* **68** 205413

[7] Takahashi H, Kota Y, Tsunoda M, Nakamura T, Kodama K, Sakuma A and Takahashi M 2011 *J. Appl. Phys.* **110** 123920

[8] Chen C T, Idzerda Y U, Lin H -J, Smith N V, Meigs G, Chaban E, Ho G H, Pellegrin E and Sette F 1995 *Phys. Rev. Lett.* **75** 152

[9] Eimüller T, Kato T, Mizuno T, Tsunashima S, Quitmann C, Ramsvik T, Iwata S and Schütz G 2004 *Appl. Phys. Lett.* **85** 2310

[10] Letellier F, Lechevallier L, Lardé R, Le Breton J M, Akmaldinov K, Auffret S, Dieny B and Baltz V 2014 *J. Appl. Phys.* **116** 203906

IOP Conf. Series: Journal of Physics: Conf. Series **903** (2017) 012062 doi:10.1088/1742-6596/903/1/012062

On Berezinskii-Kosterlitz-Thouless transition in monoaxial chiral helimagnets

Igor Proskurin[1,2], Alexander S Ovchinnikov[2] and Jun-ichiro Kishine[3]

[1] Faculty of Science, Hiroshima University, Higashi-Hiroshima, Hiroshima 739-8526, Japan
[2] Institute of Natural Sciences, Ural Federal University, Ekaterinburg 620083, Russia
[3] Division of Natural and Environmental Sciences, The Open University of Japan, Chiba 261-8586, Japan

E-mail: iprosk@ouj.ac.jp, alexander.ovchinnikov@urfu.ru, kishine@ouj.ac.jp

Abstract. We analyze the possibility of Berezinskii-Kosterlitz-Thouless (BKT) transition in XY-model with in-plane Dzyaloshinskii-Moriya (DM) interaction. It is demonstrated that under standard duality transformation of XY-model into the model of two-dimensional Coulomb gas of magnetic vortices, DM interaction is mapped to the effective electric field perpendicular to the original DM vector. Since the electrostatic energy in the constant electric field becomes dominating over the logarithmic Coulomb attraction at sufficiently large length scales, BKT transition is ruled out by the presence of DM interaction. This behavior is confirmed in the renormalization group analysis, which shows that for finite effective electric field the system always flows towards the high temperature phase.

1. Introduction

The possibility of topological phase transitions in two-dimensional (2D) systems was first anticipated independently by Berezinskii [1], Kosterlitz and Thouless [2]. Since that time, this concept found numerous applications in different fields of condensed matter physics including hydrodynamics, superconductors, and magnetic systems [3]. A classical Berezinskii-Kosterlitz-Thouless (BKT) scenario states that there exists a certain transition temperature such as that below this temperature the topological defects – vortices are bound and hidden from observation, while above it free defects can proliferate into the system. This transition is accompanied by characteristic features in macroscopic properties like, for example, in current-voltage characteristics of semiconductors [3].

During the last decade, a considerable progress has been made in physics of chiral magnets where the existence of asymmetric Dzyaloshinskii-Moriya (DM) exchange interaction leads to a rich variety of magnetic phases (see e. g. [4] and references therein). In monoaxial chiral magnets, an important shift was marked recently by experimental discovery of chiral soliton lattice structure [5] originally predicted by Dzyaloshinskii [6]. With thin films of chiral magnets becoming available, realization of BKT scenario in such materials is an intriguing question.

In this paper we consider BKT transition in 2D monoaxial chiral magnet by mapping the correspondent XY-model into 2D Coulomb gas of magnetic vortices. We found that under such transformation in-plane DM interaction is mapped into an effective electric field in the direction perpendicular to DM vector. Interestingly, similar result was obtained in 2D semiconductors

Content from this work may be used under the terms of the Creative Commons Attribution 3.0 licence. Any further distribution of this work must maintain attribution to the author(s) and the title of the work, journal citation and DOI.
Published under licence by IOP Publishing Ltd

where applied current density is mapped into topological electric field [7]. Note that locked-unlocked transition in the sine-Gordon model, where the misfit parameter plays the role of electric field in the equivalent Coulomb gas model, was also studied in [8].

Our discussion goes along the following line. First, we consider general formulation of the model. After that, the model is subjected to the duality transformation, and for the resulting effective model the renormalization group (RG) equations are derived. Conclusions based on numerical solution of the RG equations are given in the last part. For completeness, some technical details are specified in Appendices.

2. Formulation

We consider 2D chiral helimagnet withing the framework of XY-model on a square lattice with the Hamiltonian given by

$$\mathcal{H} = -J\sum_{\langle ij\rangle} \boldsymbol{S}_i \cdot \boldsymbol{S}_j - \boldsymbol{D}\cdot\sum_{\langle ij\rangle} \boldsymbol{S}_i \times \boldsymbol{S}_j - \tilde{H}_x \sum_i S_{ix}, \tag{1}$$

where $J > 0$ is the ferromagnetic exchange constant, $\boldsymbol{D} = D\boldsymbol{e}_z$ is DM vector taken along the z-axis, and $\tilde{H}_x$ is the magnetic field applied along the x-direction, see Fig. 1. All the classical spin variables $\boldsymbol{S}_i$ are confined in the basal xy-plane and parametrized as $\boldsymbol{S}_i = S(\cos\theta_i, \sin\theta_i, 0)$, where $0 \leq \theta_i < 2\pi$ is the polar angle at the site i. Angular brackets denote summation over the nearest neighboring sites.

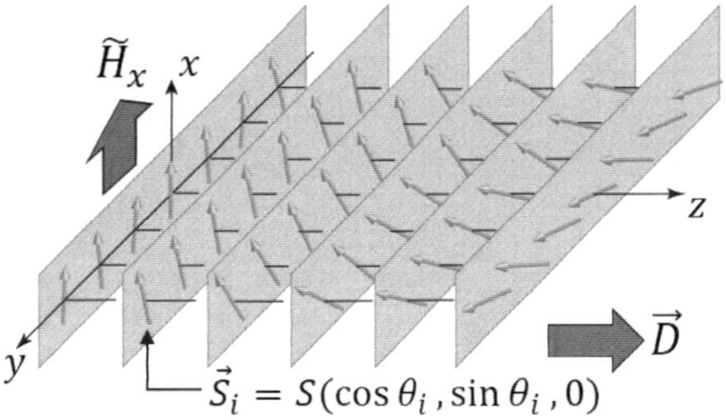

Figure 1. Schematic picture of the monoaxial 2D chiral helimagnet. The classical spins $\boldsymbol{S}_i$ rotate in the basal xy-plane, the DM vector is along the z-axis, and in-plane magnetic field $\tilde{\boldsymbol{H}}$ is applied along the x-direction.

The thermodynamic properties of the model with the Hamiltonian in (1) are described by the partition function

$$\mathcal{Z} = \int \prod_i d\theta_i \exp\left[\beta\tilde{J}\sum_{\langle ij\rangle}\cos\left(\theta_i - \theta_j - \alpha_{ij}\right) + \beta h\sum_i \cos\theta_i\right], \tag{2}$$

where $\beta = (k_B T)^{-1}$ is the inverse temperature, $\tilde{J} = S^2\sqrt{J^2 + D^2}$ is the effective exchange parameter, $h = \tilde{H}_x S$, and $\alpha_{ij} = \tan^{-1}(D/J)$ for the nearest neighbor ij-bond along the z-direction and zero otherwise.

2.1. *Effective model*

In the low temperature limit, the spin-wave contribution to the partition function (2) can be integrated out using the duality mapping technique [9]. The essential steps of the duality transformation are summarized in Appendix A. The resulting effective model takes into account vortex configurations where θ_i field, while still being slowly varying form site to site over a closed

IOP Conf. Series: Journal of Physics: Conf. Series **903** (2017) 012062 doi:10.1088/1742-6596/903/1/012062

loop, acquires total contribution proportional to the multiply of 2π. After the transformation partition function takes the form

$$\mathcal{Z} \propto \sum_{\{l_\mu(\boldsymbol{r})\}} \exp\left[-\sum_{r\mu}\left(\frac{l_\mu^2(\boldsymbol{r})}{2\beta\tilde{J}} + i\alpha_\mu l_\mu(\boldsymbol{r})\right)\right] \times \prod_{\boldsymbol{r}}\left\{\sum_{\kappa=-\infty}^{\infty} I_\kappa(\beta h)\delta\left[\sum_\mu (l_\mu(\boldsymbol{r}) - l_\mu(\boldsymbol{r}-\boldsymbol{a}_\mu)) - \kappa(\boldsymbol{r})\right]\right\}, \quad (3)$$

where a particular configuration is expressed by an integer vector field $l_\mu(\boldsymbol{r})$ defined on the bond between two lattice sites with radius vectors $\boldsymbol{r}$ and $\boldsymbol{r}+\boldsymbol{a}_\mu$ where $\boldsymbol{a}_\mu$ ($\mu = y, z$) is the lattice vector, and $\alpha_\mu = \alpha\delta_{\mu z}$ with $\alpha = \tan^{-1}(D/J)$. The effect of the magnetic field h is to generate magnetic charges $\kappa(\boldsymbol{r})$ at the center of each lattice cell with the amplitudes of such contributions proportional to the modified Bessel functions $I_\kappa(\beta h)$. The total summation is over all possible configurations with $l_\mu(\boldsymbol{r})$ running form $-\infty$ to ∞ on each bond.

The physical meaning of the constraint imposed by δ-function in equation (3) can be elucidated through the splitting of $l_\mu(\boldsymbol{r})$ into the longitudinal and transverse parts [10]

$$l_\mu(\boldsymbol{r}) = m(\boldsymbol{r}) - m(\boldsymbol{r}+\boldsymbol{a}_\mu) + \sigma(\boldsymbol{r}) - \sigma(\boldsymbol{r}+\boldsymbol{a}_\mu) + \varepsilon_{\mu\nu}\left[n(\boldsymbol{r}) - n(\boldsymbol{r}-\boldsymbol{a}_\nu)\right], \quad (4)$$

where $m(\boldsymbol{r})$ and $n(\boldsymbol{r})$ are integers, $|\sigma(\boldsymbol{r})| < 1$ and $\varepsilon_{\mu\nu}$ denotes the second rank Levi-Civita tensor. The longitudinal part satisfies lattice form of the Poisson equation

$$\sum_\mu \left[2m(\boldsymbol{r}) + 2\sigma(\boldsymbol{r}) - m(\boldsymbol{r}+\boldsymbol{a}_\mu) - \sigma(\boldsymbol{r}+\boldsymbol{a}_\mu) - m(\boldsymbol{r}-\boldsymbol{a}_\mu) - \sigma(\boldsymbol{r}-\boldsymbol{a}_\mu)\right] = \kappa(\boldsymbol{r}), \quad (5)$$

while the transverse part is a pure rotation over the unit cell with zero lattice divergence

$$\sum_\mu \varepsilon_{\mu\nu}\left[n(\boldsymbol{r}) - n(\boldsymbol{r}-\boldsymbol{a}_\nu)\right] = 0. \quad (6)$$

By substituting (4) in (3) and applying the Poisson formula to the summation over $n(\boldsymbol{r})$, we obtain

$$\mathcal{Z} \propto \sum_{\{\kappa(\boldsymbol{r})\}}\sum_{\{q(\boldsymbol{r})\}} \int \prod_{\boldsymbol{r}} \varphi(\boldsymbol{r}) I_{\kappa(\boldsymbol{r})}(\beta h) \times \exp\left\{-\sum_{r\mu}\left[\frac{\left(\hat{\Delta}_\mu\varphi(\boldsymbol{r})\right)^2}{2\beta\tilde{J}} + i\alpha_\mu\hat{\Delta}_\mu\varphi(\boldsymbol{r})\right] + 2\pi i\sum_{\boldsymbol{r}} q(\boldsymbol{r})\varphi(\boldsymbol{r})\right\}, \quad (7)$$

where

$$\hat{\Delta}_\mu\varphi(\boldsymbol{r}) = \varphi(\boldsymbol{r}+\boldsymbol{a}_\mu) - \varphi(\boldsymbol{r}) + \varepsilon_{\mu\nu}\left[m(\boldsymbol{r}+\boldsymbol{a}_\nu) + \sigma(\boldsymbol{r}+\boldsymbol{a}_\nu) - m(\boldsymbol{r}) - \sigma(\boldsymbol{r})\right]. \quad (8)$$

In what follows, we only consider the case of vanishing magnetic fields, since in this paper we are mainly focused on the effect of DM interaction. The model that includes magnetic charges will be analyzed in further publications. In the $\beta h \to 0$ limit, $I_\kappa(\beta h)$ is replaced by $\delta_{\kappa 0}$, and equation (5) has a trivial solution $m(\boldsymbol{r}) + \sigma(\boldsymbol{r}) = 0$. Taking into account that $\alpha_\mu = \alpha\delta_{\mu z}$ and making the shift of variables $\varphi(\boldsymbol{r}) \to \varphi(\boldsymbol{r}) + i\alpha\beta\tilde{J}y$, we obtain an effective model of a 2D Coulomb gas of magnetic vortices with integer charges $q(\boldsymbol{r})$ in the external electric field along the y-axis proportional to the strength of DM interaction

$$\mathcal{Z}_{\text{eff}} \propto \sum_{\{q(\boldsymbol{r})\}} \exp\left\{-2\pi^2\beta\tilde{J}\sum_{\boldsymbol{r}\boldsymbol{r}'} q(\boldsymbol{r})G(\boldsymbol{r}-\boldsymbol{r}')q(\boldsymbol{r}') + 2\pi\alpha\beta\tilde{J}\sum_{\boldsymbol{r}} y\, q(\boldsymbol{r})\right\}, \tag{9}$$

where

$$G(\boldsymbol{r}-\boldsymbol{r}') = \int \frac{d^2k}{(2\pi)^2}\frac{e^{i\boldsymbol{k}\cdot(\boldsymbol{r}-\boldsymbol{r}')}}{4-2\cos k_x - 2\cos k_y} \approx -\frac{1}{2\pi}\log|\boldsymbol{r}-\boldsymbol{r}'| + G(0), \tag{10}$$

is 2D Coulomb potential, and $G(0)$ contains the divergent part of the interaction potential that ensures 'electroneutrality' of the system (hereafter we take the lattice constant $a = 1$).

Equation (9) is the starting point for the RG analysis [11, 12]. It can be rewritten in a more convenient form [9, 11]

$$\mathcal{Z}_{\text{eff}} \propto \sum_{\{q(\boldsymbol{r})\}} \exp\left[2\pi K\sum_{\boldsymbol{r}\neq\boldsymbol{r}'} q(\boldsymbol{r})\log|\boldsymbol{r}-\boldsymbol{r}'|q(\boldsymbol{r}') - 2\pi\sum_{\boldsymbol{r}}\epsilon_y(\boldsymbol{r})\,q(\boldsymbol{r})\right] y_0^{\sum_{\boldsymbol{r}} q^2(\boldsymbol{r})}, \tag{11}$$

where $K = \beta\tilde{J}$, $\boldsymbol{\epsilon}(\boldsymbol{r}) = -\alpha\beta\tilde{J}\boldsymbol{r}$, and $y_0 = \exp(-\pi^2\beta\tilde{J}/2)$ is so called fugacity of the Coulomb gas. Since we consider the case $\beta\tilde{J} \gg 1$, all the contribution with large $|q(\boldsymbol{r})|$ are exponentially small, and we can restrict the summation in (11) to $q(\boldsymbol{r}) = 0, \pm 1$.

2.2. RG equations

The partition function in equation (11) can be expanded in terms of the total number of vortex pairs $N = \frac{1}{2}\sum_{\boldsymbol{r}} q^2(\boldsymbol{r})$

$$\mathcal{Z}_{\text{eff}} = \sum_{N=0}^{\infty}\frac{y_0^{2N}}{(N!)^2}\int\prod_{i=1}^{2N} d^2s_i \exp\left[2\pi K\sum_{i<j} q(\boldsymbol{s}_i)q(\boldsymbol{s}_j)\log|\boldsymbol{s}_i-\boldsymbol{s}_j| - 2\pi\sum_i \epsilon_y(\boldsymbol{s}_i)q(\boldsymbol{s}_i)\right], \tag{12}$$

where we take heed of $|q(\boldsymbol{r})| = 0, 1$. The factor $(N!)^2$ arises from independent permutations of all the vortices and antivortices. In the lowest order in y_0, the system contains only one vortex-antivortex pair with ± 1 charges at the positions $\boldsymbol{s}$ and $\boldsymbol{s}'$

$$\mathcal{Z}_{\text{eff}}^{(1)} = 1 + y_0^2\int d^2s\int d^2s' \exp\left[-2\pi K\log|\boldsymbol{s}-\boldsymbol{s}'| - 2\pi\epsilon_y(\boldsymbol{s}-\boldsymbol{s}')\right] + O(y_0^4). \tag{13}$$

In order to derive RG equations for the model parameters, we calculate effective interaction energy for the pair of ± 1 external charges, placed at the positions $\boldsymbol{r}$ and $\boldsymbol{r}'$, which includes the screening effects from the internal charges at $\boldsymbol{s}$ and $\boldsymbol{s}'$ [11]. In the lowest order expansion, the effective interaction reads

$$e^{\mathcal{S}_{\text{eff}}} = \frac{1}{\mathcal{Z}_{\text{eff}}^{(1)}}\left[1 + y_0^2\int d^2s\int d^2s' e^{\mathcal{S}_{\text{int}}(\boldsymbol{r},\boldsymbol{r}';\boldsymbol{s},\boldsymbol{s}')}\right] + O(y_0^4), \tag{14}$$

where

$$\begin{aligned}\mathcal{S}_{\text{int}}(\boldsymbol{r},\boldsymbol{r}';\boldsymbol{s},\boldsymbol{s}') = -2\pi K\,(&\log|\boldsymbol{r}-\boldsymbol{r}'| + \log|\boldsymbol{s}-\boldsymbol{s}'| + \log|\boldsymbol{r}-\boldsymbol{s}'| + \log|\boldsymbol{r}'-\boldsymbol{s}| \\ &- \log|\boldsymbol{r}-\boldsymbol{s}| - \log|\boldsymbol{r}'-\boldsymbol{s}'|) - 2\pi\epsilon_y(\boldsymbol{r}+\boldsymbol{s}-\boldsymbol{r}'-\boldsymbol{s}').\end{aligned} \tag{15}$$

Direct calculation of $\mathcal{S}_{\text{eff}}$ in equation (14) results in

$$\mathcal{S}_{\text{eff}} = -2\pi K_{\text{eff}}\log|\boldsymbol{r}-\boldsymbol{r}'| - 2\pi\epsilon_{\text{eff}}(\boldsymbol{y}-\boldsymbol{y}'), \tag{16}$$

where the screening from internal charges is absorbed in the effective model parameters

$$K_{\text{eff}} = K - 4\pi^3 K^2 y_0^2 \int_1^\infty dx x^{3-2\pi K} + O(\epsilon^2, y_0^4), \tag{17}$$

$$\epsilon_{\text{eff}} = \epsilon - 4\pi^3 K \epsilon y_0^2 \int_1^\infty dx x^{3-2\pi K} + O(\epsilon^2, y_0^4). \tag{18}$$

Here, $\epsilon = \alpha\beta\tilde{J}$ is the effective electric field strength. Some technical details of the derivation are reproduced in Appendix B. Giving the cut-off parameter $b = e^\ell$ in the integration over x, we eventually obtain the following system of RG equations [7]

$$\frac{dK}{d\ell} = -4\pi^3 y_0^2 K^2, \tag{19}$$

$$\frac{dy_0}{d\ell} = (2 - \pi K + \pi\epsilon) y_0, \tag{20}$$

$$\frac{d\epsilon}{d\ell} = \epsilon + 4\pi^3 y_0^2 K \epsilon. \tag{21}$$

3. Discussion and summary

Below, we discuss the system of RG equations (19)–(21). For $\epsilon = 0$, these equations describe BKT transition at the temperature $T_C = \pi\tilde{J}/2$ between the underscreened and overscreened regimes. In the underscreened regime realized at $T < T_C$ and $\pi^2 y_0 < |K^{-1} - \pi/2|$, all the vortices in the system are coupled into dipoles resulting in zero effective screening. In RG language, it means that $K(\ell)$ and $y_0(\ell)$ flow towards the $y_0 = 0$ line that corresponds to $K_{\text{eff}} = K$ according to equation (17). In contrast, in the overscreened phase, $T > T_C$ or $\pi^2 y_0 > |K^{-1} - \pi/2|$, the dipoles disperse into free charges and the screening appears. In this scenario, the model parameters flow towards the strong coupling fixed point at $K = 0$ and $y_0 = \infty$ [11].

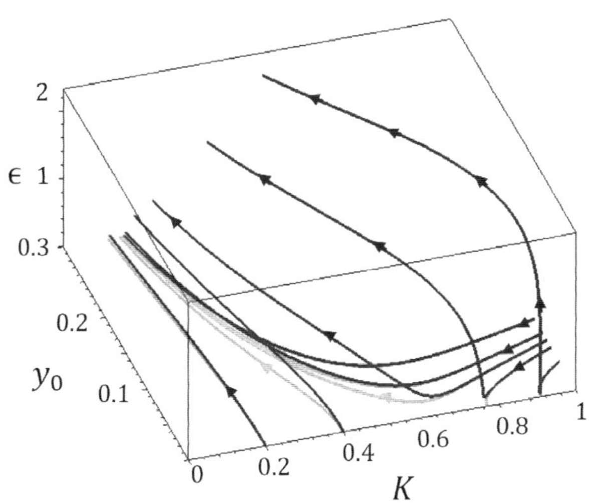

Figure 2. RG flow for the Coulomb gas in finite electric field. The flow in $\epsilon = 0$ plane is shown by green color. The BKT transition point is at the position $(2/\pi, 0, 0)$.

The presence of the electric field drastically changes the above picture. Even infinitesimal ϵ lifts BKT transition in favor of the overscreened scenario. Figure 2 show RG flow in the axes K, y_0, and ϵ calculated numerically from equation (19)–(21). In the $\epsilon = 0$ plane, RG flow is consistent with BKT transition behavior at the point $(2/\pi, 0, 0)$. However, above the transition point the electric field pulls positive and negative charges apart breaking the dipoles and redirecting the flow towards the strong screening fixed point, $K \to 0$ and $y_0 \to \infty$.

The absence of BKT transition in the system with finite ϵ can be understood from the following simple argument. In 2D system, the electrostatic energy in a constant electric field being linear in $\boldsymbol{r}$ becomes dominant over the logarithmic Coulomb attraction at sufficiently large distances. Therefore, to gain the total energy, it is always favorable to break the dipole pairs apart for any values of K and y_0.

By noting that in our model the effective electric field is proportional to the strength of DM interaction, we make the conclusion that DM interaction is relevant. The presence of in-plane DM exchange in magnetic thin film should remove BKT transition and assist the observation of free magnetic vortices.

In summary, we considered duality transformation and RG analysis in 2D chiral magnets with in-plane DM interaction. It was demonstrated that this system can be mapped into the model of 2D Coulomb gas of magnetic vortices with DM interaction playing the role of the effective electric field directed perpendicularly to the original DM vector. By applying RG analysis, we showed that the effective electric field greatly effects on the vortex-antivortex pairs, prevents the formation of magnetic dipoles, and destroys BKT transition.

Acknowledgments
We acknowledge support by the Government of Russian Federation Program 02.A03.21.0006, by the Ministry of Education and Science of the Russian Federation, projects Nos. 1437 and MK-6230.2016.2, and by Japan Society for the Promotion of Science Grants-in-Aid for Scientific Research (S) Nos. 25287087 and 25220803.

Appendix A. Duality transformation

In the low temperature limit $\beta\tilde{J} \gg 1$, one can approximate periodic function in equation (2) defined on the bond between ith and jth sites by a series of its expansions near each minimum labeled by the integer m_{ij}

$$e^{\beta\tilde{J}(\cos\Theta_{ij}-1)} \approx \sum_{m=-\infty}^{\infty} \exp\left[-\frac{\beta\tilde{J}}{2}\left(\Theta_{ij}-2\pi m_{ij}\right)^2\right], \tag{A.1}$$

where $\Theta_{ij} = \theta_i - \theta_j - \alpha_{ij}$. The summation over m_{ij} is transformed into the integration over ϕ_{ij}-field using the Poisson formula

$$\sum_{m=-\infty}^{\infty} f(m) = \sum_{l=-\infty}^{\infty} \int_{-\infty}^{\infty} d\phi\, f(\phi) e^{2\pi i l\phi}, \tag{A.2}$$

and after integrating out ϕ, we arrive to the following expression for the partition function

$$\mathcal{Z} \propto \sum_{\{l_{ij}\}} \int \prod_i d\theta_i \exp\left[-\sum_{\langle ij\rangle}\left(\frac{l_{ij}^2}{2\beta\tilde{J}} - i l_{ij}\Theta_{ij}\right) + \beta h\sum_i \cos\theta_i\right]. \tag{A.3}$$

We replace $\theta_i \to \theta(\boldsymbol{r})$ and $l_{ij} \to l_\mu(\boldsymbol{r})$ where $\mu = y, z$, and take the advantage of infinite lattice

$$-i\sum_{\langle ij\rangle} l_{ij}\left(\theta_i - \theta_i\right) \to -i\sum_{r\mu} l_\mu(r)\left[\theta(\boldsymbol{r}) - \theta(\boldsymbol{r}+\boldsymbol{a}_\mu)\right] = -i\sum_{r\mu}\left[l(\boldsymbol{r}) - l(\boldsymbol{r}-\boldsymbol{a}_\mu)\right]\theta(\boldsymbol{r}), \tag{A.4}$$

which gives

$$\mathcal{Z} \propto \sum_{\{l_\mu(\boldsymbol{r})\}} \int \prod_{\boldsymbol{r}} d\theta(\boldsymbol{r})$$
$$\times \exp\left[-\sum_{r\mu}\left(\frac{l_\mu^2(\boldsymbol{r})}{2\beta\tilde{J}} - i\left[l(\boldsymbol{r}) - l(\boldsymbol{r}-\boldsymbol{a}_\mu)\right]\theta(\boldsymbol{r}) + i\alpha_\mu l_\mu(\boldsymbol{r})\right)\right] \exp\left[\beta h\sum_{\boldsymbol{r}}\cos\theta(\boldsymbol{r})\right]. \tag{A.5}$$

Now it is possible to integrate over θ from 0 to 2π using the expansion

$$e^{z\cos\theta} = I_0(z) + 2\sum_{k=1}^{\infty} I_k(z)\cos(k\theta), \tag{A.6}$$

that gives equation (3).

Appendix B. Effective model parameters

Here we highlight crucial steps in the derivation of RG equation following the discussion in [11]. First, we rewrite (14) in the following form

$$\begin{aligned} e^{\mathcal{S}_{\rm eff}} &= e^{-2\pi K\log|\boldsymbol{r}-\boldsymbol{r}'|-2\pi\epsilon_y(\boldsymbol{r}-\boldsymbol{r}')} \\ &\times\left[1+y_0^2\int d^2s\int d^2s' e^{-2\pi K\log|\boldsymbol{s}-\boldsymbol{s}'|-2\pi\epsilon_y(\boldsymbol{s}-\boldsymbol{s}')}\left(e^{2\pi K D(\boldsymbol{r},\boldsymbol{r}';\boldsymbol{s},\boldsymbol{s}')}-1\right)\right]+O(y_0^4), \end{aligned} \tag{B.1}$$

where

$$D(\boldsymbol{r},\boldsymbol{r}';\boldsymbol{s},\boldsymbol{s}') = \log|\boldsymbol{r}-\boldsymbol{s}'| + \log|\boldsymbol{r}'-\boldsymbol{s}| - \log|\boldsymbol{r}-\boldsymbol{s}| + \log|\boldsymbol{r}'-\boldsymbol{s}'|, \tag{B.2}$$

and change to the center of mass variables $\boldsymbol{X} = (\boldsymbol{s}+\boldsymbol{s}')/2$ and $\boldsymbol{x} = \boldsymbol{s}-\boldsymbol{s}'$. Second, we expand over $|\boldsymbol{x}|$ keeping only terms up x^2

$$e^{2\pi K D(\boldsymbol{r},\boldsymbol{r}';\boldsymbol{s},\boldsymbol{s}')} - 1 = 2\pi K\boldsymbol{x}\cdot\boldsymbol{\nabla}_{\boldsymbol{X}} f(\boldsymbol{X}) + 2\pi^2K^2\left(\boldsymbol{x}\cdot\boldsymbol{\nabla}_{\boldsymbol{X}} f(\boldsymbol{X})\right)^2 + O(x^3), \tag{B.3}$$

$$e^{-2\pi\epsilon_y(\boldsymbol{s}-\boldsymbol{s}')} = 1 - 2\pi\epsilon\left(\boldsymbol{x}\cdot\boldsymbol{e}_y\right) + O(x^2), \tag{B.4}$$

where $f(\boldsymbol{X}) = \log|\boldsymbol{r}-\boldsymbol{X}| - \log|\boldsymbol{r'}-\boldsymbol{X}|$. Substitution of these equations in (B.1) gives

$$\begin{aligned} e^{\mathcal{S}_{\rm eff}} &= e^{-2\pi K\log|\boldsymbol{r}-\boldsymbol{r}'|-2\pi\epsilon_y(\boldsymbol{r}-\boldsymbol{r}')} \\ &\times\left[1-4\pi^2K\epsilon y_0^2\int d^2x\int d^2X e^{-2\pi K\log|x|}(\boldsymbol{x}\cdot\boldsymbol{e}_y)\,\boldsymbol{x}\cdot\boldsymbol{\nabla}_{\boldsymbol{X}} f(\boldsymbol{X})\right. \\ &\left.+2\pi^2K^2\int d^2x\int d^2X e^{-2\pi K\log|x|}\left(\boldsymbol{x}\cdot\boldsymbol{\nabla}_{\boldsymbol{X}} f(\boldsymbol{X})\right)^2\right]+O(\epsilon^2,y_0^4). \end{aligned} \tag{B.5}$$

In the first integral we perform integration over the angular part of $\boldsymbol{x}$ and use the identity $\boldsymbol{e}_y = \boldsymbol{\nabla}_{\boldsymbol{X}}(\boldsymbol{X}\cdot\boldsymbol{e}_y)$ that provides the answer

$$\int d^2x\int d^2X e^{-2\pi K\log|x|}(\boldsymbol{x}\cdot\boldsymbol{e}_y)\,\boldsymbol{x}\cdot\boldsymbol{\nabla}_{\boldsymbol{X}} f(\boldsymbol{X}) = -2\pi^2\boldsymbol{e}_y\cdot(\boldsymbol{r}-\boldsymbol{r}')\int_1^{\infty} dx x^{3-2\pi K}, \tag{B.6}$$

where we introduced an infrared cut-off at the lattice spacing distances. The second integral is calculated using 2D form of the Poisson equation

$$\nabla^2_{\boldsymbol{X}}\log|\boldsymbol{r}-\boldsymbol{X}| = 2\pi\delta(\boldsymbol{r}-\boldsymbol{X}), \tag{B.7}$$

that provides

$$\int d^2x\int d^2X e^{-2\pi K\log|x|}\left(\boldsymbol{x}\cdot\boldsymbol{\nabla}_{\boldsymbol{X}} f(\boldsymbol{X})\right)^2 = (2\pi)^2\log|\boldsymbol{r}-\boldsymbol{r}'|\int_1^{\infty} dx x^{3-2\pi K}. \tag{B.8}$$

After substitution of (B.6) and (B.8) in (B.5), we obtain equations (17, 18).

References

[1] Berezinskii V 1971 *Sov. Phys. JETP* **32** 493

[2] Kosterlitz J M and Thouless D J 1973 *Journal of Physics C: Solid State Physics* **6** 1181

[3] José J V 2013 *40 years of Berezinskii-Kosterlitz-Thouless theory* (World Scientific)

[4] Kishine J and Ovchinnikov A S 2015 Theory of monoaxial chiral helimagnet (*Solid State Physics* vol 66) ed Camley R E and Stamps R L (Academic Press) pp 1 – 130

[5] Togawa Y, Koyama T, Takayanagi K, Mori S, Kousaka Y, Akimitsu J, Nishihara S, Inoue K, Ovchinnikov A S and Kishine J 2012 *Phys. Rev. Lett.* **108**(10) 107202

[6] Dzyaloshinskii I 1964 *Sov. Phys. JETP* **19** 960

[7] Sujani S, Chattopadhyay B and Shenoy S R 1994 *Phys. Rev. B* **50**(22) 16668–16678

[8] Puga M W, Simanek E and Beck H 1982 *Phys. Rev. B* **26**(5) 2673–2676

[9] Nagaosa N 1999 *Quantum Field Theory in Condensed Matter Physics* Texts and monographs in physics (Springer)

[10] Polyakov A M 1987 *Gauge Fields and Strings* Contemporary concepts in physics (Taylor & Francis)

[11] Altland A and Simons B 2006 *Condensed Matter Field Theory* (Cambridge University Press)

[12] Chaikin P and Lubensky T 2000 *Principles of Condensed Matter Physics* (Cambridge University Press)

IOP Conf. Series: Journal of Physics: Conf. Series **903** (2017) 012063 doi:10.1088/1742-6596/903/1/012063

Correlation between Gilbert damping and electric resistivity in $Fe_{94}Co_6$/GaAs single crystal thin films: A role of electron scattering

Akihiro Iwasaki[1], Takahiko Suzuki[1], Nobuyuki Inaba[1], Yutaka Takahashi[1], Fumiyoshi Kirino[2], Mitsuru Ohtake[3] and Masaaki Futamoto[4]

[1] Graduate School of Science and Engineering, Yamagata University, Yonezawa, Yamagata, 992-8510, Japan
[2] Graduate School of Fine Arts, Tokyo University of the Arts, Tokyo, 110-8714, Japan
[3] Faculty of Engineering, Kogakuin University, Hachioji, Tokyo, 192-0015, Japan
[4] Faculty of Science and Engineering, Chuo University, Tokyo, 112-8551, Japan

takahasy@yz.yamagata-u.ac.jp

Abstract. Relation between Gilbert damping and scattering of conduction electrons in 3*d* transition metals is investigated. $Fe_{94}Co_6$ thin films were epitaxially deposited on GaAs(001) by rf magnetron sputtering. Some films were annealed *in situ*, and compared with the as-deposited films. The electric resistivity was smaller in the annealed films both at room and low temperatures, indicating that the scattering rate of conduction electrons was reduced by annealing. Gilbert damping was evaluated from the peak width of ferromagnetic resonance spectra. We have found that at room temperature the damping is smaller in the annealed sample (with smaller scattering rate), while at low temperature the damping is enhanced when the scattering rate is decreased. The opposite dependencies of Gilbert damping on the electron scattering rate at low and room temperatures agree with the theoretical studies based on the linear response.

1. Introduction

Gilbert damping is recognized as one of the key magnetic properties for designing spintronic devices. Unravelling its microscopic origin is of pressing importance, and many theoretical approaches to this problem have been put forward [1,2]. Several factors are pointed out that determine the damping such as spin-orbit interaction and density-of-states at the Fermi energy. For very thin films (few monolayers) interfacial effects and the coupling to adjacent layers also influence the damping [1].

One of the important theoretical findings is the role of electron scattering [3,4]. It affects Gilbert damping constant α through two distinct mechanisms: In the intraband contribution α *decreases* with *increasing* scattering rate Γ_e, and in the interband contribution α *increases* with *increasing* Γ_e. The damping is dominated by the intraband contribution in the weak scattering region (α behaves like $\alpha \propto 1/\Gamma_e$) while it is dominated by the interband contribution in the strong scattering region. ($\alpha \propto \Gamma_e$ in this case) When Γ_e in a sample could be reduced slightly, α would increase if the sample were in the weak scattering region while α would decrease if it were in the strong scattering region.

For bulk 3*d* transition metals the experimental temperature variations of the damping were reported in Ref. 5 (down to liq. He temp.). The explicit relation between the temperature and the scattering rates is not given, but their results can be understood that the damping constants drop gradually with

Content from this work may be used under the terms of the Creative Commons Attribution 3.0 licence. Any further distribution of this work must maintain attribution to the author(s) and the title of the work, journal citation and DOI.
Published under licence by IOP Publishing Ltd

IOP Conf. Series: Journal of Physics: Conf. Series **903** (2017) 012063 doi:10.1088/1742-6596/903/1/012063

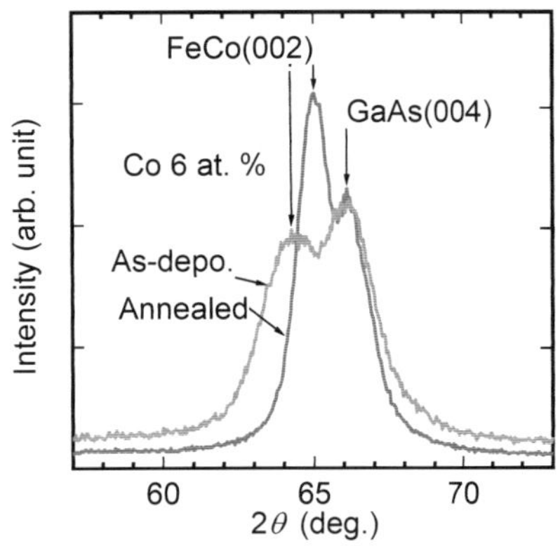

Figure 1. XRD diffraction patterns of 100-nm thick samples.

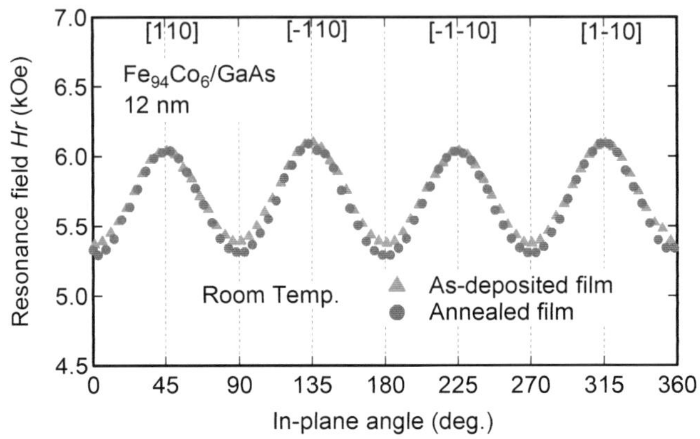

Figure 2. In-plane angle dependence of the resonance field measured by Q-band FMR.

the reduction of temperature as the phonon scattering dwindles, then upturn below 100 K with weak residual scatterings. Recently Ma *et. al.* reported the relationship between the antisite disorder and the Gilbert damping in $L1_0$ FePt thin films (17 and 22 nm) at 200 K [6]. They found that the damping was enhanced as the antisite disorder percentage increased.

In order to study the relationship between Γ_e and α experimentally we prepared single-crystal thin films with different Γ_e (as-deposited and annealed films), evaluated Γ_e by the in-plane resistivities, and measured α by ferromagnetic resonance (FMR) at room and low temperatures.

2. Growth and structure

Thin films of single crystal bcc $Fe_{94}Co_6$ were deposited on GaAs(001) substrates by rf magnetron sputtering. The base pressure of the growth chamber was less than 0.8×10^{-4} Pa, and the Ar gas pressure during the deposition was kept at 0.7 Pa. Co concentrations were confirmed by energy-dispersive x-ray analyses. Two samples with the same thickness were deposited successively, and the second one was annealed at 300°C *in situ*. The x-ray diffraction (XRD) patterns from 100-nm thick films are shown in Fig. 1. Diffraction peaks from bcc $Fe_{94}Co_6$(002) were observed, and a slight narrowing of peak width by annealing was confirmed. (The details of XRD analysis in Ref. 7)

Magnetization curves of samples were measured by a vibrating sample magnetometer with the external field in the plane of the film. The saturation magnetization reached 1750 emu/cm^3, and the in-plane magnetic anisotropy with a $\langle 100 \rangle$ ($\langle 110 \rangle$) direction being an easy (hard) direction was observed. The anisotropy was also confirmed by FMR measurements. Angular dependencies of the resonance fields (H_r) for 12-nm thick films are shown in Fig. 2. Four-fold variations of H_r caused by the in-plane magnetic anisotropy were observed. The estimated anisotropy constant K_1 increased from 3.51×10^5 erg/cm^3 in the as-deposited film to 3.91×10^5 erg/cm^3 in the annealed film. Magnetic and XRD analyses confirm that bcc $Fe_{94}Co_6$ films were grown epitaxially on GaAs(001) with cube-on-cube relationship. The peak narrowing in XRD and the anisotropy enhancement in FMR indicate that the crystallographic alignment was improved by annealing.

3. Electron scattering and Gilbert damping

Electron scattering was evaluated by measuring in-plane resistivity of films using the four-point probe method at room and low (liquid nitrogen) temperatures. (No magnetic field was applied.) The resistivities of as-deposited and annealed films are compared in Fig. 3. The resistivity dropped approximately by 0.09 μΩ·m by cooling (△→▲ or ○→●) which was the contribution from electron-phonon collisions. The residual resistivity is due to the collisions of electrons with impurities, crystal imperfections and interface roughness which depend on sample preparation, *e. g.*, on annealing. The

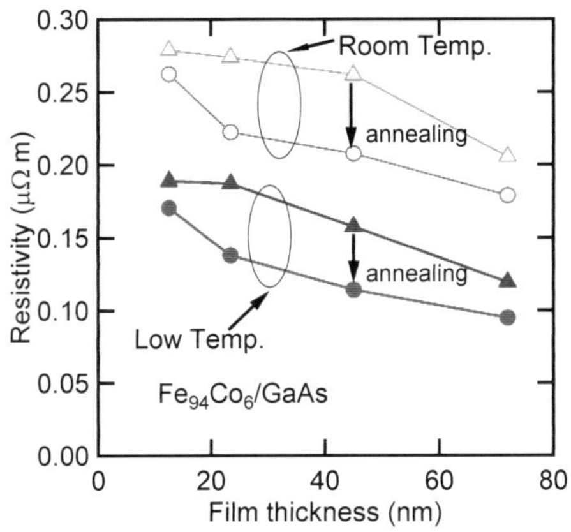

Figure 3. Reduction of resistivities by annealing, measured at room and low temperatures.

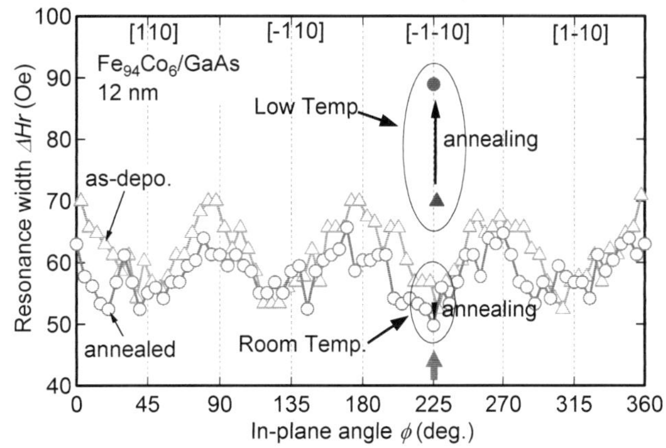

Figure 4. FMR linewidths (ΔH_r) change by annealing: ΔH_r was measured for $0° < \phi < 360°$ at room temperature, and ΔH_r was measured only in 225° at low temperature.

resistivities decreased by annealing at both temperatures, showing that scattering rates Γ_e were smaller in annealed films than in as-deposited films due to the reduction of residual scattering.

Magnetization dynamics was investigated by FMR using a conventional, field-swept spectrometer with the in-plane external magnetic field. At the microwave frequency of 35 GHz (Q-band), the resonance field (5.3-6.0 kOe as in Fig. 2) was strong enough to saturate in-plane magnetization, ensuring that magnetization was parallel to the external field. Damping constants were evaluated by FMR linewidth ΔH_r. Several factors contribute to ΔH_r, which are distinguished by in-plane variations of $\Delta H_r(\phi)$ [8]. The angular variation of Gilbert damping is small for $3d$ transition metals, while that of two-magnon scattering (TMS), which leads to non-Gilbert type damping, shows four-fold symmetric angular variation in bcc Fe-based thin films on GaAs [9].

ΔH_r of the 12-nm thick films at room temperature are plotted in Fig. 4. (△ for as-deposited and ○ for annealed films) ΔH_r consists of a four-fold symmetric contribution from TMS superposed on an angle-independent component due to the Gilbert damping. ΔH_r at $\langle 110 \rangle$ directions, at which TMS shuts off, corresponds to the upper bound of the Gilbert damping. The data points of room temperature ΔH_r scattered (we estimated the experimental uncertainty in an individual data point of ΔH_r was 5 Oe), but the angular distribution of ΔH_r as whole for the annealed film (red lines) was slightly narrower than that for the as-deposited film (green lines). The angle-averaged ΔH_r reduced from 62 Oe to 58 Oe by annealing. Linewidth was also measured at liq. N_2 temperature with the same samples at the fixed angle $[\bar{1}\,\bar{1}\,0]$ (due to the instrumental limitation of our spectrometer), as plotted in Fig. 2. (▲ for as-deposited and ● for annealed films) ΔH_r *increased* from 70 Oe to 89 Oe by annealing. The present result supports the theoretical conjecture, i. e., when the sample is in the strong scattering region due to the frequent electron-phonon collisions at room temperature, the reduction of the residual scattering by annealing leads to the slight decrease of damping. While at low temperature when the sample is in the weak scattering region, the damping increases as the electron scattering rate is reduced by annealing.

4. Conclusions

Our experimental results for $Fe_{94}Co_6$ thin films are in favour of the theoretical predictions that the damping α is affected by the electron scattering rate Γ_e as $\alpha \propto 1/\Gamma_e$ or $\alpha \propto \Gamma_e$ in weak and strong scattering region, respectively. However, our experiment was performed on the single material with limited data points, and further study is required to make clear whether the relation between α and Γ_e

holds generally in other materials or how the individual scattering processes are involved. The present finding shows the potential way to control α in terms of transport properties of materials.

Acknowledgements
This work was supported in part by Grant-in-Aid for Scientific Research (Grant Nos. 23560384 and 26420295) from the Japan Society for the Promotion of Science.

References

[1] Mankovsky S, Ködderritzsch D, Woltersdorf G and Ebert H 2013 *Phys. Rev.* B **87** 014430
[2] Turek I, Kudrnovský J and Drchal V 2015 *Phys. Rev.* B **92** 214407
[3] Gilmore K, Idzerda Y U and Stiles M D 2007 *Phys. Rev. Lett.* **99** 027204
[4] Sakuma A 2013 *J. Mag. Soc. Jpn.* **37** 343
[5] Bhagat S M and Lubitz P 1974 *Phys. Rev.* B **10** 179
[6] Ma X, Ma L, He P, Zhao H B, Zhou S M and Lüpke 2015 *Phys. Rev.* B **91** 014438
[7] Ikeya H, Takahashi Y, Inaba N, Kirino F, Ohtake M and Futamoto M 2011 *J. Phys.: Conf. Ser.* **266** 012116
[8] Wada Y, Takahashi Y, Inaba N, Kirino F, Ohtake M and Futamoto M 2013 *J. Magn. Soc. Jpn.* **37** 166
[9] Woltersdorf G and Heinrich B 2004 *Phys. Rev.* B **69** 184417

Institute of Physics
Dirac House, Temple Back
Bristol BS1 6BE UK

ISSN: 1742-6588
ISBN 978-1-5108-5062-0

CURRAN ASSOCIATES INC.
proceedings.com